Flora of Siberia

Volume 2
Poaceae (Gramineae)

Flora of Siberia

Volume 2
Poaceae (Gramineae)

Volume Editors
Prof. L.I. Malyschev
and
G.A. Peschkova

CRC Press
Taylor & Francis Group
Boca Raton London New York

CRC Press is an imprint of the
Taylor & Francis Group, an **informa** business
A SCIENCE PUBLISHERS BOOK

CRC Press
Taylor & Francis Group
6000 Broken Sound Parkway NW, Suite 300
Boca Raton, FL 33487-2742

© 2001 by Taylor & Francis Group, LLC
CRC Press is an imprint of Taylor & Francis Group, an Informa business

First issued in paperback 2019

No claim to original U.S. Government works

ISBN 13: 978-0-367-44729-8 (pbk)
ISBN 13: 978-1-57808-101-1 (hbk)

**Visit the Taylor & Francis Web site at
http://www.taylorandfrancis.com**

**and the CRC Press Web site at
http://www.crcpress.com**

PREFACE

The family of grasses (*Poaceae*) represents one of the largest of Siberian flora. The well-known taxonomists who devoted themselves to the study of this family include K.B. Trinius, A. Grisebach, R.Yu. Roshevits, S.A. Nevski, V.V. Reverdatto and others. The classic work of N.N. Tzvelev *Zlaki SSSR*[1] [Grasses of the USSR] represents a major contribution to understanding grasses.

Grasses enjoy immense economic importance. Some of them are dominant and codominant in marshy, meadow, and steppe vegetation used as hay and pasture resources.

The present volume was prepared by specialists at the Siberian Central Botanical Garden, Siberian Division of USSR Academy of Sciences[2], as well by botanists from the other institutions: E.B. Alekseev (Moscow State University) on genus *Festuca*, and M.V. Olonova (Tomsk State University) on genus *Poa*.

A thorough revision and several years of study of the entire herbarium material established that the family comprises 72 genera and 440 species and subspecies. In this book 27 new species and subspecies (7 of them with Latin diagnoses) are described; 17 varieties have been elevated to the rank of species and their distribution ranges identified; more than 10 taxa regarded as synonyms by former researchers are restored to the rank of species; 14 species detected in the herbarium material are published for the first time for the Siberian territory. The somatic chromosome number, mainly based on the work of R.E. Krogulevich and T.S. Rostovtseva[3], is cited for several species and subspecies.

For compiling chorological data, apart from the M.G. Popov Herbarium (NSK) and General Herbarium at the Central Siberian Botanical Garden (NS), Siberian Division of USSR Academy of Sciences, collections of the following herbaria were also utilized: Siberian Herbarium of V.L. Komarov Botanical Institute (A.A. Korobkov, curator); P.N. Krylov Herbarium at Tomsk State University (A.V. Polozhii, curator); D.P.

---

[1] N.N. Tzvelev. 1976. Zlaki SSSR [Grasses of the USSR]. Nauka, Leningrad. 788 pp.

[2] Since this book was published in Russian in 1990, the abbreviation USSR has been retained (instead of the present RF)—General Editor.

[3] R.E. Krogulevich and T.S. Rostovtseva. 1984. Khromosomnye chisla tsvetkovykh rastenii Sibiri i Dal'nego Vostoka [Chromosome Numbers of Flowering Plants of Siberia and the Far East]. Nauka, Siberian Division, Novosibirsk.

vi

Syreishchikov Herbarium at Moscow State Univeristy (I.A. Gubanov, curator), and partly the Herbarium at the Yakutsk Institute of Biology, Siberian Division of USSR Academy of Sciences, (late V.N. Andreev, curator).

The drawings presented in the volume are original and were prepared respectively by: genus *Festuca*—E.B. Alekseev, *Poa*—M.V. Olonova, and *Puccinellia*—S.V. Bubnova. Distribution maps were mainly compiled by V.S. Dyudzhok. N.M. Mal'tseva much assisted in manuscript preparation. The compilers are grateful to them all.

The following abbreviations are used in describing the diagnostic features of plants:

auct. non ... —auctores, non ... (authors, not ...)
class. hab.—classic habitat (locus classicus)
comb. nova—combinatio nova (new combination)
s.l.—sensu lato (in a broad sense)
sp.—species
s. str.—sensu stricto (in a narrow sense)
subsp.—subspecies
syn.—synonym
var.—varietas (variety)

The distribution range of plants depicts in coded form the 28 administrative divisions or the effective floristic areas of Siberia (see Fig. 1) as follows:

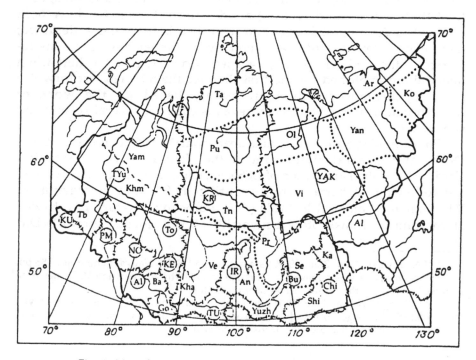

Fig. 1: Map showing nominal floristic regions of Siberia

West. Sib.: TYU—Yam, Khm, Tb, KU, OM, TO, NO, KE, AL—Ba, Go
Cent. Sib.: KR—Ta, Pu, Tn, Kha, Ve, TU
East. Sib.: IR—An, Pr, BU—Se, Yuzh, ChI—Ka, Shi, YAK—Ar, Ol, Vi, Al, Yan, Ko.

Here,

West. Sib.—Western Siberia
TYU—Tyumensk Province
  Yam—Yamal-Nenets Autonomous Territory
  Khm—Khanty-Mansi Autonomous Territory
  Tb—Tobol Floristic District
KU—Kurgan Province
OM—Omsk Province
TO—Tomsk Province
NO—Novosibirsk Province
KE—Kemerovo Province
AL—Altay Territory
  Ba—Barnaul Floristic District
  Go—Gorno-Altay Autonomous Territory
Cent. Sib.—Central or Middle Siberia
KR—Krasnoyarsk Territory
  Ta—Taimyr Floristic District
  Pu—Putoran Floristic District
  Tn—Tunguska Floristic District
  Kha—Khakassia Autonomous Territory
  Ve—Verkhneyenisei (Upper Yenisey) Floristic District
TU—Tuva Autonomous Republic
East. Sib.—Eastern Siberia
IR—Irukutsk Province
  An—Angara-Sayan Floristic District
  Pr—Fore Lena-Katanga Floristic District
BU—Buryat Autonomous Republic
  Se—Severoburyat (Northern Buryat) Floristic District
  Yuzh—Yuzhno-Buryat (Southern Buryat) Floristic District
CHI—Chita Province
  Ka—Kalar Floristic District
  Shi—Shilka-Argun Floristic District(Dauria)
YAK—Yakut Autonomous Republic
  Ar—Arctic Floristic District
  Ol—Olenek-Nizhnelensk (Lower Lena) Floristic District
  Vi—Vilyui-Verkhnelensk (Upper Lena) Floristic District
  Al—Aldan Floristic District
  Yan—Yano-Indigirka Floristic District
  Ko—Kolyma Floristic District

G.A. Peschkova

CONTENTS

x

KEY TO GENERA[1]

1.   Spikelets usually 2–35 mm long, in various types of inflorescences but not forming umbellate clusters .... 2.

+   Spikelets small (up to 1.2 mm long), arranged in umbellate clusters on main rachis and lateral branches, forming narrow paniculate inflorescences; glumes lacking. Annuals, 1–6 cm tall .... 49. **Coleanthus.**

2.   Spikelets without scabrous bristles at base, sometimes with hairs .... 3.

+   Spikelets surrounded at base by fairly long (longer than half spikelet), scabrous, stiff bristles persisting in inflorescence even after spikelets fall .... 71. **Setaria.**

3.   Keel of palea without awnlike appendages .... 4.

+   Keel of palea with 2–4 awnlike appendages, arising distinctly below paleate tip .... 51. **Pleuropogon.**

4.   Inflorescence a true spike; spikelets sessile or on very short, thick stalks, arranged singly or in groups of 2–4 (6) on edge of spike in regular rows .... 5.

+   Inflorescence paniculate, racemose or capitate, sometimes very dense, spicate, but inflorescence rachis invariably highly branched; spikelets not in regular rows .... 15.

5.   Ovary and caryopsis glabrous at tip .... 6.

+   Ovary and caryopsis pilose at tip .... 8.

6.   Spikelets with 2 or more florets. Glumes usually present (sometimes only 1 or 2) .... 7.

+   Spikelets single-flowered. Glumes absent; only their connate rudiments forming leptodermatous, 0.5 mm broad rim encircling base of spikelet .... 55. **Nardus.**

7.   All spikelets with 2 glumes, their broad side turned toward rachis; sometimes lemma very small, barely visible. Sheath without auricles .... 63. **Tripogon.**

+   All spikelets, except uppermost, with single glume; rib of glume (narrow side) turned toward rachilla. Sheath tip with lanceolate auricles .... 34. **Lolium.**

8.   Spikelets usually 2 or 3 together on rachis with single, rarely

---

[1] Patterned after N.N. Tzvelev, *Zlaki SSSR* [Grasses of the USSR].

2

        2, fully developed florets on rachis; lateral spikelets in groups of 3, often underdeveloped .... 9.

+   Spikelets usually with more than 2 well-developed florets, occasionally with 1 or 2 (but then present singly on edges of spikes) .... 10.

9.   All spikelets sessile, equally developed, 1–2 (3)-flowered .... 9. **Psathyrostachys**.

+   Spikelets in groups of 3, lateral ones on short (0.8–1.5 mm long) stalks, usually underdeveloped or with single staminate floret, central ones sessile, with single bisexual floret .... 10. **Hordeum**.

10.   Glumes without keel or keel visible only in upper half of glume. Lemma not keeled on back .... 11.

+   Glumes with nerve prominently keeled throughout length. Back of lemma slightly keeled ....6. **Agropyron**.

11.   Glumes subulate, without distinct nerves or linear-lanceolate with 1–3 (5) faint nerves; spikelets in groups of 2–4; leaves grayish-green, very stiff, with thick veins .... 12.

+   Glumes oblong-lanceolate, lanceolate or elliptical, with 3–7 (9) distinct nerves. Spikelets more often singly on edges of spikes; if in groups of 2–4, leaves soft, green, with slender veins .... 13.

12.   Glumes from subulate to oblong-lanceolate, usually normally developed, occasionally underdeveloped (plants then some-what caespitose with stiff, grayish-green leaves) ...... 8. **Leymus**.

+   Glumes in form of short, 1–7 mm long bristles, often under-developed or absent. Rhizomatous plants with delicate green leaves .... 7. **Hystrix**.

13.   Spikes with several (more than 10) spikelets, latter sessile or on very short (not more than 0.8 mm) stalks. Lemma with 5 nerves, distinct only in upper quarter .... 14.

+   Spikes with few (not more than 10) spikelets, lower spikelets on 1–2 mm long thick stalks. Lemma with 7 distinct nerves .... 3. **Brachypodium**.

14.   Glumes glabrous or subglabrous. Spikelets sessile. Callus of lemma broadly rounded. Anthers 4.5–7 mm long, more than half length of palea .... 5. **Elytrigia**.

+   Glumes rather scabrous along nerves. Spikelets on very short (up to 0.8 mm) but distinct stalks. Callus of lemma rather broadly cuneate. Anthers 1.5–3 (4) mm long, as long or less than half length of palea .... 4. **Elymus**.

15 (4).   Leaf invariably with ligule, sometimes only in form of barely visible transverse rim or transverse row of hairs .... 16.

+   Leaf without ligule; even without transverse row of hairs at its place .... 68. **Echinochloa**.

16. Spikelets without pinched off pulvinate ridge at base ....17.

+ Spikelets with pinched off pulvinate ridge at base shed together with spikelet. Rachis and spicate branches of panicles with clusters of long lustrous hairs .... 69. **Eriochloa**.

17. Lateral branches not digitate. Usually shorter than main rachis .... 18.

+ Spicate branches digitate or subdigitate at tip of stem. Main rachis usually shorter than lateral ones .... 70. **Digitaria**.

18. Glumes not inflated, saccate on back .... 19.

+ Glumes inflated, saccate on back; spikelets rounded, in 2 close rows along one side of spicate branches, in narrow elongated panicle .... 28. **Beckmannia**.

19. Branches of inflorescence not disintegrating into segments. Spikelets not enclosed in long hairs .... 20.

+ Branches of panicle in fruit disintegrating into segments. Spikelets in groups of 2 or 3 on branches; of these 1 sessile, rest stalked and enclosed in long (more than half length of spikelet) hairs diverging from stalks of spikelets and lower part of glumes .... 72. **Spodiopogon**.

20. Spikelet with one fully developed floret. Axis of spikelet extended above it in form of rachilla .... 21.

+ Spikelets 2 to many-flowered, rarely single-flowered; in latter case, above with row of scales with reduced terminal florets .... 39.

21. All spikelets bisexual, some underdeveloped sometimes ...... 22.

+ Spikelets unisexual: upper ones in panicle pistillate; lower ones staminate. Glumes usually reduced .... 2. **Zizania**.

22. Lemma and palea smooth, rather scabrous or pilose but not diffusely covered with stiff bristles. Palea usually with 2 keeled nerves, occasionally reduced .... 23.

+ Lemma and palea diffusely covered with stiff bristles, changing into spinules. Palea with 3 nerves not keeled. Glumes reduced; lemma and palea of reduced lower florets replace them .... 1. **Leersia**.

23. Lower glumes less than half length of lemma, sometimes lacking .... 24.

+ Lower glumes as long or 1.5–2 times shorter than spikelets .... 25.

24. Spikelets usually with 2 or 3 florets, occasionally single-flowered. Glumes invariably present. Stamens 3 .... 40. **Catabrosa**.

+ Spikelets invariably single-flowered. Glumes often lacking or reduced. Stamens 1 or 2 .... 42. **Phippsia**.

25. Ligule usually glabrous or with very short (much shorter than ligule) cilia along margin, occasionally transformed into row of hairs but then lemma with fairly long awns .... 26.

+ Ligule transformed almost from base into row of dense hairs, longer than ligules; lemma awnless, rarely with short cusp. Inflorescence very dense, forming cylindrical, ovate or capitate, spicate panicles .... 65. **Crypsis.**

26. Callus of lemma glabrous or with short (not less than 1/4 shorter than flower) hairs (then, rachilla glabrous or absent) .... 27.

+ Callus of lemma densely pilose, hairs as long as floret or not more than 1/6 shorter (then, axis of spikelet extended above base of flower like a crinite rachilla) .... 21. **Calamagrostis.**

27. Glumes as well as lemma and palea free down to base, their margins not intergrown. Inflorescence of diverse shape, often cylindrical or spicate .... 28.

+ Glumes intergrown in lower part, margins of lemma and palea rather intergrown at base. Inflorescense a very dense spicate panicle .... 31. **Alopecurus.**

28. Lemma awnless or short-awned (awn up to 1.5 mm long) .... 29.

+ Lemma with awn longer than 1.8 mm .... 34.

29. Inflorescence spreading or rather dense, sometimes spicate panicles, without regular cylindrical shape. Glumes often unequal, shorter than spikelets .... 30.

+ Inflorescence very dense spicate panicles of regular cylindrical shape. Glumes similar, as long as spikelets .... 29. **Phleum.**

30. Branches of panicle rather scabrous due to spinules, sometimes with only isolated spinules, subglabrous .... 31.

+ Branches of panicle totally glabrous .... 41. **Paracolpodium.**

31. Florets with 3 stamens. Spikelets not laterally flattened, sometimes flattened; in latter case, panicles not spreading, their branches rather thick, obliquely ascending .... 32.

+ Florets with single stamen. Spikelets strongly laterally flattened (appear flat), rather deltoid. Panicles spreading, with slender branches, nutant .... 47. **Cinna.**

32. Spikelets nearly not laterally flattened. Lemma as long as lower glume or shorter .... 33.

+ Spikelets strongly laterally flattened. Lemma distinctly longer than glumes, especially lower ones. Panicles rather dense, often spicate .... 48. **Arctagrostis.**

33. Lemma leptodermatous, glabrous and lustrous, significantly different in texture from coriaceous-membranous glumes .... 20. **Milium.**

+	Lemma coriaceous-membranous, rather scabrous in upper part, not lustrous, almost similar in texture to glumes ... 23. **Agrostis**.
34 (28).	Awn subterminal on lemma .... 35.
+	Awn terminal on entire or bidentate tip of lemma .... 37.
35.	Spikelets not laterally flattened. Glumes not shedding in fruit, persisting on branches of panicles. Rachilla jointed above spikelet .... 36.
+	Spikelets strongly laterally flattened; glumes shedding in fruit. Rachilla not jointed .... 30. **Limnas**.
36.	Perennial. Awn of lemma usually geniculate, arising from middle or slightly below .... 23. **Agrostis**.
+	Annual. Awn of lemma erect, arising near apex ... 22. **Apera**.
37 (34).	Callus of lemma somewhat obtuse or subacute; awn up to 25 mm long, fairly incurved or erect; slightly convoluted in lower part .... 38.
+	Callus of lemma usually acuminate; awn longer than 35 mm, geniculate once or twice; highly convoluted in lower part .... 58. **Stipa**.
38.	Awns of lemma scabrous due to spinules ..... 56. **Achnatherum**.
+	Awns of lemma plumose .... 57. **Ptilagrostis**.
39 (20)	Spikelets in inflorescence rather similar, few sometimes underdeveloped .... 40.
+	Spikelets in second spicate inflorescence, of 2 types: fruiting ones usual in shape; barren ones surrounding them aggregate, resemble a ridge .... 45. **Cynosurus**.
40.	Lemma with 1 (3) glabrous or scabrous awn(s) or awnless .... 41.
+	Lemma with 9 long terminal or subterminal awns, plumose in lower part, scabrous in upper part .... 61. **Enneapogon**.
41.	Callus of lemma short, glabrous or with relatively short hairs surrounding base. Plants of different shapes, terrestrial as well as coastal-aquatic .... 42.
+	Callus of lemma highly elongated, covered with long lustrous hairs enveloping lemma and palea. Coastal-aquatic plant with long thick rhizome .... 59. **Phragmites**.
42.	Spikelets with 2–3 florets, only uppermost bisexual and fertile; lower ones staminate or reduced (in latter case, bisexual floret with additional small bracts (lemma and palea) of reduced florets at base in addition to glumes) .... 43.
+	Spikelets with 2 or more bisexual florets, very rarely with only single developed floret; in latter case, rachilla extended and bearing bracts (lemma and palea) reduced florets at tip with 2 (1) glumes at base .... 48.
43.	Spikelet usually with 1 or 2 fully developed florets, greenish-

yellow .... 44.

+ Spikelet with three fully developed florets: upper bisexual, 2 lower ones staminate. Glumes membranous, semitransparent; lemma and palea coriaceous, yellowish-brown, lustrous .... 24. **Hierochloë**.

44. Developed florets with 3 stamens .... 45.

+ Developed bisexual floret 1, upper with 2 stamens; 2 lower reduced and represented only by lemma with geniculate awn .... 25. **Anthoxanthum**.

45. Glumes as long as spikelet and exceeding lemma and palea in spikelet. Lemma and palea of fertile floret leptodermatous, white or yellowish, rather pilose, lustrous .... 46.

+ Lower glume invariably shorter than spikelet .... 47.

46. Perennials, with long decumbent subsurface shoots. Keels of glumes wingless or narrow-winged .... 26. **Phalaroides**.

+ Annuals, without decumbent subsurface shoots. Keels of lemma broad-winged .... 27. **Phalaris**.

47. Lemma of fertile floret scabrous, coriaceous, its callus with small tufts of short hairs along sides. Perennials, with long decumbent subsurface shoots .... 66. **Arundinella**.

+ Lemma of fertile floret glabrous, lustrous, cartilaginous, its callus glabrous. Annuals, with fibrous roots .... 67. **Panicum**.

48. (42). Ovary and tip of caryopsis densely covered with short hairs .... 49.

+ Ovary and caryopsis glabrous, rarely (in some species of *Festuca*) with several hairs .... 55.

49. Annuals, without long-persisting shortened vegetative shoots .... 50.

+ Perennials, with long rhizome or densely caespitose, with long-persisting shortened vegetative shoots .... 51.

50. Sheath closed almost throughout length. Glumes with 1–5 (7) nerves; upper glume considerably shorter than spikelet .... 12. **Bromus**.

+ Sheath laciniate almost to base. Glumes with (3) 5–9 (11) nerves; upper glume usually as long as spikelet .... 13. **Avena**.

51. Awns of lemma geniculate, convolute in lower part .... 52.

+ Awns of lemma erect or somewhat curved, terminal or sub-terminal; sometimes absent .... 53.

52. Leaf blades flat, linear, with highly developed whitish sclerenchymatous tissue along margin and midrib .... 14. **Avenula**.

+ Leaf blades setaceous along margin, without whitish sclerenchymatous tissue .... 15. **Helictotrichon**.

53. Sheath of cauline leaves closed for not more than half length

from base .... 54.
+ Sheath of cauline leaves closed almost throughout length. Spikelets 12–30 mm long, with 3–10 florets, awnless or short-awned .... 11. **Bromopsis.**
54. Lemma with 3–5 nerves keeled or not, subobtuse, acute or with awn but without teeth, terminating in awn ..... 33. **Festuca.**
+ Lemma with 5–7 nerves, without keel, with 3 teeth, terminating in short cusps. Large coastal-aquatic plant with long decumbent subsurface shoots .... 32. **Scolochloa.**
55 (48). Lemma with awn diverging distinctly below whole or bidentate tip .... 56.
+ Lemma awnless or awned, emerging from whole, occasionally bidentate tip .... 59.
56. Lemma with (3) 5 poorly distinct nerves. Sheath closed for less than third of length from base .... 57.
+ Lemma with 7–9 distinct nerves, without keel, bidentate at tip, with erect awn emerging between tooth bases. Sheaths of all leaves closed almost throughout length ...... 52. **Schizachne.**
57. Lemma without keel, awn usually erect, emerging near middle. Panicle branches glabrous, rather scabrous or smooth. Leaf blades invariably glabrous .... 58.
+ Lemma keeled, awn rather curved, emerging from upper third of back. Panicle branches densely pilose or scabrous. Leaf blades usually pilose .... 16. **Trisetum.**
58. Leaf blades flat or longitudinally convolute; deep longitudinal furrows and prominent scabrous nerves on upper surface. Spikelets with 2–3 distant florets. Awns erect or slightly curved .... 18. **Deschampsia.**
+ Leaf blades longitudinally folded bristlelike, covered on upper surface with extremely tiny papillae, without distinct nerves. Spikelets with 2 proximate florets. Awns geniculate, exceeding spikelet .... 19. **Lerchenfeldia.**
59 (55). Ligule represented by rows of dense hairs. Sheaths of all leaves laciniated almost to base .... 60.
+ Ligule scarious, glabrous or with short cilia along margin; leaf sheaths closed almost throughout length .... 63.
60. Axils of few upper cauline leaves without cleistogamous florets. Lemma awnless .... 61.
+ Axils of upper cauline leaves of shortened branches with cleistogamous spikelets under sheaths. Lemma awned, occasionally awnless. Caespitose perennials with or without short decumbent shoots .... 62. **Cleistogenes.**
61. Leema with 3–5 nerves. Surface shoots, usually erect, occasionally ascending. Inflorescence not secured. Spikelets 3–14

mm long .... 62.

+ Lemma with 7–9 nerves. Perennials with procumbent surface shoots. Inflorescence secund, very dense panicles comprising several alternating spicate branches. Spikelets 2.2–5 mm long .... 60. **Aeluropus**.

62. Perennials. Stems with tuberous thickened lower internode. Lemma 3–5 mm long. Spikelets with few (2–4) florets .... 54. **Molinia**.

+ Annuals. Stems with distant nodes; internode not thickened. Lemma 1.5–3 mm long. Spikelets with many (3–20 to 40) florets .... 64. **Eragrostis**.

63 (59). Lemma lanceolate, truncate at tip, not cordate at base ... 64.

+ Lemma broadly ovate (up to suborbicular), obtuse at tip, cordate at base .... 46. **Briza**.

64. Annuals, easily removable with roots, without shortened vegetative shoots .... 65.

+ Perennials, usually with shortened vegetative shoots forming mats or with long decumbent subsurface shoots .... 66.

65. Lemma lanceolate-ovate or ovate, pilose near base along nerves. Panicle branches glabrous or scabrous .... 35. **Poa**.

+ Lemma lanceolate, glabrous. Panicle branches scabrous .... 36. **Eremopoa**.

66. Panicles spreading or compressed, usually spicate but not secund; spikelets not aggregated into dense clusters .... 67.

+ Panicles secund, with rather thick branches bearing large clusters of closely aggregated spikelets. All glumes keeled .... 44. **Dactylis**.

67. Panicle branches rather scabrous or glabrous, rarely (in some species of *Festuca*) rather pilose but then panicles not spicate. Palea of same texture as lemma .... 68.

+ Branches of dense spicate panicles densely covered with very short hairs. Palea membranous, distinctly different in texture from coriaceous-membranous lemma .... 17. **Koeleria**.

68. Lemma with well-developed keel throughout length .... 69.

+ Back of lemma rounded or with very weak keel, distinct only in upper part .... 72.

69. Tip of ovary glabrous. Lemma without cusp; very rarely with short cusp at tip (*Hyalopoa*) .... 70.

+ Tip of ovary rather pilose, sometimes with sparse hairs. Tip of lemma with short awn or cusp .... 33. **Festuca**.

70. Palea without spinules along keels but usually puberulent. Panicle branches glabrous .... 71.

+ Palea with spinules (sometimes only few) throughout length or only in upper part of keel; rarely without spinules but then

plants annual or biennial. Panicle branches rather scabrous or
glabrous .... 35. **Poa**.
71. Spikelets 1.5–4 mm long, with (1) 2 (3) florets. Lemma up to
3.5 mm long, its callus glabrous. Coastal-aquatic plant without decumbent subsurface shoots but with procumbent sur-
face or submerged shoots .... 40. **Catabrosa**.
+ Spikelets 5–10 mm long, with 2–4 florets. Lemma 3.5–5 mm
long, its callus with several flexuous hairs. Surface plants with
decumbent subsurface shoots .... 37. **Hyalopoa**.
72 (68). Lemma scarious for most part, lustrous, glabrous (rarely, rather
pilose), with corona of fairly stiff hairs on callus, sometimes
with only few such hairs (in *Arctophila*) .... 73.
+ Lemma not lustrous, not scarious or with narrow scarious
margin, its callus glabrous .... 74.
73. Glumes rather obtuse, considerably shorter than spikelet and
adjoining lemma. Spikelets with 2–6 developed florets ....
.... 38. **Arctophila**.
+ Glumes subacute, more or less as long as spikelet and longer
than adjoining lemma. Spikelets with 1 (2) developed florets,
rest underdeveloped .... 39. **Dupontia**.
74. Sheaths of all leaves closed almost throughout length. Lemma
with many (7–9, upto 13) nerves, awnless .... 75.
+ Sheaths of cauline leaves closed for not more than half length.
Lemma with 3–5 (7) nerves, awned or awnless .... 76.
75. Glumes with single nerve. Spkiletes with 3–15 (20) fully developed and 1 or 2 upper underdeveloped florets. Plants of
coastal-aquatic or humid habitats .... 50. **Glyceria**.
+ Glumes with 3–7 (9) nerves. Spikelets with 1–3 (5) developed
florets; upper underdeveloped florets in spikelet form whitish
clavate lump. Plants of rather arid habitats .... 53. **Melica**.
76. Lemma without keel or weakly keeled, usually rather pilose
near base, occasionally glabrous, obtuse or subacute at tip but
invariably without cusp or awn .... 43. **Puccinellia**.
+ Lemma keeled, glabrous, but rather scabrous or pilose throughout surface, acute at tip, usually with cusp or awn ....
.... 33. **Festuca**.

1. Leersia Sw.

1. **L. oryzoides** (L.) Sw. 1788, Nov. Gen. Sp. Pl.: 21.
Fairly large rhizomatous perennial with stems rooting at lower nodes.
Leaves 6–10 mm broad, broadly linear, scabrous along margin and midrib.
Ligules short (0.7–2 mm) and obtuse. Panicles lax, with scabrous flexuous
branches. Spikelets 5–6 mm long, strongly compressed, on very short

stalk. Glumes reduced. Lemma on back with stiff bristles and, along keel, with cristate-ciliate hairs.

Wet places along banks of brooks. **West. Sib.**: AL—Ba (Novo-Belokurikha village). —Europe, Caucasus, Asia, North America. Described from North America (Virginia state).

2. Zizania L.

1. **Z. latifolia** (Griseb.) Stapf 1909 in Kew Bull.: 385.

Very large (up to 3 m tall) perennial with thick decumbent subsurface shoots and glabrous stems rooting at lower nodes. Leaves broadly linear, 10–25 mm broad, ligules 7–15 mm long. Panicles up to 55 cm long, branches with diffuse spinules; staminate spikelets concentrated in lower part of panicle and pistillate in upper. Lemma of staminate floret with 5 nerves and short (2–3 mm) awn; in pistillate floret, awn almost as long as spikelet.

Banks of slow rivers and meanders. **East. Sib.**: ChI—Shi (Boldurui II and Argunsk villages). —East. Asia, introduced in European USSR. Described from Dauria (between Shilka and Argun').

3. Brachypodium Beauv.

1. Plant with long rhizome not forming mat. Lemma of all florets with terminal awn not longer than 6 mm .... 1. *B. pinnatum*.
+ Plant cespitose or with very short rhizome. Lemma of central and upper florets in spikelet with long awn (7–12 mm), almost as long as lemma .... 2. *B. sylvaticum*.

1. **B. pinnatum** (L.) Beauv. 1812, Ess. Agrost.: 101, 155.

Stems erect, fairly stout. Leaves rather short, flat, quite frequently light or yellowish-green, scabrous on lower surface, with diffuse long hairs along veins on upper surface. Spikes erect, comprise 7–15 fairly large (up to 3 cm long) spikelets. Glumes unequal: lower with 3 nerves, narrowly lanceolate, upper with 7 nerves, 2–3 mm longer than lower. Lemma usually pilose throughout surface; hairs on back of lemma very short (sometimes absent), longer on sides. Anthers 3–4.5 mm long.

Grasslands, coniferous as well as deciduous forests, forest glades, exposed slopes, dry meadows and scrubs. **West. Sib.**: TYu—Tb, KU, OM, TO, NO, KE, AL—Ba, Go. **Cen. Sib.**: KR—Tn, Kha, Ve, TU, **East. Sib.**: IR—An, Pr, BU—Se, Yuzh, ChI—Shi (Vozdvizhenka village, Argunsk and Yamarovka settlements), YAK—Vi (Peledui river). —Europe, Caucasus, Asia. Described from Europe. Map 1.

2. **B. sylvaticum** (Hudson) Beauv. 1812, Ess. Agrost.: 101, 155.

Stems relatively weak. Leaves long, flat, usually dark green, scabrous on both surfaces; sometimes with additional sparse long hairs on upper surface. Spikes inclined or nutant with 2–7–10 spikelets 2–2.5 cm long. Glumes acute, upper 2–3 mm longer than lower, short-awned. Back of lemma glabrous, covered with more or less dense, short, subulate hairs on sides. Anthers 2–3 mm long.

Humid, birch, asp, fir and asp-fir tall grass forests. **West. Sib.**: NO (Maskalikha area in Iskitimsk region), KE, AL—Ba, Go, **Cen. Sib.**: KR—Ve (Amyl river near Tuva estuary, Mozhar Lake). **East. Sib.**: BU—Yuzh (Vydrino station). —Europe, Caucasus, Asia. Described from Great Britain. Map 2.

4. Elymus L.

1. Spikes with spikelets in groups of 2 or 3 along margin throughout length or only in middle part .... 2.
+ Spikes with spikelets occurring singly throughout length; very rarely, in some species, spikelets in groups of 2 along margin of spike in lower part .... 4.
2. Spikes stout, erect. Glumes almost as long as lower floret in spikelet; nerves 3–7 .... 3.
+ Spikes weak, nutant. Glumes 2 or 3 times shorter than lower floret in spikelet; nerves 1–3 .... 23. *E. sibiricus*.
3. Awns of lemma erect, appressed. Glumes almost as long as adjacent lemma, gradually narrowing into 2–7 (10) mm long awn at tip. Awns of glumes and adjacent lemma arising at same level .... 3. *E. dahuricus*.
+ Awns of lemma laterally rather steeply recurved. Glumes distinctly shorter (by 1–3 mm) than adjacent lemma, sharply narrowed into short up to 3 mm long cusp or awn. Awn of lemma arising at level of awn tip of adjacent glumes ...... 4. *E. excelsus*.
4 (2). Awns of lemma longer (more than 15 mm), laterally steeply recurved .... 5.
+ Awns of lemma very short, usually not more than 15 mm, erect or slightly flexuous, ascending, or lacking .... 10.
5. Spikes dense, erect (sometimes slightly inclined at tip), with stout ascending rachis. Glumes nearly as long as lower floret or less than 1.5 times shorter .... 6.
+ Spikes lax, weak, with highly flexuous or nutant rachis. Glumes 1.5–2 times shorter than lower floret .... 7.
6. Glumes nearly as long as lower flower, broadly lanceolate. Palea with crowded short spinules along keels ....
.... 5. *E. fedtschenkoi*.
+ Glumes distinctly shorter than lower floret (sometimes up to 1.5 times), narrowly lanceolate or lanceolate. Palea with long divergent stiff spines or cilia along keels .... 7. *E. gmelinii*.
7. Spikes short (5–12 cm long), dense, with approximate spikelets. Plant of high altitudes, often at moderate heights .... 8.

+   Spikes very long (10–20 cm), with loosely arranged spikelets. Tall plants of forest-mountain belt .... 9.

8.   Leaf blades with dense and short hairs on both surfaces .... 18. *E. pamiricus.*

+   Leaf blades glabrous on lower surface and with sparse, divergent hairs on upper along veins .... 22. *E. schrenkianus.*

9.   Rachilla subglabrous, with extremely short spinules. Lemma glabrous, somewhat longer than palea. Anthers 2–3 mm long .... 2. *E. confusus.*

+   Rachilla puberulent. Lemma scabrous or pilose on back, nearly as long as palea. Anthers 1–1.5 (1.8) mm long .... 20. *E. pubiflorus.*

10 (4).   Rachilla segments subglabrous, covered with extremely short, rather appressed bristles, visible only under high magnification .... 11.

+   Rachilla segments covered with more or less divergent hairs throughout surface or only in uper part .... 12.

11.   Back of lemma glabrous .... 12. *E. kronokensis.*

+   Back of lemma scabrous due to short, diffuse spinules .... 29. *E. zejensis.*

12.   Lemma of all florets in spikelet awnless or with awns not longer than 6 mm .... 13.

+   Lemma of all or only uppermost florets in spikelet with erect awns longer than 6 mm .... 24.

13.   Glumes few, usually less than 1.5 times (not more than by 2.5 mm) shorter than adjacent lower floret, gradually narrowed into short cusp or awn .... 14.

+   Glumes 1.5–2 times (by 3–5 mm) shorter than lemma, usually abruptly acute, forming short cusp or awn .... 17.

14.   Glumes lanceolate, with 5–7 thick nerves, nearly as broad as spaces between nerves. Spikes relatively narrow, 4–6 mm thick .... 15.

+   Glumes broadly lanceolate, with (5) 7–9 slender nerves, much narrower than spaces between nerves. Spikes fairly thick (6–10 mm), sometimes branched at base .... 17. *E. nevskii.*

15.   Back of lemma more or less scabrous or pilose .... 16.

+   Back of lemma glabrous, sometimes with few spinules only in upper part .... 25. *E. trachycaulus* subsp. *novae-angliae.*

16.   Glumes more or less pilose within (especially at base). Spikelets not glaucescent. Anthers 1.5–2.5 mm long ... 16. *E. mutabilis.*

+   Glumes glabrous or scabrous within. Spikelets usually glaucescent. Anthers 1–1.8 mm long ..... 26. *E. transbaicalensis.*

17 (13).   Back of lemma glabrous, pilose on callus and near base along sides, with very short spinules near tip .... 18.

+ Back of lemma covered with diffuse or more or less dense hairs or elongated spinules .... 19.

18. Glumes usually with 3 nerves, pilose within, half as long as lower floret. Lemma, except for callus, glabrous, awnless .... 6. *E. fibrosus.*

+ Glumes usually with 3–5 nerves, glabrous within, 1.5 times shorter than lower floret. Base of lemma along sides more or less pilose, awnless or with up to 2 mm long awn .... 24. *E. subfibrosus.*

19. Glumes of all spikelets 1.5–2 times shorter than lower floret. Spikes greenish or gray-violet, with rather separated spikelets; lowermost rachis segments more than 7 mm long. Plants of river valleys in forest and forest-tundra zone .... 20.

+ Glumes of variable length; glumes in lower spikelets nearly equaling lower floret; up to 1.5–2 times shorter in upper spikelets. Spikes pink-violet, erect, fairly dense; lowermost rachis segments up to 7 mm long. Compactly caespitose, arctic, and bald-peak plant .... 22.

20. Leaf blade glabrous on upper surface. Glumes with narrow membranous margin .... 21.

+ Leaf blade diffusely pilose on upper surface. Glumes with fairly broad membranous margin .... 15. *E. macrourus* subsp. *turuchanensis.*

21. Stems glabrous under spikes .... 13. *E. macrourus* s. str.

+ Stems scabrous under spikes .... 14. *E. macrourus* subsp. *neplianus.*

22. Glumes and lemma dingy, usually with glaucescent bloom. Lower nodes of stems sometimes puberulent. Plants of tundra zone .... 23.

+ Glumes and lemma somewhat lustrous, without glaucescent bloom. Nodes of stems usually glabrous. Plants of hill tundra .... 21. *E. sajanensis.*

23. Leaf blades glabrous on both surfaces but somewhat scabrous due to short spinules .... 27. *E. vassiljevii.*

+ Leaf blades puberulent on both surfaces ... 8. *E. hyperarcticus.*

24 (12). Palea with tiny, crowded spinules along keels (more than 35 per keel). Spikes erect or slightly inclined .... 25.

+ Palea with fairly long sparse cilia along keels (less than 35 per keel). Spikes nutant .... 19. *E. pendulinus.*

25. Glumes gradually narrowed toward tip; along edge, with membranous fringe of uniform breadth or slightly broadened near center and narrowed toward tip. Spikelets on spike somewhat uniformly arranged on spike. .... 26.

+ Glumes abruptly narrowed toward tip, with membranous

14

fringe along margin, broader near tip (under awn). Lower spikelets 1.5–2 cm apart .... 10. *E. jacutensis.*

26. Glumes small, 1.5–2 times shorter than lower floret with 3–5 nerves .... 27.

+ Glumes nearly as long as lower floret or slightly shorter, with 5–7 (9) prominent nerves .... 29.

27. Lemma scabrous or pilose on back. Glumes usually glabrous within; very rarely puberulent .... 28.

+ Lemma glabrous, smooth on back. Glumes with short pubescence within, sometimes subglabrous .... 1. *E. caninus.*

28. Rachilla segments subglabrous in lower part, with extremely short appressed bristles; puberulent only under floret. Back of lemma slightly scabrous .... 29. *E. zejensis.*

+ Rachilla segments pilose throughout surface. Back of lemma puberulent .... 9. *E. ircutensis.*

29. Glumes broadly lanceolate, nearly as long as lower floret, with 5–7 (9) thick nerves, awned and frequently with additional tooth at tip; broadly white membranous along margin ....
.... 11. *E. komarovii.*

+ Glumes narrowly lanceolate, somewhat shorter than lower floret, with (3) 5–7 nerves, awned, without additional teeth, narrowly membranous along margin, weakly pilose within ......
.... 28. *E. viridiglumis.*

1. **E. caninus** (L.) 1755, Pl. Suec. ed. 2: 39.—*Agropyron caninum* (L.) Beauv. —*Roegneria canina* (L.) Nevski.

Plant with short rhizome. Leaves up to 1.5 cm broad, flat, with diffuse long hairs on upper surface; glabrous, rather scabrous on lower surface. Spikes relatively long, slightly nutant. Glumes with 3 (5) nerves scabrous due to spinules, with short pubescence within. Back of lemma glabrous, very rarely with isolated spinules in upper part, with erect awn as long as lemma or longer. Palea with crowded short spinules along nerves; scabrous in spaces between nerves.

Fir, spruce, aspen, and mixed forests, birch groves, forest glades, overflow meadows, riparian scrubs, near thermal springs. **West. Sib.**: TYU—Khm, Tb, KU (Chesnokovka village), OM, TO, NO, KE, AL—Ba, Go. **Cen. Sib.**: KR—Kha, Ve, TU. **East. Sib.**: IR—An, Pr, BU—Se, Yuzh, YAK—Vi (Peledui village). —Europe, Caucasus, Mid. Asia, Mediterranean, West. Asia, West. China (East. Tien Shan). Described from Europe. Map 4.

2. **E. confusus** (Roshev.) Tzvelev 1968 in Rast. Tsentr. Azii 4: 221. — *Agropyron confusum* Roshev.

Tall, laxly cespitose plant with narrow (3–6 to 9 mm broad) flat leaves, glabrous beneath, covered on upper surface with long slender hairs. Spikes nutant, lax, with flexuous rachis. Glumes usually unequal in length, 2 or 3 times shorter than lower floret (upper with 3–5, lower with 1–3 nerves), narrowly lanceolate, sharply narrowed at tip into cusp or short awn. Lemma lanceolate, glabrous or with isolated short spinules along sides, with long (2–3 cm) recurved awn at tip. Rachilla subglabrous, with extremely short spinules. Palea distinctly shorter than lemma, truncate at tip. Anthers (1.8) 2–2.5 (3) mm long. In East. Sayan (Tunkinsk mountain range), $2n = 28$.

Mountain-forest belt on open turf-covered meadow slopes, forest fringes, forests, sometimes forest roadsides, rock debris, sometimes coastal meadows. Common at 1000–2000 m elevations but descending much lower. **Cen. Sib.**: TU (Sangilen upland, Tsagan-Shibetu mountain range). **East. Sib.**: IR—An, BU—Se, Yuzh, ChI—Ka (Kodar mountain range), Shi. —Far East, Mongolia, NE China (Manchuria). Described from Forebaikal, probably from Khamar-Daban moutain range. Map 3.

3. **E. dahuricus** Turcz. ex Griseb. 1852 in Ledeb., Fl. Ross. 4: 331, emend. Peschkova. —*Clinelymus dahuricus* (Turcz. ex Griseb.) Nevski.

Cespitose perennial with erect, stout stems and flat leaves glabrous on upper surface. Spikes erect, dense, green or slightly violet. Glumes lanceolate, with 3–5 (7) scabrous nerves, with 3–7 (10) mm long awn at tip. Lemma almost always scabrous all over back, rarely glabrous, with long (up to 2.5 cm), erect (not recurved) scabrous awn. Rachilla with short spinules or hairs. Anthers 2–3 mm long.

Forest-steppe and steppe belt in dry meadows, meadow steppes, sometimes rocky steppe slopes, quite often fallow lands and roadsides. **East. Sib.**: ChI—Shi. —Far East (upper course of Amur river), NE. China (Manchuria), NE Mongolia. Described from Nerchinsk Dauria. Map 17.

4. **E. excelsus** Turcz. ex Griseb. 1852 in Ledeb., Fl. Ross. 4: 331, emend. Peschkova. —*E. dahuricus* auct. fl. sib. non Turcz. nec Griseb.

Cespitose plant. Leaves flat or convolute, usually with diffuse squarrose long hairs on upper surface; occasionally glabrous. Spikes erect or slightly curved, green or with violet tinge. Glumes linear-lanceolate, as long as lower floret, with 3 (5) scabrous nerves, gradually narrowing into short cusp or 1–3 mm long awn. Lemma with 10–20 mm long awn, laterally sharply recurved, glabrous near base, scabrous on back, spinules longer toward pilose tip. Rachilla with very short spinules, not pilose. Anthers 1.5–2.5 mm long. Altay (Katun Belki—snow-covered, flattened moutain summits in Siberia), $2n = 42$.

Forest-steppe and steppe belts, on sand and pebble beds along river valleys; adventitious in steppified forest land, forest meadows, steppe slopes, rocks, and debris. **West. Sib.**: AL—Go. **Cen. Sib.**: KR—Kha, Ve, TU. **East. Sib.**: IR—An, Pr (Kachug town, Makarova village), BU—Yuzh (Kharatsai village—class. hab.—and others), ChI—Shi (Tolbaga village on Khilok river).—Mid. Asia, West. Mongolia, West. China. Not east of Yablonov mountain range. Map 6.

5. **E. fedtschenkoi** Tzvelev 1973 in Novosti sist. vyssh. rast. 10: 21. —*Roegneria curvata* (Nevski) Nevski. —*Agropyron macrolepis* Drobov.

Compactly cespitose plant with convolute or flat leaves covered with dense and very short hairs on upper or both surfaces, sometimes mixed with much longer hairs. Spikes erect or slightly inclined, light green. Glumes broadly lanceolate, broadly white membranous along margin in upper half, sharply narrowing at tip, nearly as long as lower floret, with 5–7 nerves bearing very short and slender spinules. Lemma with rather short spinule-hairs on back and long (10–25 mm) slender scabrous recurved terminal awn. Rachilla pilose. Palea along keels shortly ciliate with short slender hairs in-between. Anthers 2.7 mm long.

Subalpine meadows and meadow slopes in upper mountain belt; valley forests. **West. Sib.**: AL—Go (Belyi Anui village, Dzhosator river, Tueryk gorge). —Mid. Asia, West. China (Junggar), Mongol. Altay, western Himalyas. Described from Nor. Tien Shan (northern slope of Ketmen mountains).

6. **E. fibrosus** (Schrenk) Tzvelev 1970 in Spisok rast. Gerb. fl. SSSR 18: 29. —*Agropyron fibrosum* (Schrenk) Candargy. —*Roegneria fibrosa* (Schrenk) Nevski.

Cespitose plant with glabrous and scabrous, more or less (3–7 to 10 mm) leaves. Spikes narrow, relatively not dense, inclined, usually violet. Glumes half as long as lemma, with 3 (5) nerves, fine pubescence within. Lemma glabrous (very rarely, slightly scabrous), awnless, callus crinite, rachilla pilose. Palea with short and dense spinules along keels, with sparse pubescence-between, especially along sides of keels. Anthers 1.2–1.8 mm long.

Exposed slopes of meadows in forests, valley meadows, sometimes in forests, on sand and along roadsides. **West. Sib.**: TYU—Yam (Polui river near Sukhoi Polui river estuary), Khm, Tb, KU (Borodino village), TO, NO (Karachi station), KE (Pod"yakovo village). **Cen. Sib.**: KR—Pu, Tn (Baikit village), Ve (Sutyaga village, Kezhma settlement). **East. Sib.**: IR—An, Pr.—Nor. Europe, Urals, Nor. Kazakhstan. Described from Kazakhstan (Karkaralinsk hills). Map 5.

7. **E. gmelinii** (Ledeb.) Tzvelev 1968 in Rast. Tsentr. Azii 4: 216. — *Agropyron turczaninovii* Drobov. —*Roegneria turczaninovii* (Drobov) Nevski.

Coarse, laxly cespitose perennial with stout stems. Leaves convolute or flat (4–10 mm broad); squarrose long hairs on upper surface (sometimes, on both surfaces), occasionally glabrous. Spikes erect, narrow, violet. Glumes with 5–7 coarse and stiff scabrous nerves, gradually acuminate, but awnless, distinctly shorter (sometimes almost 1.5 times) than lower floret. Lemma with stiff strong spinules, along sides, often even on back, with long (2–4 cm) terminal awn; awn strongly recurved in fruit, stiff and scabrous. Palea with strong and long, uniformly spaced spinules along keels; coarse and scabrous between keels, especially near tip. Rachilla covered with dense but short spinules. Anthers 2–3 mm long, yellow.

Mountain-steppe and forest-steppe belts in steppe and dry meadows, forest fringes, dry thin forests; forest belt, exposed southern rocky steppe slopes; adventitious sometimes on fallow lands. **West. Sib.**: TO, NO, KE, AL—Ba, Go. **Cen. Sib.**: KR—Kha, Ve, TU. **East. Sib.**: IR—An, Pr, BU—Se, Yuzh, ChI—Shi, YAK—Vi. Mid. Asia, Far East, Nor. Mongolia, China. Described from Altay hills. Map 7.

Hybrid *E. gmelinii* × *E. sibiricus* has been reported for the northwestern bank of Baikal (Muzhinai bay).

8. **E. hyperarcticus** (Polunin) Tzvelev 1972 in Novosti sist. vyssh. rast. 9: 61. —*Roegneria hyperarctica* (Polunin) Tzvelev. —*Elymus sajanensis* subsp. *hyperarcticus* (Polunin) Tzvelev.

Low (up to 30 cm tall) caespitose plant with stems puberulent on lower nodes. Leaves narrow (1–2 mm), with dense and short hairs on both surfaces. Spikes short, erect, greenish or with pink-violet tinge. Glumes 1.5–2 times shorter than lemma, with 3–5 nerves, relatively broad and abruptly acuminate in upper part, sometimes with 1 or 2 lateral teeth, rather pilose, especially along nerves. Lemma more or less pilose, with 1–5 mm long awn. Rachilla pilose.

Turf-covered slopes of knolls in Arctic tundra. **Cen. Sib.**: KR—Ta (Yenisey Lake on Gydansk peninsula, Pyasina river near Tareya river estuary). —Far East (Chukchi, Kamchatka), North America, Greenland. Described from Baffin Bay (American Arctic).

9. **E. ircutensis** Peschkova, sp. nova. —*Agropyron turczaninovii* auct. non Drobov: Popov in Spisok rast. Gerb. fl. SSSR, No. 4008. —*Elymus magadanensis* auct. non Khokhr: Peschkova, 1985, Novosti sist. vyssh. rast. 22: 42.

Gramen magnum (ad m et altius). Folia 3–8 mm lata, plana, glabra vel supra longe et patenter pilosa, subtus interdum breviter pilosa. Spica subdelflexa vel subnutans, angusta haud densa. Glumae lemmate accumbente sesquiduplo breviores, nervis 3–5 plus minusve scabris in mucronem et aristam brevem sensim angustatae, margine angusto membranaceo aequilato vel ad medium latissimo cinctae, interdum dente accessorio sub apice praediatae. Lemma dorso breviter pilosum, apice aristatum, arista lemmati subaequilonga. Rachilla breviter pilosa. Antherae 1.8–2.2 mm longae.

Typus: Burjatia, districtus Tunkinskij, pagum Mondy, laricetum harum in valle fl. Irkut, 12 IX 1966, no. 2655. G. Peschkova (NSK).

Affinitas. Species *E. jacutensi* et *E. komarovii* affinis, sed a priore glumis sensim acutatis, margine angusto membranaceo aequilato vel medio latissimo (nec glumis ad apicem subito angustatis margine membranaceo superne latissimo cinatis), a posteriore vero glumis lemmate accumbente sesqui-duplo brevioribus, nervis 3–5 (nec 5–7) percursis differt.

Large (up to 1 m tall or more) plant with flat, 3–8 mm broad leaves glabrous or covered on upper surface with long, soft, separated hairs. Spikes slightly inclined or weakly nutant, not dense. Glumes 1.5–2 times shorter than lower floret, with 3–5 scabrous nerves, gradually narrowing into cusp of awn at tip; along margin, with narrow membranous fringe of uniform breadth or broader near center; sometimes with additional tooth at tip. Lemma pilose on back, with long terminal awn, nearly as long as lemma or shorter. Rachilla puberulent. Anthers 1.8–2.2 mm long.

Mixed, birch and larch forests, forest meadows, glades, borders, sometimes forest roadsides. **West. Sib.**: KE (between Ailskoe village and Taradanovo; Mal. Elan' village on Kondoma river), AL—Go. **Cen. Sib.**: KR—Ve (Mal. Kamalinskoe village, Kansk region, Amyl river below Bes' river estuary), TU. **East. Sib.**: IR—An, Pr (Tutura village, Orlinga river upper course; Chikan river near Bushulka brook, Zhigalovsk region), BU—Se, Yuzh (Mondy village, Tunkinsk region—class. hab., Turuntaeva settlement), YAK—Al. —Endemic. Map 10.

10. **E. jacutensis** (Drobov) Tzvelev 1972 in Novosti sist. vyssh. rast 9: 61. —*Agropyron jacutense* Drobov. —*Agropyron pubescens* (Trin.) Schischkin. —*Roegneria jacutensis* (Drobov) Nevski.

Caespitose plant with narrow (2–4 mm) leaves glabrous on both surfaces, sometimes with distant hairs on upper surface. Spikes fairly long, scattered, erect, with separated lower spikelets. Glumes 1.5–2 times shorter than lower floret, usually with 3 (5) nerves, glabrous or rather scabrous along nerves, glabrous within. Lemma pilose throughout length or only in lower part; awn terminal nearly as long as or longer than lemma. Rachilla covered with long hairs. Altay (Katun' river and Ust'-Sema settlement), $2n = 28$.

Sand-pebble debris along river valleys and lake banks, occasionally in tundra belt or zone along rocky slopes and valley meadows. **West. Sib.**: AL—Ba (Srostki village, Biisk region), Go. **Cen. Sib.**: KR—Pu, Tn, Ve. **East. Sib.**: IR—An (environs of Nizhneudinsk), Pr, BU—Se (along Bombuika and Vitimakan rivers), YAK—Ol, Vi (Bilyuchai river—class. hab.), Al, Yan. —Northern Far East, Alaska, NW Canada. Map 9.

11. **E. komarovii** (Nevski) Tzvelev 1968 Rast. Tsentr. Azii 4: 216. — *Agropyron komarovii* Nevski. —*Elymus uralensis* subsp. *komarovii* (Nevski) Tzvelev.

Plant with short rhizome. Leaves up to 1 cm broad, flat, with long, soft, sparse hairs (sometimes without them), on upper surface along veins. Spikes fairly dense, green, nearly one-sided, erect or inclined. Glumes with broad white membranous margin, slightly asymmetrical, with 5–7 (9) scabrous nerves, steeply acuminate forming short cusp. Lemma lanceolate, rather pilose on back, with erect awn as long as or longer than lemma. Palea short-ciliate along keels.

Birch, asp, larch and Siberian pine-larch forests, forest glades, exposed meadow slopes, forest and floodplain meadows, scrubs; ascends pebble beds up to high altitudes. **West. Sib.**: AL—Go. **Cen. Sib.**: KR—Kha, TU. **East. Sib.**: IR—An, Pr, BU—Yuzh (Khorok river—class. hab.—and others). —Mid. Asia, Mongolia, West, China (Junggar). Map 8.

12. **E. kronokensis** (Kom.) Tzvelev 1968 in Rast. Tsentr. Azii 4: 216. —*Roegneria. borealis* (Turcz.) Nevski. —*R. scandica* Nevski. —*R. kronokensis* (Kom.) Tzvelev. —*Agropyron boreale* (Turcz.) Drobov. —*A. lenense* M. Popov.

Caespitose plant with narrow (up to 6 mm) leaves, glabrous on both surfaces (var. *scandica* (Nevski) Tzvelev), pilose only on upper surface (var. *borealis* (Turcz.) Tzvelev) or on both surfaces (var. *kronokensis*). Spikes erect or slightly inclined, at first green, turning violet later, with crowded spikelets. Glumes shorter than lemma, sometimes by 1.5 times, glabrous, scabrous only along nerves, sharply narrowed into short cusp or 1–3 mm long awn, broadly scarious along margin, especially under awn. Lemma glabrous and smooth, awnless or with 3–5 (up to 10) mm long awn. Rachilla bearing extremely short spinules. Putoran plateau, $2n = 28$.

Sand and pebble beds along river and lake valleys, exposed rocky slopes, mounds, floodplain meadows, riverbed forests, scrubs of forest and tundra belts in northern Siberia and bald-peak belt of southern Siberia. **West. Sib.**: TYU—Yam, AL—Go (Saratan village on Bashkaus river, Kurkura river, crossing of Koksu and Kara-Alakha rivers). **Cen. Sib.**: KR—Pu, Tn, Ve (Kezhma settlement). **East. Sib.**: IR—An (Ilimsk town, source of Khongorok river on Udinsk mountian range), Pr (Erbogachen and Nakanno settlements), BU—Se, ChI—Ka, YAK—Ar, Ol, Vi, Al, Yan, Ko. —Arctic and Northern Europe, Urals, Mid. Asia (Junggar Alatau), Far East, Mongolia, North America. Described from Kamchatka (Kronotsk pass). Map 11.

13. **E. macrourus** (Turcz.) Tzvelev s. str. 1970 in Spisok rast. Gerb. fl. SSSR 18: 30. —*Agropyron macrourum* (Turcz.) Drobov.—*Roegneria macroura* (Turcz.) Nevski. —*Agropyron nomokonovii* M. Popov.

Caespitose, rarely rhizomatous plant up to 90 cm tall. Leaves narrow, flat, glabrous or crinite on upper surface. Spikes narrow, erect, long (up to 30 cm), with widely separated lower spikelets. Glumes 1.5–2 times shorter than lemma, with 3 or 4 scabrous nerves, acuminate or with short awn. Lemma more or less densely pilose on back, especially in lower part, awnless or with short (up to 2–3 mm) awn. Rachilla pilose. Yakutia (Shandrin river, Ulakhan-Tas mountain range), $2n = 28$.

Sand banks of rivers, valley meadows and Chosenia-poplar groves, sometimes on rubble, turf-covered slopes of forest zone. **West. Sib.**: AL—Go (Katun' river). **Cen. Sib.**: KR—Pu. **East. Sib.**: IR—Pr, BU—Se (Verkhn. Angara river—class. hab.—and others), YAK—Ar, AL, Yan, Ko.—Arctic zone of East. Europe, Far East (northeast), North America (Alaska and Yukon river basin). Map 12.

14. **E. macrourus** subsp. **neplianus** (V. Vassil.) Tzvelev 1973 in Novosti sist. vyssh. rast. 10: 25. —*Roegneria nepliana* V. Vassil.

Differs from type subspecies only in stems scabrous (not glabrous) below spike.

Rocky slopes and pebble beds in tundra and forest tundra zones. **East. Sib.**: YAK—Ar (lower course of Kolyma river), Ko. —Northern Far East. Described from Far East (Anyui mountain range).

15. **E. macrourus** subsp. **turuchanensis** (Reverd.) Tzvelev 1971 in Novosti sist. vyssh. rast. 8: 63. —*Agropyron turuchanense* Reverd. — *Roegneria turuchanensis* (Reverd.) Tzvelev.

Rather low (up to 40 cm) caespitose plant with short (up to 10 cm) and relatively dense spikelets. Glumes more or less pilose along nerves, especially in lower part, broadly membranous along margin, steeply narrowed into short cusp at tip.

Sandy slopes and sand-pebble banks of rivers in tundra and forest tundra zones. **West. Sib.**: TYU—Yam. **Cen. Sib.**: KR—Ta, Pu (Lapkaikha river—class. hab.—and others). **East. Sib.**: YAK—Ar, Ol, Yan, Ko. — Circumpolar Urals.

16. **E. mutabilis** (Drobov) Tzvelev 1968 in Rast. Tsentr. Azii 4: 217. —*Agropyron mutabile* Drobov. —*A. angustiglume* Nevski. —*Roegneria mutabilis* (Drobov) Hyl.

Cespitose plant. Leaves flat, fairly broad, scabrous, with long sparse hairs on upper surface (hairs very rarely absent). Spikes erect or slightly inclined, dense, rather elongated, green or sometimes violet. Glumes slightly shorter than adjoining lemma, with (3) 4 or 5 thick, light-colored scabrous nerves, broadly lanceolate, gradually narrowed into short cusp or awn, sometimes with tooth near tip, scarious along margin, somewhat pilose within. Lemma scabrous or puberulent, gradually narrowed into short cusp or awn up to 5–6 mm long. Palea with frequent short spinules or ciliate along keels. Putoran plateau and Lake Baikal, $2n = 28$.

Thin coniferous and deciduous forests, largely in herbaceous vegetation, forest meadows, borders, ascending into high hills along meadow valleys and pebble riverbeds in forest and subalpine meadows, near rock debris; confined to grasslands near snowline, rocky and pebble banks of rivers, meadows on riverbeds. **West. Sib.**: TYU—Yam, Khm, Tb, OM, TO, NO, KE, AL—Ba, Go. **Cen. Sib.**: KR—Pu, Tn, Kha, Ve, TU. **East. Sib.**: IR—An, Pr, BU—Se, Yuzh, ChI—Shi (Nerchinsk Zavod settlement), YAK—Vi (Dushekan village on Czona river—class. hab.—and others), Al.—Nor. Europe, Mid. Asia, northern part of Far East, West. China, Mongolia, North America (Cordilleras). Map. 13.

17. **E. nevski** Tzvelev 1970 in Spisok rast. Gerb. fl. SSSR 18: 27. — *Agropyron ugamicum* Drobov, non *Elymus ugamicus* Drobov. —*Roegneria ugamica* (Drobov) Nevski. —*Elymus dentatus* subsp. *ugamicus* (Drobov) Tzvelev.

Lax, cespitose plant with thick geniculate stems in lower nodes. Leaves flat, glabrous, smooth or slightly scabrous on both surfaces. Spikes erect, relatively broad, dense, green or grayish-violet. Glumes slightly shorter than lower florets, broadly lanceolate, with (5) 7–9 slender and finely scabrous nerves (space between them broader than nerves themselves), white membranous along margin, terminating in short cusp or awn (1–2 mm), often with teeth. Lemma scabrous or puberulent, with short (2–7 mm) awn. Palea with slender short spinules crowded along keels. Rachilla pilose. Anthers 1–1.5 mm long.

High-altitude belt (above 2,000 m) along rocky slopes and pebble beds. **West. Sib.**: AL—Go (Yazulu village on Chulyshman river, Dzhulu-Kul' Lake, Tueryk gorge on Kuraisk moutain range). —Mid Asia. Described from West. Tien Shan (Urgam river near Tashkent).

18. **E. pamiricus** Tzvelev 1960 in Bot. mat. (Leningrad) 20: 425.—*E. schrenkianus* subsp. *pamiricus* (Tzvelev) Tzvelev.

Closely related to *E. schrenkianus*, differing in dense leaves puberulent on both surfaces. Spikes, compared to those of *E. schrenkianus*, somewhat lax, nutant, with flexuous awn. Glumes somewhat longer than in *E. schrenkianus*, gradually narrowed into short cusp or 1–3 mm long awn. Lemma subglabrous, with poorly distinct and sparse spinules. Palea nearly as long as lemma; keels with short spinules recurved on one side. Rachilla with very short and dense hairs. Anthers 1.5–2 mm long, greenish-violet.

High mountains along rocky slopes and talus. **West. Sib.**: AL—Go (Ak-Alakha river near Karabulak river estuary, Bogoyash hill in Chulyshman mountain range, upper course of Buguzun river in Chikhachev mountain range). —Mid. Asia, Tibet , Mongolian Altay. Described from Pamir.

19. **E. pendulinus** (Nevski) Tzvelev 1968 in Rast. Tsentr. Azii 4: 217. —*E. brachypodioides* (Nevski) Peschkova.—*Roegneria pendulina* Nevski. —*R. brachypodioides* Nevski.—*Agropyron brachypodioides* (Nevski) Serg. —*Elymus yezoensis* (Honda) Worosch. non Honda, 1930.

Laxly cespitose plant with flat leaves glabrous on both surfaces; scabrous along veins. Spikes rather loose, sometimes with greatly separated lower spikelets, slightly inclined. Glumes with 3–5 scabrous nerves, shorter than adjoining lower floret, sometimes by 1.5 times. Lemma glabrous or with sparse semiappressed bristles on sides, sometimes pilose on back, with fairly long (1.5–2 times longer than glume) erect scabrous awn. Altay (Ust'-Sema settlement), $2n = 28$.

Birch-spruce valley forests, floodplain scrubs, pine groves, rocky slopes under elm bushes as well as cliffs and rocks. **West. Sib.**: AL—Go (Chemal village, Ust'-Sema settlement and along Katun' river). **Cen. Sib.**: KR—Kha (Sartykovo village), Ve (Sayanskoe and Potroshilovo villages, Veselen'kaya hill), TU (Cherbi settlement, Derzig community farm). **East.**

22

Sib.: BU—Yuzh, ChI—Shi. —Far East, Mongolia, East. Asia. Described from Amur. Map 19.

20. **E. pubiflorus** (Roshev.) Peschkova 1985 in Novosti sist. vyssh. rast. 22: 41. —*Agropyron confusum* var. *pubiflorum* Roshev. —*Elymus confusus* var. *pubiflorus* (Roshev.) Tzvelev.

Laxly caespitose plant with narrow flat leaves, glabrous beneath, scabrous or with squarrose hairs on upper surface. Spikes lax, pendant. Glumes usually 2 or 3 times shorter than lower floret, almost equal in length, 3-nerved, gradually narrowed into cusp or short awn. Lemma linear-lanceolate, pilose on back or scabrous due to short spinules, with long, strongly recurved terminal awn. Palea nearly as long as lemma, sometimes barely shorter or longer. Rachilla covered with short spinules or pilose. Anthers 1–1.5 (1.8) mm long.

Sand-pebble riverbeds, floodplain forests, ascending into bald-peak belt along river valleys. **Cen. Sib.**: KR—Pu, TU (Toora-Khem village, Todzhinsk region). **East. Sib.**: IR—An (Murino and Slyudyanka stations), Pr, BU—Se, Yuz (Khoitogol ulus—nomad tent village), ChI—Ka, Shi, YAK—Vi (environs of Yakutsk), Al (Timpton river between Georgievsk village and Kavykty river—class. hab.—and others), Yan, Ko. —Northern part of Far East. Map 14.

21. **E. sajanensis** (Nevski) Tzvelev 1972 in Novosti sist. vyssh. rast. 9: 61.—*Roegneria sajanensis* Nevski. —*Agropyron sajanense* (Nevski) Grubov.

Rather small caespitose plant with stems foliated almost up to top. Leaves flat and glabrous on both surfaces, sometimes barely scabrous leaves like nodes. Spikes short and fairly compact, more or less violet. Glumes shorter (sometimes by 1.5–2 times) than lemma, with 3–5 rather pilose nerves, with broadened membranous margin at tip and sharply narrowed into short cusp. Lemma rather pilose (hair longer along nerves), aristate (awn 1–3 to 5 mm long). Rachilla covered with short spinules or bristles. Altay (Chui steppe), $2n = 28$.

Alpine mountains on dry rocky, usually carbonate slopes and debris, sometimes in sand and pebble riverbeds. **West Sib.**: AL—Go (upper course of Ak-Kol river on Ukok plateau). **Cen. Sib.**: KR—Ve, TU (Sol'bel'der river in Sangilen upland, Bura hill in Uyuksk mountain range). **East. Sib.**: IR—An, BU—Yuzh (Naryn-Daban area—class. hab. —and others). Endemic. Map 18.

22. **E. schrenkianus** (Fischer et Meyer) Tzvelev 1960, Bot. mat. (Leningrad 20: 428. —*Agropyron schrenkianum* (Fischer et Meyer) Candargy. —*Roegneria schrenkiana* (Fischer et Meyer) Nevski.

Low, laxly caespitose plant with narrow (3–5 mm) leaves, glabrous or slightly scabrous beneath, with diffuse divergent hairs on upper surface along veins. Spikes relatively short, compact, inclined. Spikelets on short stalks. Glumes 2 or 3 times shorter than lower floret, lanceolate, with 3 scabrous nerves, steeply acuminate into up to 10 mm long awn. Lemma rather scabrous on back due to short spinules, broadened in midsection and sharply narrowed into long (up to 3 cm) recurved awn. Palea slightly shorter than lemma, ciliate along keels. Rachilla crowded with short spinules (sometimes transformed into hairs). Anthers 1.5–2 mm long, greenish-violet.

Rocky slopes, debris, river sand and grassland in belt below bald peaks. **West. Sib.**: AL—Go. **Cen. Sib.**: TU (upper course of Balyktyg-Khem river and water divide of Barlyk and Khemchegeilik-Khem rivers). **East. Sib.**: BU—Yuzh (Uburt-Khongoldoi river, Il'chir Lake). —Mid. Asia, Western China, the Himalayas, Mongolian and Gobi Altay, Described from Tarbagatai.

23. **E. sibiricus** L. 1753, Sp. Pl.: 83. —*Clinelymus sibiricus* (L.) Nevski.

Cespitose plant with nutant spikes. Leaves 3–13 mm broad, flat, less often convolute, scabrous beneath, glabrous or covered with scattered long hairs on upper surface. Spikes rather lax, sometimes quite dense, with weak flexuous rachis. Spikelets, especially in midportion, paired on spike edge. Glumes 2 or 3 times shorter than floret, with (1) 3 (5) scabrous nerves, linear to lanceolate, and 1–10 mm long awn at tip. Lemma usually scabrous, occasionally with spinules only along nerves, gradually narrowed into long (10–25 mm), rather strongly recurved awn. Palea with short spinules along keels. Rachilla with very short spinules, sometimes subglabrous. Siberia, $2n = 28$ (Lake Baikal, Kurtushibinsk mountain range in West. Sayan, Kuraisk mountain range in Altay), 42 (environs of Kyzyl).

Floodplain and dry meadows, sand and pebble beds, forests and scrubs, forest glades, borders, sometimes as weed on fallow land, railroad embankments, farm boundaries and roadsides. **West. Sib.**: TYU—Tb, OM, TO, NO, KE, AL—Ba, Go. **Cen. Sib.**: Kr—Pu, Tn, Kha, Ve, TU. **East. Sib.**: IR—An, Pr, BU—Se, Yuzh, ChI—Ka, Shi, YAK—Vi, Al. — Europe, Urals, Far East, Alaska (only as escape west of Urals). Described from Siberia. Map 15.

24. **E. subfibrosus** (Tzvelev) Tzvelev 1970 in Spisok rast. Gerb. fl. SSSR 18: 30. —*Roegneria subfibrosa* Tzvelev. —*Elymus fibrosus* subsp. *subfibrosus* (Tzvelev) Tzvelev.

Cespitose perennial with stout stems and flat leaves (up to 1 cm broad), galbrous or diffusely pilose in upper part. Spikes fairly long, erect, with separated spikelets in lower part green or slightly violet. Spikelets large, 2–7-flowered. Glumes with 3–5 nerves, 1.5 times shorter than lower florets, glabrous, scabrous along nerves, acuminate or with short cusp. Lemma large (up to 18 mm long), glabrous, with a few rather long hairs only at base, sometimes scabrous on sides, acuminate or short (up to 2–6 mm) and aristate at tip. Northern Yakutia, $2n = 28$.

Sand-pebble beds along river valleys, floodplain meadows, occasionally on exposed dry slopes. **West. Sib.**: TYU—Yam (Tazovsk bay; Nadym village), KE (Ordzhonikidze settlement). **Cen. Sib.**: KR—Pu. **East. Sib.**: IR—Pr, YAK—Ar, Ol, Vi, Al (Balaganakh—class. hab.—and others), Yan, Ko. —Endemic. Map 16.

25. **E. trachycaulus** (Link) Gould ex Shinners subsp. **novae angliae** (Scribner) Tzvelev 1973 in Novosti sist. vyssh. rast. 10: 23. —*Agropyron peschkovae* M. Popov. —*Agropyron tenerum* auct. non Vasey.

Relatively tall (50–100 cm) cespitose plant with several nodes, grayish-green leaves flat or recurved along margin and scabrous on upper or both surfaces. Spikes narrow, long (up to 20 cm), rather lax, erect or slightly inclined. Glumes as long as lower floret, with 4–7 scabrous nerves, acuminate or with short cusp; glabrous within. Lower lemma glabrous and smooth; slightly scabrous near apex, acuminate or with short awn (up to 1.5 mm). Palea with short spinules along keels.

Cultivated as an edible plant, sometimes escapes and found on farm boundaries, fallow lands, roadsides, waste dumps near mines, very rarely on steppe slopes, steppified terraces above floodplains and in dry pine groves. **West. Sib.**: OM, NO, AL—Ba, Go. **Cen. Sib.**: KR—Kha, Ve, TU. **East. Sib.**: IR—An, BU—Yuzh, ChI—Shi. —North American species, introduced in many extratropical countries of both hemispheres. Described from North America (Vermont state, USA).

Type subspecies *trachycaulus* is distinguished by subglabrous (with extremely fine, short spinules) spikelet rachilla; sometimes grown in nurseries (reported from environs of Karachi Lake in Chanovsk region of Novosibirsk province). Not reported yet outside cultivated fields.

26. **E. transbaicalensis** (Nevski) Tzvelev 1968 in Rast. Tsentr. Azii 4: 219. —*Agropyron transbaicalense* Nevski. —*Roegneria transbaicalensis* (Nevski) Nevski. —*Agropyron pallidissimum* M. Popov. —*Roegneria buriatica* Sipl. —*E. mutabilis* subsp. *transbaicalensis* (Nevski) Tzvelev.

Highly similar to *E. mutabilis* but differing in glumes glabrous within (sometimes covered with extremely short spinules) and leaf blades usually glabrous on upper surface (sometimes rather pilose—var. *burjaticus* (Sipl.) Tzvelev). Pubescence of lemma ranges from abundant divergent to diffuse appressed spinules only along nerves. Awns of lemma short (1–3 mm). Spikelets often covered in grayish bloom. Eastern Siberia (Lake Baikal, Tunksnsk mountain range in Eastern Sayan), $2n = 28$.

Sand and pebble riverbeds, rock and sand, usually carbonate steppified slopes toward river valleys, in poplar, Chosenia and spruce forests in floodplains; ascends along river valleys into high hills (up to 2,000 m or more). **West. Sib.**: AL—Ba, Go. **Cen. Sib.**: KR—Pu (environs of Noril'sk, Khantaisk Lake, Loganchi hill), Tn, Ve, TU. **East. Sib.**: IR—An, Pr, BU—Se, (Verkhnyaya Angara river—class. hab.—and others), Yuzh, ChI—Ka (Syul'ban river), YAK—Vi, Al. —Northern Mongolia. Map 21.

27. **E. vassiljevii** Czer. 1981, Sosudistye Rasteniya SSSR: 315. — *Roegneria villosa* V. Vassil. —*Elymus sajanensis* subsp. *villosus* (V. Vassil.) Tzvelev.

Closely related to *E. sajanensis* and *E. hyperarcticus*. Differs from former in narrower leaf blades, puberulent lower nodes of stems, and much smaller anthers on average; from latter, in glabrous leaf blades scabrous only to some extent.

Exposed rocky slopes in arctic tundra zone, pebble and sand river-beds, sometimes on rocks. **Cen. Sib.:** KR—Ta. **East. Sib.:** YAK—Ar. — Arctic section of Far East. Described from Chukchi (Chegitun' river valley). Map 20.

28. **E. viridiglumis** (Nevski) Czer. 1981, Sosudistye Rasteniya SSSR: 351. —*Roegneria viridiglumis* Nevski. —*R. taigae* Nevski. —*Elymus uralensis* subsp. *viridiglumis* (Nevski) Tzvelev.

Large (up to 1 m or more tall) plant with coarse stems and broad flat leaves, scabrous and with long divergent hairs on upper surface. Spikes erect, thick, with crowded large violet spikelets. Glumes slightly shorter than lower floret, with (3) 5–7 thick nerves, with short pubescence within, gradually acuminate into awn, with practically no membranous margin on sides. Lemma on back with dense short spinules or coarsely scab-rous, long terminal awn, generally as long as lemma, slightly recurved on sides. Rachilla puberulent. Anthers about 2 mm long.

Mixed forests, glades and fringes. **West. Sib.:** TO (in Mariinsko-Chulymsk taiga, Bubeev area and Kozevinkoya village). —Urals. De-scribed from Southern Urals (Argayashsk canton, Bashkir ASSR).

29. **E. zejensis** Probat. 1984 in Bot. zhurn. 69 (2): 257.

Plant with short rhizome up to l m tall forming lax mats. Stems thin and slender. Leaf blades 2–5 mm broad, flat, with slightly convoluted margin and long, diffuse hairs on upper surface. Spikes about 10 cm long, erect or slightly curved, weak, loose. Glumes 1.5–2 times shorter than lower florets, narrowly lanceolate, with 3–5 scabrous nerves, acute or with short (up to 2 mm) terminal cusp, narrowly scarious along margin. Lemma on back with short diffuse hairs, denser on margin, with erect or slightly curved 5–10 mm long terminal awn. Palea with dense short spinules along keels. Anthers 2.2 mm long.

Black birch forests. **East. Sib.:** ChI—Shi (Vozdvizhenka village, Nerchinsko-Zavodsk region). —Far East. Described from Amur prov-ince (Novovysokoe village on Zeya river).

× Elyhordeum Mansf. ex Cziczin et Petrov

1. × **E. arachleicum** Peschkova sp. hybr. nov. (*Elymus sibiricus* L. × *Hordeum brevisubulatum* (Trin.) Link.

Species *E. sibirico* affinis. Spica nutans sed densior, spiculis ad excavationes racheos binis-ternis, subviolaceis. Glumae lineari-setiformes, longae, spiculae aequilongae vel eam superantes, rectae; lemmata aristis brevioribus haud reflexis. Antherae 1.5–2.2 mm longae.

Typus: Transbaicalia orientalis opp. Czita, lacus Arachlei, in prato 17 VII 1964, G. Peschkova, L. Turova (NS).

Outwardly similar to *E. sibiricus*. Spike nutant but denser, spikelets in groups of 2 or 3 along margin, slightly violet, glumes linear-setaceous, long as long or longer than spikelet, erect. Awns of lemma very short, not recurved. Anthers 1.5–2 mm long.

Type: Eastern Transbaikal, Chita town, Lake Arakhlei, meadow, July 17, 1964, G. Peschkova and L. Turova (NS).

2. × **E. chatangense** (Roshev.) Tzvelev (*Elymus macrourus* (Turcz.) Tzvelev × *Hordeum jubatum* L.). Quite common in northeastern Siberia. Reported from Khatanga, Aldan and Kolyma river basins.

3. × **E. kolymensis** Probat. (*Elymus pubiflorus* (Roshev) Peschkova × *Hordeum jubatum* L.).

Hybrid, possibly fertile, frequent in Kolyma river basin (environs of Verkhnekolym and others) as well as near Khandyg settlement in Tompo region.

4. × **E. pavlovii** (Nevski) Tzvelev (*E. mutabilis* (Drobov) Tzvelev s.l. × *Hordeum turkestanicum* Nevski). Found in Altay (near Bogoyash river estuary in Altay sanctuary).

5. Elytrigia Desv.

1. Lemma glabrous and smooth on back .... 2.
+ Lemma more or less pilose on back .... 6.
2. Lemma with long (8–20 mm) awn strongly laterally recurved ..... 3.
+ Lemma awnless or with short erect awn not longer than 5 mm
.... 4.
3. Upper surface of leaf puberulent or covered with very short spinules. Spikes with crowded spikelets, lower spikelets 0.7–1 cm apart. Palea glabrous, smooth along keels near base, with short crowded spinules above. Stocky plant, rarely taller than 50 cm ....
.... 2. *E. gmelinii*.
+ Upper surface of leaf covered with long squarrose hairs along with velutinous short pubescence, rarely long hairs absent, sometimes glabrous on both surfaces. Spikes usually scattered; lower spikelets 2–3.5 cm apart. Palea with crowded short spinules along keels, usually up to base. Large plant 80–100 cm tall .... 3. *E. jacutorum*.
4. Upper surface of leaf with greatly thickened veins projecting like thick ridges. Glumes (1.5) 2–3 times shorter than lower floret ..... 5.
+ Upper surface of leaf with slender veins, not prominent like thick ridges. Glumes usually as long as lower floret, or not more than 1.5 times shorter .... 6. *E. repens*.
5. Stems numerous, forming compact mat with tufts of radical leaves

at base. Leaf usually appressed to stem. Leaf blades more or less densely puberulent on upper surface; long hairs usually absent .... 1. *E. geniculata.*
+ Tall rhizomatous plant. Stems single. Radical leaves poorly developed, cauline leaves mostly laterally recurved. Leaf blades densely puberulent interspersed with long sparse hairs on upper surface; very rarely pubescence absent .... 5. *E. lolioides.*
5 (2). Lemma uniformly covered on back with disperesed, rather short appressed hairs; terminal erect awn up to 6 mm long ....
.... 4. *E. kaachemica.*
+ Lemma on back villous due to dense long hairs, usually glabrous near very tip, obtuse or with short (not more than 1.5 mm) cusp ....
.... 7. *E. villosa.*

1. **E. geniculata** (Trin.) Nevski 1936 in Tr. Bot. in-ta AN SSSR 1 (2): 82. —*Agropyron geniculatum* (Trin.) Koch. —*A. repens* var. *geniculatum* Krylov. —*Elytrigia geniculata* subsp. *geniculata.* —*Elymus bungeanus* (Trin.) Meld.

Glaucescent or glaucescent green plant with numerous, usually geniculate stems and abundant tufts of radical leaves forming fairly dense mat, sometimes with short decumbent shoots. Leaves convoluted or flat, narrow, densely covered with very short hairs on upper, and sometimes on lower surface as well; cauline leaves usually appressed to stem. Spikes slender, with interrupted lower spikelets. Glumes 1.5–2 times shorter than lower floret, obtuse or acuminate, sometimes with cusp, 3–5 thick nerves, broadly white scarious along margin. Lemma glabrous, acuminate or with cusp at apex, frequently with straight, not recurved, up to 5–6 mm long awn. Palea with short cilia along keels up to base. Anthers 4 or 5 mm long. Altay (Terektinsk mountain range), $2n = 28$.

Hill steppes, rocky southern slopes, debris, marble outcrops, sometimes ascending to 1,750 m above sea level along pebble scarps. **West. Sib.**: AL—Go (Charysh river—class. hab.—and others). **Cen. Sib.**: KR—Kha, Ve, TU. —Northern Mongolia. Map 25.

2. **E. gmelinii** (Trin.) Nevski 1936 in Tr. Bot. in-ta AN SSSR 1 (2): 78. —*Agropyron aegilopoides* Drobov. —*A. gmelinii* (Trin.) Candargy. —*A. propinquum* Nevski. —*A. roshevitzii* Nevski. —*A. strigosum* auct. non Boiss. —*Elytrigia strigosa* subsp. *aegilopoides* (Drobov) Tzvelev.

Not very tall, 20–70 (90) cm, cespitose, glaucescent or yellowish-green plant, frequently forming short decumbent shoots. Leaves narrow, convolute, occasionally flat, glabrous beneath, densely puberulent on upper surface. Spikes straight, slender, glaucescent-green or violet tinged, slightly scabours along nerves and awn. Glumes 1.5–3 times shorter than lower floret, slightly uneven, glabrous, with 3–5 (7) nerves, obtuse or acuminate, sometimes with short cusp. Lemma lanceolate or linear-lanceolate, glabrous, with 5 nerves, distinctly visible in upper part; midrib extended into scabrous, strongly recurved awn. Palea with short dense spinules along keels, especially in upper half; glabrous in lower part. Anthers 4–5 mm long, yellow or violet.

Hilly rocky-rubble steppe and semisteppe associations, rocks and precipitous, often carbonate, southern slopes, ascending to forest and bald-peak belts. **West. Sib.**: KE, AL—Ba, Go. **Cen. Sib.**: KR—Kha, Ve (environs of Minusinsk and Krasnoyarsk), TU. **East. Sib.**: BU—Yuzh (environs of Selenginsk—class. hab.—and others), ChI—Shi. —Mid. Asia, West. China (Junggar), Mongolia. Map 24.

3. **E. jacutorum** (Nevski) Nevski 1933 in Tr. Bot. in-ta AN SSSR 1 (1): 24. —*Agropyron jacutorum* Nevski. —*Elytrigia strigosa* subsp. *jacutorum* (Nevski) Tzvelev.

Tall, glaucescent cespitose plant. Leaves long, narrow, more often convolute, upper surface densely puberulent and, additionally, fairly densely covered with long interrupted hairs (hairs rarely absent). Spikes long (up to 20 cm), slender, very lax (lower spikelets 2–3.5 cm apart and upper 1–1.5 cm apart). Glumes slightly or up to 1.5 times shorter than lower floret, glabrous, with 5 or 6 nerves, slightly unequal, somewhat obtuse, sometimes with indistinct terminal teeth. Lemma up to 1 cm long, glabrous, with short pubescence only at bases, 5 nerves of which middle one extending into scabrous 1–2 cm long terminal awn, strongly laterally recurved. Palea as long as or slightly longer than lemma, covered with very short cilia or spinules along keels, up to base. Anthers yellow, about 4 mm long.

Dry, often carbonate southern slopes and clayey tall cliffs on river terraces. **East. Sib.**: IR—Pr, YAK—Vi, Al (environs of Amginsk village on Aldan upland—class. hab.—and others), Yan (Sobolookh settlement), Ko. —Far East. Map 22.

4. **E. kaachemica** Lomonosova et Krasnob. 1982 in Bot. zhurn. 67 (8): 1138.

Grayish-green plant 30–45 cm tall with numerous stems together with radical leaves forming small compact mats. Leaves flat, about 2 mm broad, glabrous and smooth on both surfaces, rarely slightly scabrous on upper surface, cauline leaves usually obliquely upturned. Spikes slender, sparse, erect or slightly declinate. Glumes 1.5–2 times shorter than lower floret, with 3 thick nerves, sharply narrowed into short cusp; broad scarious fringe along margin. Lemma with sparse and short semiappressed hairs throughout surface, gradually transformed into 4–6 mm long terminal awn. Palea with short spinules-cilia along keels disappearing in lower third. Anthers 1.5–2 mm long.

Rocky slopes and cliffs of rivers. **Cen. Sib.**: KR—Ve (Bol. Ur river estuary), TU (Bel'bei settlement—class. hab.). —Endemic.

5. **E. lolioides** (Kar. et Kir.) Nevski 1934 in Tr. Sredneaz. un-ta, ser. 8c, 16: 61. —*Agropyron lolioides* (Kar. et Kir.) Candargy. —*A. intermedium* var. *angustifolium* Krylov. —*Elymus lolioides* (Kar. et Kir.) Meld.

Plant with long rhizome, stems few (12) 30–90 cm tall; cauline leaves flat, convolute, standing subperpendicularly; radical leaves absent or poorly developed; cauline as well as radical leaves with dense short hairs on upper surface; hairs sparser and long along veins. Spikes lax, with interrupted spikelets, especially in lower part. Glumes 1.5–2 times shorter than lower floret, obtuse or subobtuse, sometimes acute, slightly keeled, with (3) 5–7 nerves, white scarious margin, usually glabrous, with long sparse cilia along midrib occasionally. Lemma glabrous, with 5 nerves, distinct near tip, midrib extended into thick, short cusp, barely longer than lemma. Palea with sparse, fairly long and rather thick spinules along keels. Anthers 4–5 mm long.

Flat, more often sandy steppes, fallow land and dry pine forests. **West. Sib.**: TYU—Tb (environs or Tobol'sk), KU (Ukrainets village, Zverinogolovsk region), NO, KE (environs of Kemerovo town, Lachinovsk kurya—long narrow oxbow detached from river at upper end—in Krapivinsk region), AL—Ba, Go. **Cen. Sib.**: KR—Kha, Ve.—European part of USSR and northern Kazakhstan. Described from Eastern Kazakhstan (Sukhaya Rechka river near Semipalatinsk). Map 26.

6. **E. repens** (L.) Nevski 1933 in Tr. Bot. in-ta AN SSSR 1 (1): 14. — *Agropyron repens* (L.) Beauv.

Fairly large plant with long rhizome. Leaves usually flat, rarely convolute, green or gray, glabrous or pilose on upper surface, with slender veins. Spikes erect, bilateral, with ciliate or pubescent rachis. Glumes almost as long as lemma, glabrous and smooth, with 5–7 weakly protruding nerves, acuminate or short-awned. Lemma glabrous, smooth, somewhat obtuse to aristate (with awn up to 5–8 mm long), obtuse, acute or aristate. Rachilla subglabrous, with very short appressed spinules. Anthers 3.5–5 mm long. Altay (Chui steppe), $2n = 28, 42$.

Steppes, meadows, fields, fallow land, sand-pebble beds of river valleys; descends into forest; scrubs. **West. Sib.**: TYU—Khm, Tb, KU, OM, NO, KE, AL—Ba, Go. **Cen. Sib.**: KR—Tn, Kha, Ve, TU. **East. Sib.**: IR—An, Pr, BU—Se, Yuzh, ChI—Ka, Shi, YAK—Ol, Vi, Al, Yan, Ko. — Eurasia, North America. Described from Europe.

7. **E. villosa** (Drobov) Tzvelev 1964 in Arkt. fl. SSSR 2: 247. — *Brachypodium villosum* Drobov. —*Agropyron karawaewii* P. Smirnov. — *Roegneria karawaewii* (P. Smirnov) Karav.

Glaucous plant with long rhizome, 30–85 cm tall, with few stems and few radical leaves. Leaves flat or convolute, up to 5 mm broad, with veins of unequal thickness, scabrous on upper surface due to short spinules or puberulent, sometimes interspersed with long diffuse hairs. Spikes bilateral, somewhat lax, with fairly large declinate spikelets.

Glumes lanceolate, 1.5 times shorter than lower floret, glabrous, with 3–5 somewhat thick nerves, white scarious margin. Lemma except for uppermost part, pubescent with long dense hairs, with 5 nerves distinctly visible only near glabrous tip, obtuse or with short cusp not longer than 1.5 mm. Palea with short, thick, sparse cilia along keels. Anthers 3.5–5 mm long.

On steppified southern slopes, partially fixed sands, pine groves and their fringes, sand steppes, along borders. **East. Sib.**: YAK—Ol (30 km away from Menkere river estuary), Vi (Sangar settlement, environs of Yakutsk), Al (Okhotsk ferry), Yan (Meginsk ulus—nomad village, Tulaginsk column—class. hab.—and others). Endemic. Map 23.

6. Agropyron Gaertner

1. Plant with long rhizome and decumbent subsurface shoots. Stems usually single. Radical leaves absent or very few ...... 2.
+ Plant cespitose or with short rhizome (sometimes with decumbent subsurface shoots). Stems usually few or several. Radical leaves present, frequently forming dense mats .... 4.
2. Spikes dense, largely cristate, with spikelets declinate from inflorescence rachis; spikelets proximate. Glumes densely pilose or ciliate along keel. Lemma with dense long hairs ... 3.
+ Spikes loose, not cristate; spikelets separated randomly and appressed to inflorescence rachis. Glumes glabrous or with cilia along keel. Lemma with sparse hairs of varying size ....
.... 11. *A. pumilum.*
3. Stem glabrous under spike. Spikes broad- or oblong-linear, indistinctly cristate. Upper surface of leaf blades densely puberulent (velutinous pubescence), sometimes interspersed with very long hairs .... 8. *A. michnoi.*
+ Stem sparsely pilose under spike. Spikes oblong-ovate, cristate. Upper surface of leaf blades diffusely scabrous or sparsely covered with fine spinules .... 9. *A. nataliae.*
4. Spikes ovate, oblong or broadly linear, (1.5) 2–6 cm long, 8–25 mm broad, cristate, with parallel spikelets, strongly declinate from inflorescence rachis .... 5.
+ Spikes linear, 2–7 cm long, 5–10(15) mm broad, subcylindrical, not cristate, sometimes indistinctly cristate with superposed spikelets, mostly appressed to inflorescence rachis .... 9.
5. Stem, rather densely pubescent with soft hairs under spike, more often throughout length, rarely glabrous; in latter case, with hairy joints .... 6.
+ Stem glabrous under spike, rarely scabrous or with short, diffuse, rigid hairs .... 7.
6. Sheath and blades of lowerst (outer) radical leaves densely

short pilose on both surfaces .... 5. *A. erickssonii.*

+ Leaf sheath and blade usually glabrous on lower (outer) surface, puberulent (subvelutinous) or glabrous, sometimes with long diffuse hairs on upper (inner) surface ..... 2. *A. cristatum.*

7. Radical leaves few and usually developing in autumn. Leaf blades flat (rarely convolute), scabrous on upper surface, sometimes with long isolated hairs .... 8.

+ Radical leaves usually developed, often very few, rigid, convolute, filiform, sometimes arcuate, aggregated into dense mats; upper (inner) surface of leaf blade densely pilose .... 7. *A. kazachstanicum.*

8. Spikes large, fairly broad, oblong-ovate, dense, without space between spikelets. Glumes and lemma dense and crinite, with up to 5–6 mm long terminal awn. Leaf blades subhorizontally recurved from stem, galbrous or diffusely puberulent on upper surface .... 4. *A. distichum.*

+ Spikes elongate- or ovate-oblong, slightly narrowed toward tip, diffuse, with distinct spaces between spikelets. Glumes and lemma glabrous or diffusely pilose, with 2–3 (4) mm long terminal awn. Leaf blades usually scabrous on upper surface, usually with interrupted diffuse long hairs, slightly declinate from stem .... 10. *A. pectinatum.*

9 (5). Lemma with (1) 2–4 mm long terminal awn .... 10.

+ Lemma obtuse, with short, thick, cusp or terminal, awn not longer than 1 mm .... 6. *A. fragile.*

10. Radical leaves numerous, convolute, rigid, liliform, fairly long, forming mat. Long vegetative shoots absent. Leaf blades densely puberulent on upper (inner) surface .... 3. *A. desertorum.*

+ Radical leaves poorly developed. Several shoots and oblong foliated vegetative shoots forming mat. Leaf blades flat or convolute; upper surface with scattered spinules interspersed with long isolated hairs .... 1. *A. angarense.*

1. **A. angarense** Peschkova 1984 in Bot. zhurn. 69 (8): 1088.

Cespitose plant. Stems 50–75 cm tall, numerous, glabrous or diffusely puberulent under spike. Radical leaves nearly absent but leafy vegetative shoots developed. Cauline leaves usually convolute, less often flat, appressed to stem. Leaf blades covered on upper (inner) surface with dense, short and very long divergent hairs, sometimes puberulent even on outer surface. Spikes 2.5–6 cm long, cylindrical, linear, not cristate, with superposed spikelets. Glumes glabrous or long-ciliate along keel, gradually acuminate into 2–4 mm long awn. Lemma mainly pilose, sometimes subglabrous, gradually narrowed into up to 4 mm long terminal awn. Palea with spinules of different size, along keels, thickened at base.

Exposed turf-covered carbonate slopes in southern taiga belt. **Cen. Sib.**: KR—Kha (Barezovka village, Altay region), Ve (Noshino village, Ust'yanskii settlement). **East. Sib.**: IR—An (environs of Ust'-Ilimsk—class. hab.—Danilova village on Nizh. Tungusk river). —Endemic.

2. **A. cristatum** (L.) Gaertner 1770 in Novi Comment. Sci. Petropol. 14 (1): 540.

Stems 20–80 cm tall, numerous, erect or slightly geniculate at base, pubescent in upper part with long, entangled, largely appressed hairs, particularly dense under spike. Radical leaves usually few and variable in length, sometimes absent; cauline leaves recurved from stem, densely puberulent on upper surface and, along veins, additionally with long, sparse, interrupted hairs, flat or convolute. Glumes and lemma rather pubescent (up to densely pilose). Spikes dense, 1.5–4 cm long, 1–2 cm broad. Rachilla covered with very short spinules. East. Siberia, $2n = 14$ (Irkana Lake in Nor. Baikal region), 28 (Tunkinsk moutain range in Eastern Sayan).

Steppes, quite often carbonate steppe slopes of forest belt, dry terraces in and above floodplains and steppified forests. **Cen. Sib.**: KR—Ve, TU (West. Sayan). **East. Sib.**: IR—An, Pr, BU—Se, Yuzh, ChI—Shi, YAK—Ar (Tit-Ary settlement), Vi, Al, Yan (environs of Verkhoyansk and Olekminsk). —Far East (introduced), East. Mongolia, NE China. Described from East. Siberia (exact location not indicated). Map 29.

3. **A. desertorum** (Fischer ex Link) Schultes 1824 in Schultes et Schultes fil., Mant. 2: 412. —*A. sibiricum* var. *desertorum* (Fischer ex Link) Boiss.

Glaucescent green plant, usually with numerous slender, glabrous stems or slightly scabrous under spike and convolute filiform leaves forming compact mats. Leaf blades glabrous and smooth on lower surface and densely puberulent on upper. Spikes short-linear, not cristate, narrow (usually not more than 1 cm broad), grayish-green. Glumes subovate, broadly scarious along margin, terminating in short awn, scabrous or ciliate along keel. Lemma glabrous or pilose, terminating in 2–2.5 (3) mm long awn. Palea with short thick spinules or long cilia along keels, falling off later.

Rocky slopes in steppes, desertified sand steppes, sometimes on sand dunes; enters arid pine groves. **West. Sib.**: KU (Lopatinsk and Kureinsk villages), OM (Elizavetinka village in Cherlaksk region and Maksimovka village in Shcherbakul'sk region; Poltavka settlement), AL—Ba. **Cen. Sib.**: TU.—SE European USSR, Caucasus, Mid. Asia, West. China, Mongolia (western). Described from Nor. Caucasus (Kuma river basin). Map 27.

4. **A. distichum** (Georgi) Peschkova 1985 in Novosti sist. vyssh. rast. 22: 37. —*Bromus distichus* Georgi. —*Agropyron cristatum* var. *macrantha* Roshev.—*A. cristatum* subsp. *baicalense* Egor. et Sipl.

Glaucescent green cespitose plant with several coarse, erect stems 40–110 cm tall, usually glabrous or slightly scabrous under spike.

Radical leaves absent or very few. Cauline leaves flat, up to 8 mm broad (rarely convolute), subhorizontally recurved from stem, scabrous or subglabrous on upper surface. Spikes large and broad, (1.5) 3–6 cm long and 1.2–2.5 cm broad. Glumes ovate-lanceolate, pilose, gradually narrowed into fairly long (up to 5–6 mm) awn. Lemma also long-awned, rather densely pilose, narrowly membranous along margin. Forebaikal (NE Baikal coast), $2n = 28$.

Steppe slopes, forest fringes, rock debris, coastal pebble beds and thin pine groves in Baikal coastal belt. **East. Sib.:** IR—An, BU—Se. — Endemic. Described from Priol'khon on Baikal. Map 28.

5. **A. erickssonii** (Meld.) Peschkova, status novus. —*A. cristatum* var. *erickssonii* Meld. in Norlindh 1949, Fl. mong. steppe 1: 118.

Cespitose plant. Stems 20–45 cm tall, numerous, erect or strongly geniculate in lower nodes, rather densely pubescent with long flexuous, somewhat divergent hairs throughout length or in upper part. Radical leaves several, nearly equal in length, many times shorter than stem, convolute, occasionally flat, densely puberulent on both surfaces, sometimes insignificantly interspersed with long interrupted hairs. Cauline leaves slightly recurved from stem, flat or convolute, glabrous or slightly scabrous beneath, densely puberulent on upper surface (velutinous pubescence). Glumes and lemma with long, densely entangled hairs. Spikes dense, 1.5–4 cm long, 1–1.5 cm broad, slightly narrowed toward tip. Rachilla with sparse, very short spinules visible only under magnification.

Hill-steppe belt, rocky-rubble steppes, steppified meadows, sand knolls. **West. Sib.:** AL—Go. **Cen. Sib.:** TU (far west). —Mongolia. Described from Inner Mongolia (Naiman-Ul ulus).

6. **A. fragile** (Roth) Candargy 1901 in Arch. Biol. Vég. (Athénes) 1: 58. —*A. sibiricum* Willd.

Compactly cespitose plant. Stems rather slender, glabrous, few. Leaves narrow, flat or convolute, glabrous and smooth on lower surface and extremely puberulent or scabrous on upper surface. Spikes 3.5–7 cm long, narrow (about 1 cm, sometimes up to 1.5 cm broad), narrowed, cristate, sometimes with superposed indistinctly cristate spikelets toward tip. Glumes ovate, carinate, scabrous or ciliate along keel. Lemma glabrous or pilose, obtuse, with short cusp (less than 1 mm long). Palea with several (12–40) spinules along keels.

Sandy and loamy soils in flat steppes. **West. Sib.:** AL—Ba. —SE European USSR, Caucasus, Mid. Asia, West. China, Mongolia (western). Described from specimens of unknown origin found in a field.

7. **A. kazachstanicum** (Tzvelev) Peschkova 1985 in Novosti sist. vyssh. rast. 22: 37. —*A. cristatum* subsp. *kazachstanicum* Tzvelev. —*A. cristatum* auct. p.p.

Stocky (15–50 cm tall, sometimes more), compactly cespitose plant, usually with few stems, glabrous throughout length, rarely puberulent in upper part and geniculate in lower part. Radical leaves numerous, many times (rarely only up to 2 times) shorter than stem, nearly same in length, rigid, convolute, often arcuate, glabrous and smooth on outer surface, puberulent and veined on inner. Spikes ovate or oblong-ovate, cristate, usually dense but with distinct spaces between spikelets. Glumes glabrous or with long cilia along keel and with 2–4 mm long awn. Lemma pilose, occasionally scabrous or glabrous, terminating in 2–4 mm long awn. Altay (Kosh-Agach settlement), $2n = 28$.

Rocky-rubble steppes, rocks, steppe scrubs, desert associations, less often on sand and sandy steppes, ascending to 2,000 m in hills. **West. Sib.**: KE (Novopesterovo settlement), AL—Go. **Cen. Sib.**: KR—Kha, Ve, TU. —Kazakhstan, West. Mongolia, West. China. Described from Kazakhstan (Kyzyl-Rai hills). Map 32.

8. **A. michnoi** Roshev. 1920 in Izv. Glavn. bot. sada RSFSR 28: 384.

Rhizome long, decumbent. Stems 30–90 cm tall, glabrous throughout length. Radical leaves few or absent, cauline leaves 2 or 3, recurved from stem, convolute, occasionally flat, glabrous and smooth on outer surface, short- and densely velutinously pubescent on upper surface; usually without long divergent hairs. Spikes 2–10 cm long and 1–1.5 cm broad, broad- or oblong-linear, with spikelets dense and cristate or superposed with barely distinct ridge formation. Glumes ciliate along keel, aristate. Lemma more or less crinite or with short spinules on back and 2 mm long terminal awn. Buryat, $2n = 28$.

Shifting or loose sand and sandy, weakly turfed steppes. **East. Sib.**: BU—Se (Ulan-Burga and Khuterkhei ulus in Verkh. Kuitun area), Yuzh (Peschanoe Lake beyond Kumyn hill—class. hab.—and others), ChI—Shi (Kuranzha village on Onon river; Onon station; Barun-Torei Lake). —NE Mongolia and NE China (Manchuria). Map 33.

9. **A. nataliae** Sipl. 1968 in Novosti sist. vyssh. rast. 5: 13. —*A. michnoi* subsp. *nataliae* (Sipl.) Tzvelev. —*A. michnoi* auct. non Roshev.

Plant with long rhizome. Stems 40–70 cm tall, glaucescent, usually single, sparingly pilose under spike. Radical leaves developing in late autumn; cauline leaves flat or convolute, glabrous and smooth on lower surface and covered with diffuse slender spinules or very short hairs on upper surface. Spikes 3–5 cm long and 1.5–2 cm broad, oblong-ovate, cristate, with dense contiguous spikelets, compact. Glumes densely pilose, aristate. Lemma also densely squarrose pilose, with up to 3.5 mm long awn.

Sand dunes. **East. Sib.**: BU—Se (Kharkhushun village on Argode river), Yuzh (Onokhoi station), ChI—Ka (Chara river near its confluence with Srednii Sakukan river—class. hab.). —Endemic.

10. **A. pectinatum** (Bieb.) Beauv. 1812, Ess. Agrost: 146. —*A. cristatum* var. *pectinatum* (Bieb.) Krylov.—*A. pectiniforme* Roemer et Schultes. —*A. cristatum* subsp. *pectinatum* (Bieb.) Tzvelev.

Fairly large (50–75 cm tall) plant with numerous leafy stems and poorly developed radical leaves. Stems usually scabrous or puberulent under spike, occasionally glabrous. Leaves flat, sometimes convolute, scabrous on upper surface with long diffuse divergent hairs, occasionally glabrous, slightly recurved from stem. Spikes elongate- or ovate-oblong, narrowed toward tip, cristate, rather dense but with distinct spaces between spikelets. Glumes and lemma more often glabrous; occasionally glumes with long cilia along keel but lemma sparsely pilose; in both cases, drawn into 2–3 (up to 4 mm) long awn. Palea with few short spinules along keels.

Rocky steppe slopes, meadow glades, dry valley meadows and sand as well as farm hedges, in crops, fallow land, nurseries, dumps along railroads, and roadsides. **West. Sib.**: KU, OM, NO, AL—Ba, Go. **Cen. Sib.**: KR—Kha, Ve, TU. More eastward, only introduced. **East. Sib.**: IR—An, Pr (Mamakan settlement), BU—Yuzh.—Europe, Mid. Asia, Caucasus, Mediterranean, West. Asia, West. China (Junggar), Nor. Mongolia. Described from Crimea. Map 31.

11. **A. pumilum** Candargy 1901 in Arch. Biol. Vég. (Athénes) 1 (29): 49. —*A. krylovianum* Schischkin. —*A. ciliolatum* auct. non Nevski. — *Elytrigia kryloviana* (Schischkin) Nevski.

Rhizomatous plant with decumbent subsurface shoots and few stems, 15–100 cm tall. Radical leaves few, cauline leaves rather declinate, flat or convolute, scabrous or puberulent on upper surface, sometimes interspersed with long divergent hairs. Spikes oblong-linear, with interrupted, slightly declinate spikelets. Glumes lanceolate, keeled, acuminate, with cilia of different size along nerves or glabrous. Lemma covered with sparse hairs of unequal length, with 5 nerves, obtuse or with short cusp. Palea with short spinules along keels, sometimes interspersed with very long cilia. Anthers 3–3.5 mm long.

Steppes on sandy soil, pine groves, thin larch forests, steppe scrubs, rocks and rock slopes. **West. Sib.**: AL—Ba (between Pavlovka and Shadrukha settlements), Go. **Cen. Sib.**: KR—Kha, Ve (environs of Krasnoyarsk—class. hab.—and others). **East. Sib.**: IR—An (Balturino hill in Chunsk region). Mid. Asia (Balkhash region). Map 34.

7. Hystrix Moench

1. **H. sibirica** (Trautv.) Kuntze 1891, Rev. Gen.: 778. —*Asperella sibirica* Trautv.

Rhizome decumbent, branched. Stems 25–75 cm tall, glabrous or pubescent under spike. Leaves 2–7 mm broad, flat, upper surface glabrous or with diffuse hairs, sometimes scabrous. Spikes 5–13 cm long, linear, narrow. Spikelets about 15 mm long, appressed at base with joint, violet or greenish-violet, singly or paired on spike margin, each on short stalk with corona of short hairs above. Glumes 1–5 (7) mm long, setaceous, sometimes absent. Lemma 7–9 mm long, narrowly lanceolate, slightly pilose or subglabrous, with 3–5 mm long terminal awn. Palea with short and dense (especially in upper part) spinules along keels. Anthers 5 mm long.

Below belt of bald peaks and tundra zone, riverbed sand and pebble beds, rocky slopes, rock debris, sometimes descending into forest belt along river valleys. **West. Sib.:** AL—Go (Ulandryk river on Sailyugem mountain range). **Cen. Sib.:** KR—Pu, Ve (along Bol. Pit river between Sukhoi Pit and Shirokoe settlement). **East. Sib.:** IR—Pr (source of Khomolkho river in Bodaibinsk region), BU—Se, ChI—Ka (Kemen river basin 25 km east of Naminga settlement), YAK—Ol (Olenek river, near Alakit river estuary—class. hab.—Daldyn river, tributary of Markha river, Arga-Sala river), Vi (Mogdy river), Al, Yan (Khastakha river in Momsk region). —Far East (Dzhugdzhur and Pribrezhnyi mountain ranges). Map 30.

8. Leymus Hochst.

1. Glumes lanceolate, with (1) 2–3 (5) nerves, 1.5–3 mm broad, narrowed gradually toward tip, awnless or with short awn not longer than third of glume .... 2.
+ Glumes linear-subulate or subulate from narrow-lanceolate base, with 1 (3) nerves, up to 1–1.5 (2) mm broad, narrowed at tip into awn longer than third of glume .... 5.
2. Large (more than 1 m tall) plant with thick stems glabrous and smooth under spike. Spikes 10–30 cm long, rachis with spinules or hairs only along nerves. Glumes glabrous and smooth, usually longer than adjoining floret .... 3.
+ Stems usually less than 1 m high, slender, covered to some extent with hairs or spinules under spike. Spikes 5–10 cm long, rachis with short dense pilosity. Glumes pilose on back, shorter, as long as or slightly longer than adjoining floret ... 4.
3. Palea glabrous and smooth. Glumes with 1–3 nerves. Spikes farily broad, with spikelets in groups of 2 to 4 on edge ....
.... 14. *L. racemosus* subsp. *crassinervius*.
+ Palea with numerous short spinules in upper part of keels. Glumes with single nerve. Spikes narrow, with spikelets located in groups of 2 or 3 on edge ....
.... 15. *L. racemosus* subsp. *klokovii*.
4. Glumes 5–12 mm long, coriaceous-membranous, shorter or as long as lower floret .... 7. *L. interior.*

+   Glumes 12–22 mm long, slender-coriaceous, equal to or longer than lower floret .... 20. *L. villosissimus*.

5 (1).   Rachilla with dense rather long hairs. Callus of lemma with hairs longer than 0.3 mm at base. Plant usually with short rhizome and cespitose .... 6.

+   Rachilla glabrous or covered with sparse short, spinescent hairs. Callus of lemma glabrous or with hairs smaller than 0.3 mm long at base; rarely hairs longer. Plant with long and decumbent rhizome, rarely cespitose (*L. ordensis*) .... 18.

6.   Glumes glabrous and smooth along midrib and margin, usually surpassing lower floret in spikelet. Lemma awnless or terminating in short, thick cusp .... 3.

+   Glumes covered with short spinules or cilia along midrib and margin, at least in upper half, as long as lower floret in spikelet, slightly longer or distinctly shorter. Lemma, at least in lower florets, with awn not less than 1 mm long .... 7.

7.   Glumes somewhat superposed at base and masking base of lemma .... 8.

+   Glumes usually hot superposed; base of lemma clearly visible between them. .... 11.

8.   Glumes generally as long as lower floret, glabrous on back, with few short spinules only along midrib in upper part and along margin. Lemma crinite, occasionally glabrous. Palea glabrous along keels at base or throughout its length, sometimes with cilia or occasional spinule in upper part .... 9.

+   Glumes usually somewhat unequal in length and exceeding lower floret, scabrous or puberulent not only along midrib but also on back. Lemma covered with short appressed hairs; palea crowded with very short spinules throughout length .... 2. *L. angustus*.

9.   Spike rachis scabrous or rather pilose along convex surface. Hairy bands indistinct under spike edges, with comparatively short hairs. Palea with spinules or cilia along keels, rarely subglabrous .... 10.

+   Spike rachis glabrous or subglabrous along convex surface. Long, rigid cilia only along nerves. Band of dense and long (up to 1.5 mm) hairs under spike edges. Palea glabrous or with few short spinules near tip along keels ..... 17. *L. secalinus*.

10.   Medium-size, yellowish or brownish-green plant with short (7–12 cm) spikes, slightly thickened in middle part. Lemma smaller, sharply narrowed at tip into 1–3 mm long cusp or awn. Glumes distinctly shorter than lower floret .... 6. *L. dasystachys* (var. *mongolicus*).

+   Large, glaucescent green plant with long (10–20 cm) cylindrical

38

spikes, slightly narrowed toward tip. Lemma large, lanceolate, gradually acuminate, awnless (with short, 1–2 mm, awn only in lowermost florets). Glumes almost as long as lower floret .... 8. *L. jenisseiensis.*

11 (7). Lemma of nearly all flowers in spikelet with 1–4 mm long awn .... 12.

+ Lemma awnless or with awn not longer than 1–2 mm only in lowermost florets in spikelet .... 15.

12. Sheath and leaves glabrous and smooth on outer (lower) surface), rarely some slightly scabrous due to poorly visible spinules. Glumes linear-subulate, separated, with narrow barely visible membranous fringe along margin .... 13.

+ All leaves or only some (usually lowest or preceding year's) as well as their sheath distinctly scabrous or puberulent on outer (lower) surface. Glumes subulate from narrowly lanceolate base, sometimes linear-subulate, often contiguous, with distinct membranous fringe along margin .... 14.

13. Lemma covered with glaucous, erasable bloom (occasionally without it, brownish). Leaves glaucescent green, with glabrous, smooth lower surface. All spikelets appressed to spike rachis. Spikes fairly narrow (1–1.5 cm), uniform in thickness throughout length or slightly narrowed toward tip .... 9. *L. littoralis.*

+ Lemma yellowish-green or brown, very rarely with glaucous bloom. Leaves yellowish or grayish-green; lower surface somewhat scabrous due to minute sparse spinules. Lower spikelets declinate and upper appressed to spike rachis. Spikes large, broader at base (up to 2.5 cm), sharply narrowed toward tip .... 4. *L. chakassicus.*

14. Spikes usually brownish, sometimes with violet tinge, yellowish-green, oblong-cylindrical, not thickened in middle part. Lemma largely pilose, wtih poorly visible nerves. Glumes with single nerve .... 6. *L. dasystachys.*

+ Spikes green or yellowish-green, oblong-ovate, frequently highly enlarged in middle part. Lemma sparsely pilose or glabrous, with 5–7 distinctly visible nerves (especially in upper part). Glumes with single central and sometimes 2 lateral, poorly visible nerves .... 12. *L. ovatus.*

15 (11). Glumes subulate from linear-lanceolate base, packed at base and mask callus (but not base) of lower floret in spikelet, scabrous or rather pilose on back, almost as long as lower floret or not more than 1/4 shorter .... 16.

+ Glumes subulate or acicular, broadly separated at base, revealing callus of lower floret; often glabrous on back, barely scabrous only along margin; length of glumes highly unequal and not less than 1/3, sometimes 1/2 shorter than lower floret .... 11. *L. ordensis.*

16. Lemma more or less pilose or villous due to long dense hairs
.... 17.

+ Lemma glabrous or subglabrous (with separated short spinules along sides and back) .... 1. *L. akmolinensis.*

17. Glumes narrow, linear-subulate, scabrous or pilose throughout back, without cilia or membranous fringe along margin. Lemma villous-pilose due to dense white hairs ....
.... 13. *L. paboanus.*

+ Glumes subulate from narrow-lanceolate base, often joined at base, scabrous or glabrous on back, with narrow membranous fringe along margin; awn densely scabrous. Lower part of lemma sparsely pilose, glabrescent toward tip ....
.... 18. *L. sphacelatus.*

18 (5). Plants with long decumbent rhizome, not forming dense mats
.... 19.

+ Plants compactly cespitose .... 11. *L. ordensis.*

19. Stems arising in tufts from vertical part of rhizome obliquely erect .... 20.

+ Stems arising alternately or in groups from horizontal rhizome, ascending initially and erect later .... 22.

20. Rachilla covered with sparse short hairs or spinules. Lemma diffusely pilose or glabrous. Spikelets in groups of 2 or 3 along edge in middle part of spike or throughout its length ...... 21.

+ Rachilla glabrous or with minute spinules visible only under magnification. Lemma glabrous. Spikelets invariably single ......
.... 16. *L. ramosus.*

21. Spikelets in groups of 2 or 3 along edge throughout spike length. Glumes subulate and covered with short spinules or hairs along midrib, margin as well as inner surface. Lemma glabrous, rather obtuse, sharply narrowed into 1–4 mm long awn .... 10. *L. multicaulis.*

+ Spikelets in pairs along edge in middle part of spike. Glumes linear-lanceolate, glabrous on outer surface as well as inner. Lemma rather diffusely pilose or glabrous, gradually narrowed into short awn .... 19. *L. tuvinicus.*

22 (19). Rachilla covered with short sparse hairs. Lemma usually diffusely pilose .... 3. *L. buriaticus.*

+ Rachilla glabrous with extremely short, poorly visible spinules. Lemma glabrous .... 5. *L. chinensis.*

1. **L. akmolinensis** (Drobov) Tzvelev 1960 in Bot. mat. (Leningrad) 20: 430. —*Elymus akmolinensis* Drobov. —*Aneurolepidium akmolinense* (Drobov) Nevski.

Glaucescent green cespitose plant with glabrous and smooth stems up to 80 cm tall. Leaves rigid, convolute, sometimes flat, with alternating thick and slender somewhat scabrous veins. Spikes not broad; spikelets in pairs on edge of spike. Glumes very narrow, linear-subulate, with single faint nerve, scabrous along margin and back in upper part, without scarious margin, equal in length, as long or slightly shorter than adjoining lower floret. Lemma broadly lanceolate, glabrous and smooth, sometimes puberulent on sides, acuminate or with short (0.5–2 mm long) terminal awn. Palea with dense spinules of different size along keels. Rachilla puberulent. Callus glabrous, with corona of short hairs at base.

Saline meadows. **West. Sib.**: NO (Evsino settlement, Cherpanovsk region), AL—Go (Kulada village, Ongudaisk region). **Cen. Sib.**: TU (Erzin village, Chadan and Salchur settlements in Ovyursk region). — Transvolga, Urals, Mid. Asia. Described from Kazakhstan (Brai aul village, Akmolinsk province).

2. **L. angustus** (Trin.) Pilger 1947 in Bot. Jahrb. 74: 6. —*Elymus angustus* Trin. —*Anèurolepidium angustum* (Trin.) Nevski.

Cespitose, glaucescent green tall (up to 100 cm) plant with numerous convolute radical leaves, occasionally flat, rather scabrous on upper surface. Stems stout, glabrous, smooth, slightly scabrous only under spike. Spikes 8–25 (30) cm long, narrow, erect, with appressed spikelets arranged in groups of 2 or 3 along edge. Lower part of glumes overlapping along margins and base of lower floret in spikelet, lanceolate, asymmetrical, narrowed gradually into cusp, glabrous or slightly scabrous, nearly as long as spikelet. Lemma 7–10 (12) mm long, sparsely covered with short rigid hairs, acute. Palea with short dense spinules or hairs along keels.

Desert and sas steppes, solonetz meadows, steppe river valleys, pea tree thickets. **West. Sib.**: OM (environs of Omsk), NO (Bol. Chany Lake), AL—Ba, Go (Chuya river—class. hab.). **Cen. Sib.**: TU. —Mid. Asia, West. China (Junggar), Mongolia (western). Map 35.

3. **L. buriaticus** Peschkova 1985 in Bot. zhurn. 70, 11: 1556. —*L. chinensis* (Trin.) Tzvelev × *L. littoralis* (Pallas) Peschkova.

Rhizome long, rather thick, horizontal, decumbent, with glabrous, 35–90 cm tall stems arising alternately and singly. Leaves flat, occasionally convolute, usually declinate from stem, with relatively slender veins, densely puberulent, sometimes interspersed with long, divergent hairs. Spikes with more or less aggregate spikelets, 2 each on edge, especially in middle part. Glumes subulate from lanceolate base, glabrous, with spinules along margin. Lemma diffusely pilose, sometimes subglabrous, gradually narrowed into short cusp or awn. Spike rachis glabrous, rigid-ciliate along nerves, sometimes with band of hairs under edge of spikelet. Rachilla sparsely puberulent. Callus pilose on sides.

Sandy slopes of steppes, loamy solonetz soils in steppified river valleys. **East. Sib.**: IR—An (Ust'-Anga village in Ol'khonsk region), BU—

Se (Dushelan and Amalat villages), Yuzh (environs of Gusinoe Lake — class. hab.—and others), YAK—Vi environs of Yakutsk, Sangar settlement), Al. —Endemic. Map 37.

4. **L. chakassicus** Peschkova sp. nova. —*Elymus dasystachys* auct. non Trin. —*E. ovatus* auct. non Trin., quod pl. chakassia.

Plantae longe rhizomatosae 50–130 cm altae, culmis glabris, sub spica tantum pilosis. Folia flavido-vel fuscidulo-viridia plana, rarius convoluta, supra scabriuscula, interdum sparse longe pilosa, subtus glabra vel sparse scabra. Spicae plerumque magnae (5–20 cm longae), sub anthesi latae (basi 2–2.5 cm latae), dein angustae (ca 1.5 cm latae), rhachide pilosa vel scabra. Spiculae multiflorae, densae, inferiores patentes, superiores appressae. Glumae lineari-subulatae, margine et secus nervum medium ciliatae vel scabrae, subremotae, flore infimo subbreviores, ei aequilongae vel eo sublongiores. Lemmata flavo-viridia vel olivaceo-fusca, lanceolata, in acumen breve vel aristam 0.5–2 mm longam sensim acutata, plus minusve pilosa. Pleae secus carinam (interdum superne tantum) remote ciliatae.

Typus: Chakassia regio autonoma, distr. Altaisk, vicinitas pagi Letnik, vallis Jenissei, pratum inundatum. 28 VI 1967, E. Jerschova, T. Lamanova.

Affinitas. Species nostra *L. dasystachyi* et *L. littorali* affinis est, sed a priore glumis lineari-subulatis, angustissime (subindistincte) membranaceo-marginatis remotis necnon vaginis laminisque foliorum inferiorum extus glabris vel vix scabriusculis, a posteriore autem spica sat lata (basi praecipue) flavido-viridi epruinosa necnon foliis scabriusculis (nec laevibus) flavido-viridibus (nec glaucescentibus) differt.

Plant with long rhizome, 50–130 cm tall, with glabrous stems, pilose only under spike. Leaves yellowish or brownish-green, flat, occasionally convolute, short-scabrous on upper surface, sometimes with occasional long hairs, glabrous or diffusely scabrous beneath. Spikes usually large (5–20 cm long), broad during antheses (2–2.5 cm at base), becoming narrow thereafter (about 1.5 cm). Spike rachis pilose or scabrous. Spikelets many-flowered, dense; lower laterally declinate, upper appressed. Glumes linear-subulate; ciliate or scabrous along margin and midrib; slightly separated, somewhat short, as long or longer than lower floret. Lemma yellowish-green or olive-brown, lanceolate, gradually acuminate into short cusp or awn 0.5–2 mm long, more or less pilose. Palea with divergent cilia along keels, sometimes developed only in upper part.

Solonetz steppes and steppified meadows along river valleys and their steppe slopes. **Cen. Sib.**: KR—Kha, Ve, TU (Kara-Kol' Lake in Bai-Taiginsk region). —Endemic. Map 39.

5. **L. chinensis** (Trin.) Tzvelev 1968 in Rast. Tsentr. Azii 4: 205. — *Agropyron pseudoagropyrum* (Trin. ex Griseb.) Franchet. —*Aneurolepidium pseudoagropyrum* (Trin. ex Griseb.) Nevski.

Rhizome long, decumbent, funicular. Stems arising singly or in tufts from horizontal rhizome, initially ascending and later erect. Leaves flat or rather convolute, gradually acuminate toward tip, with thick glaucous vein covered with long stray hairs on upper surface, occasionally glabrous or subglabrous. Spikes oblong, 6–15 (20) cm long, separated in lower part; spikelets single, or an average of 2 each on spike edge. Glumes linear-subulate, with single nerve, glabrous. Lemma lanceolate, gradually narrowed into short cusp or awn 0.5–1.5 mm long, glabrous, smooth. Rachilla subglabrous, covered with extremely short, poorly distinct spinules. East. Siberia (Transbaikal), $2n = 28$.

Solonetz steppes and steppified saline meadows, often forming nearly pure herbaceous cover, abundant on fallow land. **West. Sib.**: AL (Ongudaisk region). **Cen. Sib.**: KR—Kha, Ve, TU ("Tselinnyi" State Farm, Erzin settlement). **East. Sib.**: IR—An, Pr, BU—Se, Yuzh, ChI—Ka, Shi, YAK—Vi (environs of Yakutia), Al (Chilimakh village on Amga river). —Far East, China, Mongolia. Described from China. Map 36.

6. **L. dasystachys** (Trin.) Pilger 1947 in Engler Bot. Jahrb. 74: 6. — *Elymus dasystachys* Trin. —*Aneurolepidium dasystachys* (Trin.) Nevski. — *Leymus secalinus* auct. p.p.

Loosely cespitose plant with short rhizome. Stems 35–80 cm tall, slender, glabrous or finely pubescent under spike. Radical leaves usually numerous, convolute or flat, grayish-green, with short dense pilosity on upper surface, whole lower surface or at least lowest part in mat densely scabrous or velutinous-pilose. Spike rachis scabrous or more or less with dense long hairs. Spikes brownish-violet, dark brown or occasionally yellowish-green, usually with isolated, not dense, lower spikelets, cylindrical, or slightly narrowed toward tip. Glumes linear-subulate or sharply transformed from lanceolate base into subulate cusp, sometimes slightly superposed (*L. dasystachys* (Trin.) Tzvelev var. *mongolicus* (Meld.) Peschkova comb. nov. = *Elymus dasystachys* var. *mongolicus* Meld. in Norlindh, 1949, Fl. mong. steppe 1: 128), frequently, as glume transforms into awn, with long cilla, scabrous due to short spinules on upper surface along margin and back. Lemma more or less pilose, sometimes subglabrous, sharply narrowed into 1–4 mm long awn; awn often absent in upper florets in spikelet; 5–7 poorly visible nerves on back. Palea with spinules of various size or short cilia along keels. Altay (Chui steppe), $2n = 28$.

Sand-pebble beds along mountain river valleys, sand-rocky steppes up to 2,000 m above sea level. **West. Sib.**: AL—Go (Chuya river—class. hab.—and others). **Cen. Sib.** : TU. **East. Sib.**: BU—Yuzh (Mondy village and Khuzhir settlement). —Mid. Asia, Nor. Mongolia, West. China. Map 40.

7. **L. interior** (Hultén) Tzvelev 1964 in Arkt. fl. SSSR 2: 253. —*Elymus interior* Hultén. —*L. ajanensis* auct.

Cespitose plant with short rhizome, with decumbent, later ascending surface shoots. Stems up to 80 cm tall, glabrous in lower part, densely pubescent with short hairs in upper part, especially under spike and along spike rachis. Leaves flat and convolute, with slender, rather scabrous veins on upper surface. Spikes 5–10 cm long. Spikelets 10–18 mm long, often brownish-violet. Glumes coriaceous-membranous, usually shorter than lower floret, occasionally as long, lanceolate, with 3 nerves of which midrib prominent, lateral ones barely visible, with soft hairs on back. Lemma lanceolate, awnless, villous-pilose. Palea terminating in 2 long teeth. East. Siberia (Kolyma upland), $2n = 28$.

Sand-pebble deposits in floodplains of mountain rivers, rising up to bald-peak belt. **Cen. Sib.:** KR—Ta (Taimyr Lake—class. hab.). **East. Sib.:** YAK—Ar, Al (Ytyga river, tributary of Yudoma river), Yan. —Far East. Map 38.

8. **L. jenisseiensis** (Turcz.) Tzvelev 1973 in Novosti sist. vyssh. rast. 10: 51. —*Elymus jenisseiensis* Turcz.

Grayish-green perennial with short rhizome, forming small loose mats. Stems 50–100 cm tall, scabrous or glabrous under spike. Leaves flat, sometimes with convolute margin, densely scabrous on upper surface, glabrous or subglabrous on lower surface. Spikes 10–20 cm long, dense, elongate, nearly uniform in breadth or slightly narrowed toward tip. Glumes subulate from lanceolate base, slightly superposed at base, more or less as long as lower florets, ciliate along margin, scabrous on back. Spike rachis scabrous or pilose, with barely distinguishable band of hairs under edge. Lemma rather pilose, sometimes subglabrous, gradually narrowed into short cusp or (in lowermost flowers) upto 1.5 mm long awn. Keels of palea with diffuse cilia or spinules in upper half.

Sandy steppes, shifting sands, poplar groves along river valleys. **West. Sib.:** AL—Go (Aktash village in Kosh-Agach region). **Cen. Sib.:** KR—Kha, Ve (Yenisey river near Krasnoyarsk—class. hab.—and others), TU (Tere-Khol' Lake, Bai-Khaak settlement). **East. Sib.:** BU—Yuzh (Ubur-Dzokoi area near Ust'-Kyakhta village). —Endemic. Map 41.

9. **L. littoralis** (Griseb.) Peschkova 1988 in Novosti sist. vyssh. rast. 24: 23. —*Elymus dasystachys β. littoralis* Griseb. —*Elymus dasystachys* auct. p.p. non Trin. —*Aneurolepidium dasystachys* auct. p.p. —*Leymus secalinus* auct. p.p.

Rhizome slender, decumbent. Stems 50–100 cm tall, glabrous or pilose under spike. Leaves flat, occasionally convolute, grayish or glaucescent green, short-scabrous on upper surface, sometimes interspersed with long isolated hairs; glabrous and smooth beneath. Spikes 6–10 (15) cm long, 0.7–1.3 (1.5) cm broad, grayish-green or brown, relatively loose, usually with separated lower spikelets. Spikelets in groups of 2 or 3 on edge. Spike rachis more or less ciliate on convex surface along nerves, with band of very dense long hairs under edge of spike. Glumes subulate from linear or narrowly lanceolate base acuminate, as long as adjacent lower florets; with short spinules and cilia almost from base along margins and midrib, with or without very narrow membranous fringe. Lemma more or less crinite, gradually narrowed into 0.5–4 mm long awn (only uppermost in spikelet often awnless). Palea covered almost from base with more or less long spinules or cilia along keels.

Coastal sand, solonetz and sandy steppes, pine groves on sand. **East. Sib.**: IR—An, Pr, BU—Se, Yuzh (environs of Posol'sk—class. hab.—and others) YAK—Vi. —Nor. Mongolia. Map 51.

10. **L. multicaulis** (Kar. et Kir.) Tzvelev 1960 in Bot. mat. (Leningrad) 20: 430. —*Elymus aralensis* Regel. —*Aneurolepidium multicaule* (Kar. et Kir.) Nevski.

Rhizome long, slender, decumbent. Stems 50–100 cm tall, glabrous, smooth, scabrous only under spike. Leaves flat or with convolute margin; scabrous on upper surface (sometimes also on lower surface). Spikes 5–12 cm long, linear, dense or with isolated spikelets in lower part. Spike rachis with appressed hairs along ribs. Spikelets in groups of 2 or 3 on edge, green or violet tinged. Glumes subulate, acuminulate, with single nerve; with short spinules or scabrous on back. Glumes as long as or slightly shorter or longer than lower floret. Lemma broadly lanceolate, glabrous, lustrous, sharply narrowed into short (1–3 mm) awn; palea short and thin ciliate along keels.

Solonetz meadows. **West. Sib.**: AL—Go (Chuya river basin). — Transvolga, Mid. Asia, Nor. China. Described from Tarbagatai.

11. **L. ordensis** Peschkova 1985 in Bot. zhurn. 70, 11: 1554.

Cespitose, glaucescent green plant. Stems 50–90 cm (or more) tall, stout; glabrous, slightly scabrous or puberulent only under spike. Leaves convolute or flat, glabrous, smooth, sometimes slightly scabrous on upper surface along thick veins. Spikes relatively lax, with easily shed spikelets; glumes, rachilla, and lowest floret preserved longest. Spiklets singly or in groups of up to 3 along spike edge, sometimes all spikelets arising singly. Glumes usually unequal in length (2–6 mm), occasionally equal, often glabrous, very narrow, linear-subulate, almost twice shorter than lower floret, sometimes one glume greatly reduced (about 1–2 mm long). Lemma 8–10 mm long, broadly lanceolate, sharply narrowed into short cusp or 0.5–2 mm long awn, more or less pilose, sometimes slightly glabrate on back or totally glabrous. Palea with short and rather various size crowded spinules along keels. Altay (Kuraisk mountain range), $2n = 28$.

Solonchak meadows and steppes in steppe regions. **West. Sib.**: AL—Go. **Cen. Sib.**: KR—Kha, Ve, TU. **East. Sib.**: IR—An (Ust'-Orda settlement—class. hab.—and others). —North Mongolia. Map 42.

12. **L. ovatus** (Trin.) Tzvelev 1960 in Bot. mat. (Leningrad) 20: 430.— *Elymus ovatus* Trin. —*Aneurolepidium ovatum* (Trin.) Nevski.—*Leymus secalinus* subsp. *ovatus* (Trin.) Tzvelev.

Rhizome slender, decumbent. Stems 20–100 cm tall, relatively slender, usually scabrous or puberulent under spike. Radical leaves 3–7 mm broad, usually numerous, flat or with convolute margin, commonly densely hairy on both surfaces as well as leaf sheaths. Spike rachis densely pilose. Spikes usually not long (5–10 (15) cm) but broad (0.7–2 cm), yellowish-green. Spikelets in middle part in groups of 3 or 4 along edge, resulting in thickened midsection of spike that narrows toward both ends. Glumes subulate from narrowly lanceolate base, with 1–3 faint nerves; with short spinules along margin and back, scabrous. Lemma lanceolate, gradually narrowed into short cusp or 0.5–2 mm long awn, with 5–7 nerves distinctly visible, especially in upper part; nerves glabrous or sparsely pilose, sometimes with broad membranous fringe along margin. Palea with diffuse spinules or cilia along keels.

Sand-pebble beds in valleys along middle and lower courses of mountain rivers (up to 1,600 m above sea level), in flat steppes and semideserts, under cover of valley forests, sometimes on rocky southern slopes. **West. Sib.:** AL—Go (Chulyshman river—class. hab.—and others). **Cen. Sib.:** TU.—Mid. Asia (Balkhash region), West, China (Junggar). Mongolia. Map 46.

13. **L. paboanus** (Claus) Pilger 1947 in Bot. Jahrb. 74: 7. —*Elymus paboanus* Claus.—*E. salsuginosus* (Griseb.) Turcz. ex Steudel. — *Aneurolepidium paboanum* (Claus) Nevski.

Glaucescent cespitose plant with glabrous and smooth (sometimes pilose or scabrous under spike) stems up to 90 cm tall. Leaves coarse, flat or with convolute margin and sharply distinct veins; scabrous or subglabrous on upper surface. Spikes elongated, not broad, dense. Glumes subulate-linear, with single faint nerve, without distinct scarious fringe along margin; usually pilose or scabrous on back, very rarely subglabrous; all glumes equal in length and nearly as long as (sometimes, slightly longer than) adjoining lower floret. Lemma broadly lanceolate, sharply narrowed into short cusp or up to 1.5 mm long awn; lemma with dense white hairs. Palea with crowded, very short spinules along keels.

Solonchak meadows, solonetzes, solonetz steppes. **West. Sib.:** TYU— Tb, KU, OM, NO, KE, AL—Ba, Go. **Cen. Sib.:** TU. **East. Sib.:** BU (Solenoe Lake near Sul'fat station; between Gusinoe and Solenoe Lakes; Zagustai river). —Transvolga, Urals, Mid. Asia, Mongolia, West. China (Junggar). Described from SE European USSR (Kinel' river). Map 43.

14. **L. racemosus** (Lam.) Tzvelev subsp. **crassinervius** (Kar. et Kir.) Tzvelev 1971 in Novosti sist. vyssh. rast. 8: 65. —*Elymus giganteus* auct. non Vahl.

Perennial with long rhizome. Stems 50–100 cm tall, stout, thick, glabrous and smooth under spike. Leaves flat or convolute, with thick, highly scabrous veins on upper surface and glabrous on lower surface.

Spikes usually large (15–30 cm long and 1–2 cm broad), thick, gradually narrowed toward tip, spikelets in groups of 3 to 5 along edge. Spike rachis ciliate along edge, glabrous elsewhere including under spike edges. Glumes lanceolate-subulate, glabrous and smooth on back and margin, as long as or longer than lower floret. Lemma with 5–7 distinct nerves, awnless or with short (up to 1 mm) thick cusp with fairly long dense hairs in lower part, glabrescent toward tip. Palea glabrous and smooth along keels, very rarely scabrous in upper fourth due to some very short spinules.

Sand dunes, sand steppes and semideserts, dry pine groves, thickets of steppe pea tree. **West. Sib.**: TYU—Tb (environs of Tobol'sk), KU (Yazeva village, Kurtamysh region), AL—Ba. **Cen. Sib.**: KR—Kha, Ve, TU (Tere-Khol' Lake, Khemchik river). **East. Sib.**: BU—Yuzh (Uletui area in Kyakhtinsk region, Kumyn hill near Chernoe Lake in Kudary river basin). —Mid. Asia, West. China and West. Mongolia. Described from East. Kazakhstan (near Semipalatinsk). Map 45.

15. **L. racemosus** (Lam.) Tzvelev subsp. **klokovii** Tzvelev 1971 in Novosti sist. vyssh. rast. 8: 65.

Spikes relatively slender, lax. Spikelets in groups of 2 or 3 along edge. Palea covered with short spinules in upper part along keels.

Sand and sand steppes. **West. Sib.**: KU (Ukrainets village in Pritobol'noe region). **Cen. Sib.**: KR—Kha, Ve. —SE European USSR, northern Mid. Asia. Described from South. Urals (Guberlinsk village).

16. **L. ramosus** (Trin.) Tzvelev 1960 in Bot. mat. (Leningrad) 20: 430. —*Agropyron ramosum* (Trin.) K. Richter. —*Aneurolepidium ramosum* (Trin.) Nevski.

Rhizome long slender, decumbent. Stems arising in dense tufts from vertical part of rhizome and obliquely erect. Leaves flat or convolute, gradually acuminate toward tip, glabrous beneath, frequently scabrous on upper surface along thick veins and sometimes with sparse divergent hairs. Spikes slender, relatively short (3–8 cm); spikelets one each along spike edge. Glumes linear-subulate, usually unequal, with single indistinct nerve, glabrous. Lemma glabrous, smooth, broadly lanceolate, sometimes awned in lower florets of spikelet; usually awnless or with very short awnlike cusp in upper florets. Rachilla subglabrous, with extremely short spinules visible only under high magnification. Altay (Kosh-Agach settlement), $2n = 28$.

Flat, usually solonetz steppes, solonchak meadows, marshy river valleys, steppe slopes. **West. Sib.**: OM (Pavlograd village, Kamyshlovo Lake near Elita settlement), AL—Ba. **Cen. Sib.**: KR—Kha, Ve, TU. **East. Sib.**: IR—An, BU—Yuzh (eastward up to Irkut river valley). —South European USSR, Mid. Asia, West. China (Junggar). Described from Irtysh river in Semipalatinsk province. Map 44.

17. **L. secalinus** (Georgi) Tzvelev 1968 in Rast. Tsentr. Azii 4: 209. — *Triticum secalinum* Georgi. —*T. littorale* Pall. —*Elymus giganteus* auct. — *L. racemosus* auct. p.p.

Rhizome long, decumbent. Stems 50–100 cm tall, coarse, rigid, relatively thick. Leaves glaucescent green, flat or with convolute margin, broadly sulcate and rigid-scabrous on upper surface and glabrous, smooth, lustrous on lower surface. Spikes 6–15 cm long, dense, rather thick (0.7–1.5–2 cm), rachis laterally rigid-ciliate, with band of long (up to 2 mm) dense hairs under edges, elsewhere glabrous or slightly scabrous, glaucous. Spikelets 2–4 each along edge. Glumes slightly overlapping, subulate from lanceolate base, with very slightly membranous margin, glabrous and smooth at base, scabrous along margin in upper part, nearly as long as lower floret. Lemma covered with glaucous bloom, sparsely pilose, glabrescent toward top, acute, awnless; only lower ones in spikelet with short (up to 1.5 mm) cusp. Palea glabrous along keels, nearly smooth, scabrous due to very short spinules only in upper quarter.

Dry coastal coarse sand along coasts of northern Baikal. **East. Sib.**: IR—An (Ol'khon Lake—class. hab.—and Ol'khonsk region), BU—Se. — Endemic. Map 47.

18. **L. sphacelatus** Peschkova 1985 in Bot. zhurn. 70 (11): 1955. — *Leymus dasystachys* (Trin.) Pilger × *L. paboanus* (Claus) Pilger.

Compactly cespitose plant. Stems 60–90 cm tall, slender, stout, strongly foliate, glabrous or slightly scabrous under spike. Radical leaves numerous, long and acuminate, convolute, occasionally flat, grayish-green, more or less coarsely short. Scabrous on upper surface; usually glabrous on lower surface. Spikes 5–13 cm long, about 1 cm broad, dense. Spike rachis more or less pilose, with indistinct band of long very dense hairs under edges. Glumes subulate from linear or narrow-lanceolate base, usually shorter than lower floret, with spinules or cilia along margin, and with diffuse short spinules or hairs on back and not just along midrib. Lemma obtuse, often awnless in upper florets, with short (not more than 2 mm) awn in lower florets, fairly pilose in lower part, glabrescent toward tip and brownish (as though opalescent) along sides. Upper part of palea also brownish along keels, with short crowded spinules, usually somewhat longer than lemma (awnless).

Marshy meadows and solonetz steppes. **Cen. Sib.**: TU (Ogneva area on Elegest river—class. hab.—and others). —Endemic. Map 48.

19. **L. tuvinicus** Peschkova 1985 in Bot. zhurn. 70 (11): 1557. —*L. ramosus* (Trin.) Tzvelev × *L. dasystachys* (Trin.) Pilger.

Rhizome long, slender, decumbent. Stems glabrous or slightly pilose under spike, arising in dense tufts obliquely erect from vertically ascending part of rhizome. Leaves flat, with relatively slender veins, scabrous or puberulent on upper surface, sometimes interspersed with long sparse hairs. Spikes relatively slender. Spikelets usually single; in pairs in middle part along edge. Spike rachis scabrous, frequently with band of very long and dense hairs under edge. Glumes linear-lanceolate, gradually acuminate, glabrous, scabrous along margin. Lemma lanceolate, gradually narrowed into 1–4 mm long awn; more or less pilose, scabrous or subglabrous on back. Palea with cilia along keels. Callus largely glabrous, with few short hairs on sides.

Marshy steppified meadows with loamy soil in river valleys. **West. Sib.**: AL—Ba (Georgievka village, Burlinsk region), Go (lower course of Chulyshman river). **Cen. Sib.**: KR—Kha, Ve, TU (environs of Kyzyl—class. hab.—and others). **East. Sib.**: BU—Yuzh (Khuzhir settlement, Obtoi river estuary). —Nor. Mongolia. Map 50.

20. **L. villosissimus** (Scribner) Tzvelev 1960 in Bot. mat. (Leningrad) 20: 429. —*Elymus villosissimus* Scribner. —*Leymus arenarius* auct. p.p.

Plant with slender rhizome, 25–60 cm tall. Stems under spike with soft hairs. Leaves glaucous, flat (3–8 mm broad) or with convolute margin, slightly scabrous on upper surface along relatively slender veins. Spikes 5–10 cm long with short-pubescent rachis. Spikelets 14–25 mm long, often brownish-violet. Glumes 12–22 mm long, usually longer than adjoining lower floret, lanceolate, coriaceous-membranous with faint nerves, more or less pilose. Lemma broadly lanceolate, villous-pilose, awnless. Palea with 2 long terminal teeth.

Sand-pebble beds on sea coasts. Found in lower courses of large rivers. **East. Sib.**: YAK—Ar (lower courses of Lena and Kolyma). —Far East, Kamchatka, Kuril' islands, Alaska. Described from St. Paul Island in the Bering Sea. Map 53.

× Leymostachys Tzvelev

1. × **L. korovinii** Tzvelev (*Leymus* ? *racemosus* (Lam.) Tzvelev s.l. × *Psathyrostachys juncea* (Fischer) Nevski).

Reported from Tuva territory (environs of Seserlig settlement).

× Leymotrigia Tzvelev

1. **L. wiluica** (Drobov) Tzvelev (*L. littoralis* (Griseb.) Peschkova × *Elytrigia repens* (L.) Nevski).

Collected near Nakanno settlement in Katanga region of Irkutsk province. Described from Yakutia (near Bilyuchansk village on Vilyui river).

2. **L. zarubinii** Peschkova sp. hybr. nova (*Leymus chinensis* (Trin.) Tzvelev × *Elytrigia repens* (L.) Nevski).

Gramen magnum longe rhizomatosum, spica longa. Spiculae ad excavationes racheos ut in *L. chinensi* binae, sed lemmata majora manifeste (superne praecipue) quinquenervosa, apice longe et sensim acutata.

Typus: Prov. Czita, distr. Olovjannaja, pag. Karaksar, terrassa inundata fl. Onon, 6 VII 1961 A. Zarubin.

Large plant, with long rhizome, with long spikes. Spikelets in pairs along spike margin as in *Leymus chinensis* but lemma larger, with 5 distinct (especially in upper part) nerves; long and gradually acuminate.

Very common hybrid, rather frequent along river valleys in southern Chita province.

9. Psathyrostachys Nevski

1. Spikes extremely brittle; only lower part of 1–3 (5) cm long spikelet with distinctly visible fracture site at tip preserved in herbarium. Spikelets fall together with glumes and spike rachis sections. Hairs at base of glumes usually as broad as latter or sometimes 1.5–2 times broader. Spikelets 1-, rarely 2-flowered .... 2.
+ Spikes relatively tenacious, slightly narrowed toward tip, (3) 5–12 cm long. Spikelets readily fall from glumes; spike rachis with glumes often long persisting. Hairs at base of glumes usually not exceeding breadth of latter. Spikelets 2- or 3-, rarely 1-flowered ....
.... 3. *P. juncea.*
2. Keels of palea covered with very short dense, upturned weak spinules, rarely interspersed with very long sparse hairs. Stem under spike usually covered with short dense squarrose hairs (velutinous pubescence); sometimes pubescence rather sparse and stiff. Root leaves 5–20 cm long, flat or convolute, more often erect, rarely slightly arcuate. Spikelets single-flowered .... 1. *P. caespitosa.*
+ Keels of palea covered with sparse long cilia or soft hairs. Stems glabrous or slightly scabrous under spike, occasionally with short stiff hairs. Radical leaves very short, 2–5 (10) cm long, convolute and arcuate, rarely erect or flat. Spikelets 1- or 2-flowered ....
.... 2. *P. hyalantha.*

1. **P. caespitosa** (Sukaczev) Peschkova comb. nova. —*Elymus caespitosa* Sukaczev 1913 in Tr. Bot. muz. Akad. nauk 11: 80. —*P. juncea* subsp. *hyalantha* auct. p.p. —*Elymus junceus* auct. non Fischer. —*E. junceus* var. *caespitosus* (Sukaczev) Reverd.

Compactly cespitose perennial. Stems 20–60 cm tall, scabrous or with velutinous pubescence under spike. Radical leaves profuse, 5–20 cm long, flat or convolute, erect or sometimes arcuate, both surfaces scabrous due to sparse short spinules. Spikes breaking readily; in collections, usually 1–3 (4) cm long, linear, spike rachis rather pilose, long-ciliate along ribs. Spikelets arranged in groups of 2 or 3 along edge; sometimes not developed in lower part. Spikelets 1- (rarely, 2-) flowered. Glumes subulate, more or less pilose, hairs in lower part as long or longer than glume breadth. Lemma rather uniformly covered on back with hairs of various size. Palea covered along keels with short dense hairs turned toward tip.

50

Steppes, solonetz meadows, dry and crustate solonetzes, steep carbonate slopes. **Cen. Sib.**: KR—Kha, Ve (Kubekovo village near Krasnoyarsk). **East. Sib.**: IR—An, YAK—Vi (Lena bank near Yakutsk—class. hab.—and others), Yan (Bala settlement, Dulagalakh river). —Endemic. Map 52.

2. **P. hyalantha** (Rupr.) Tzvelev 1968 in Rast. Tsentr. Azii 4: 202. — *P. juncea* subsp. *hyalantha* (Rupr.) Tzvelev.

Compactly cespitose perennial. Stems 20–40 cm tall, glabrous, smooth, slightly scabrous or with short appressed hairs only under spike. Radical leaves usually withering early, 2–5 (10) cm long, convolute, arcuate or, occasionally, erect. Spikes breaking readily and hence short, 1–3 (4) cm long, spike rachis glabrous or scabrous, ocassionally appressed-pilose. Spikelets 1- or 2-flowered. Glumes and lemma covered with long hairs in lower part, hairs greatly shortened upwards. Palea covered with long cilia or soft hairs along keels.

Desertified and solonetz steppes. **West. Sib.**: AL—Ba (Ainak station, Burlinsk region). **Cen. Sib.**: TU (environs of Kyzyl, Ust'-Elegest settlement, Khandagaity village in Ovyursk region). —Mid. Asia, West. China (Junggar, Kashgar). Described from Mid. Asia (Dzhaman-Tau hills in central Tien Shan).

3. **P. juncea** (Fischer) Nevski 1934 in Fl. SSSR 2: 714. —*Elymus junceus* Fischer.

Compactly cespitose perennial. Stems (10) 25–85 cm tall, glabrous, smooth, scabrous or with short appressed hairs under spike. Radical leaves profuse, 7–20 (35) cm long, flat, or occasionally convolute, scabrous on both surfaces. Spikes elongated, 5–12 cm long, narrowing slightly toward tip, usually not breaking at rachis joints but with readily shed spikelets; glumes persisting on spike rachis. Spike rachis slightly scabrous under edges, ciliate along ribs. Spikelets in groups of 2 or 3 along edge, 2- or 3-flowered, rarely 1-flowered. Glumes subulate, scabrous or puberulent; longest hairs usually not more than glume breadth. Lemma scabrous or with short stiff hairs; hairs sometimes on tubercles. Palea covered with dense hairs of various size along keels. Altay (Chagan-Uzun village), $2n = 14$.

Arid steppes, desert-steppe river valleys, banks of salt lakes, rocky steppe slopes up to 2,200 m above sea level. **West. Sib.**: TYU—Tb (Serebryanka village), KU (environs of Kurgan), OM, NO, AL—Ba, Go. **Cen. Sib.**: KR—Kha, TU. —Transvolga, Mid. Asia, West. Asia, West. China (Junggar), Tibet, Mongolia. Described from Lower Volga region. Map 49.

10. Hordeum L.-Barley

1. Glumes usually not longer than 1.5 cm, more often 0.4–1 cm long, subulate. Spikes narrow and uniform in breadth .... 2.

+ Glumes 3–6 cm long, setaceous. Spikes enlarging upward because of laterally deflecting glumes .... 3. *H. jubatum.*
2. Awns of lemma of central spikelets 5–10 mm long. Anthers 0.7–2 mm long .... 3.
+ Awns of lemma of central spikelets 1–4 (5) mm long. Anthers (2.5) 3–4 mm long .... 4.
3. Lemma of central spikelets surficially scabrous. Anthers 1.2–2 mm long. Spikes subtenacious. Stem joints with short pubescence ....
.... 1. *H. bogdanii.*
+ Lemma of central spikelets glabrous and smooth. Anthers 0.7–1 mm long. Spikes extremely brittle. Stem joints glabrous ....
.... 6. *H. roshevitzii.*
4. Plant loose bushy or cespitose with short rhizome. Auricles of radical and lower cauline leaves usually less than 1 mm long, scarious, readily falling and poorly visible .... 5.
+ Plant compactly cespitose. Auricles of radical and lower cauline leaves 1.5–2.5 mm long, linear-subulate, membranous and persisting even in old leaves .... 7. *H. turkestanicum.*
5. Lemma of central spike with 1–2 (2.5) mm long awn, scabrous or puberulent, rarely subglabrous. Glumes of central spikelet usually terminating 1–2 mm below awn base of lemma. Leaves scabrous or puberulent on both surfaces .... 6.
+ Lemma of central spikelet with 2–4 (5) mm long awn, glabrous, smooth, rarely scabrous or pilose. Glumes of central spikelet usually as long or longer (rarely, slightly shorter) than lemma. Leaves glabrous or slightly scabrous on both surfaces; occasionally crinite on upper surface .... 2. *H. brevisubulatum.*
6. Stem joints glabrous. Stems often geniculate at base, spikes thick ...
.... 4. *H. macilentum.*
+ Stem joints with short dense hairs. Stems often erect, spikes slender .... 5. *H. nevskianum.*

1. **H. bogdanii** Wilensky 1918 in Izv. Sarat. opyt st. 1 (2): 13.

Cespitose plant with 50–100 cm tall stems; joints with short pubescence. Leaves grayish-green, flat, scabrous on both sufaces, auricles minute or lacking. Spikes 3–10 cm long, tenacious (spike rachis with glumes persists after fall of fruit), grayish or with faint violet tinge. Glumes of central spikelet as long or slightly longer than lemma (without awn); lemma scabrous due to short spinules throughout back or only in lower part; tip extended into scabrous 6–10 mm long awn.

Saline meadows and desert solonetz steppes. **West. Sib.:** TYU—Tb (between Golyshmanovo settlement and Lake Karasul'), OM (Myasnikovo village in Tarsk region), AL—Ba (Malinovskaya village in Aleisk region). **Cen. Sib.:** TU (Ak-Chyra settlement in Ovyursk region). —Transvolga, Mid. Central and West. Asia, West. Mongolia, Tibet. Described from Lower Volga region (El'ton Lake).

2. **H. brevisubulatum** (Trin.) Link 1843 in Linnaea 17: 391.

Loosely cespitose plant with short rhizome sometimes forming mat. Stems slender, glabrous at joints, very rarely puberulent. Leaves green, mostly flat, up to 6 mm broad; narrow, semiconvolute only in clusters of vegetative shoots; usually glabrous on both surfaces or on upper surface; occasionally slightly scabrous or crinite on upper surface, auricles minute or lacking. Spikes 4–8 cm long, with brittle rachis, greenish-violet. Glumes of central spikelet as long or longer than lemma (without awn), rarely slightly shorter. Lemma of central spikelets usually glabrous, with few spinules or hairs near tip, occasionally with short hairs throughout surface, with 2–4 (5) mm long terminal awn. Anthers (2.5) 3–4 mm long. Siberia (Gornyi Altay, East. Sayan, Yakutia), $2n = 28$.

Saline meadow valleys, banks of salt lakes, marshes, sometimes in meadow steppes, steppe scrub; found in hills up to 2,000 m above sea level; in northern regions, as roadside weed, fallow land, sand-pebble riverbeds. **West. Sib.**: TYU—Tb, KU, OM, NO, KE, AL—Ba, Go. **Cen. Sib.**: KR—Tn, Kha, Ve, TU. **East. Sib.**: IR—An (environs of Irkutsk—class. hab.—and others), Pr, BU—Se, Yuzh, ChI—Ka (Chara settlement), Shi, YAK—Ol, Vi, Al, Yan. —European USSR (introduced plant), Mid. Asia, Far, East, Mongolia, China (Manchuria). Map 57.

3. **H. jubatum** L. 1753, Sp. Pl.: 85. —*Critesion jubatum* (L.) Nevski.

Perennial with numerous glabrous 10–40 cm high stems. Leaves narow (up to 4 mm broad), flat, scabrous or puberulent on both surfaces; usually also crinite on upper surface; auricles not developed. Spikes 3–7 (10) cm long (without awn), enlarging upward, very brittle, greenish-violet. Glumes of central spikelet 3–6 (7) cm long, very slender, setaceous, violet in upper part. Lemma of central spikelets glabrous, transformed at tip into long, 2–6 (8) cm, slender awn. Anthers about 1.5 mm long. Yakutia (Chersk settlement), $2n = 28$.

Valley meadows and pebble beds, often as roadside weed, near settlements, in wastelands. **West. Sib.**: KU (Vvedenskoe settlement), OM (environs of Omsk, Elita settlement), NO (introduced), KE. **Cen. Sib.**: KR—Pu (Essei Lake), Tn, Kha (Luk'yanovka village in Altay region). **East. Sib.**: IR—An (environs of Irkutsk), BU—Yuzh, YAK—all regions. —Far East, North America, as introduced plant in European USSR, Caucasus and Mid. Asia. Described from Canada.

4. **H. macilentum** Steudel 1855 Syn. Pl. Gram. 1: 352. —*H. turkestanicum* auct. non Nevski.

Grayish-green, compactly cespitose plant with glabrous stems 30–60 cm tall, glaucescent near joints due to bloom, sometimes joints pubescent. Leaves grayish, flat or convolute, scabrous or densely puberulent on both surfaces, with minute readily falling auricles. Spikes up to 5 cm long, very brittle (only some lowermost spikelets usually persisting), greenish with faint violet tinge. Glumes of middle spikelet usually 1–2 mm shorter than lemma (without awn), very rarely almost as long. Lemma puberulent or scabrous on back, rarely subglabrous, with short, 1–2 (2.5) mm long awn. Anthers 2.5–3.5 mm long.

Saline meadows and solonchaks along banks of salt lakes. **East. Sib.**: ChI—Shi (extreme south). —Mongolia, Manchuria. Described from Shilka. Map 54.

5. **H. nevskianum** Bowden 1965 in Canad. Journ. Genet. Cytol. 7 (3): 396. —*H. brevisubulatum* auct. p.p. —*H. brevisubulatum* var. *nevskianum* (Bowden) Tzvelev. —*H. brevisubulatum* subsp. *nevskianum* (Bowden) Tzvelev.

Cespitose plant with slender, almost invariably glabrous, erect, 60–80 cm tall stems, short dense-pubescence only along joints. Leaves green, rather scabrous on both surfaces, stiff, narrowly linear, flat; radical leaves sometimes convolute. Auricles poorly developed and readily shedding. Spikes 4–7 cm long, narrow, brittle (predominantly in upper half), grayish violet-green. Glumes of middle spikelet slightly shorter than lemma (without awn); lemma rather scabrous on back, occasionally subglabrous, scabrous only in upper part, with short, 1–2 (2.5) mm long awn. Anthers 3–4 mm long.

Saline meadows along river valleys. **West. Sib.**: TYU—Tb, KU, OM, NO (Gorevka village), AL—Ba. **Cen. Sib.**: TU. —Transvolga, Urals, Mid. Asia, West. Asia. Described from southern Urals. Map 55.

6. **H. roshevitzii** Bowden 1965 in Canad. Jour. Genet. Cytol. 7(3): 395. —*H. sibiricum* Roshev. non Schenk.

Cespitose plant. Stems 35–75 cm tall, numerous, very slender, glabrous, smooth, with glabrous joints, usually erect or slightly geniculate at base. Leaves flat or convolute, grayish-green or green, narrow, slightly scabrous on both surfaces, auricles not developed. Spikes 3–7 mm long, brittle in upper half dark violet. Glumes of middle spikelet 1.5–2 times longer than lemma (without awn) and usually almost as long as awn of latter. Lemma glabrous, slightly scabrous only near tip, with 6–8 mm long awn. Anthers 0.7–1 mm long.

Saline and solonetz meadows on salt lake banks. **West. Sib.**: TYU—Tb (between Bol. Sorokin and Vorsikha villages), KE, AL—Go (Koo village in Chul'cha river estuary, Chuisk steppe). **Cen. Sib.**: KR—Kha, Ve, Tu. **East. Sib.**: IR—An (Kultuk station in Slyudyansk region), BU—Se, Yuzh (Ichotoi river in Dzhida river basin—class. hab.—and others), ChI—Mid. Asia (Balkhash region), Far East (south), NE China, Mongolia. Map 58.

7. **H. turkestanicum** Nevski 1934 in Tr. Sredneaz. un-ta 8c, 17: 45. —*H. secalinum* var. *brevisubulatum* f. *puberulum* Krylov. —*H. brevisubulatum* subsp. *turkestanicum* (Nevski) Tzvelev.

Compactly cespitose plant. Stems 20–60 cm tall, more or less geniculate at base, relatively thick, glabrous, with glabrous or short-haired stems. Leaves glaucous or glaucescent green, flat or convolute, radical leaves often short, densely puberulent or scabrous on both surfaces due to short dense spinules; auricles of lower leaves 1.5–2 mm long, linear-subulate, persisting even in old leaves. Spikes 2.5–6 cm long, grayish or dark violet, relatively broad, brittle, especially in upper part. Glumes of middle spikelet nearly as long as lemma (without awn) or slightly longer. Lemma of central flowers usually densely puberulent on back, sometimes subglabrous, with short (1–2.5 mm long) awn. Anthers about 4 mm long.

Saline meadows of hill river valleys. **West. Sib.**: AL—Go (Ak-Alakh river on Ukok plateau, Bel'tir village, Kosh-Agach region). —Mid. Asia, Mongolian Altay, West. Asia (Iran), West. China, the Himalayas, Tibet. Described from Mid. Asia (Alai mountain range).

11. Bromopsis Fourr.

1. Lemma with 5–10 mm long awn. Rhizome shortened or with short decumbent shoots .... 2.
+ Lemma awnless or with short, 1–4 (6) mm long awn. Rhizome usually with long decumbent shoots .... 3.
2. Shoots covered at base with dead leaf sheaths. Sheaths undivided or slightly split into erect fibers and bearing deltoid auricles ....
   .... 4. *B. benekenii*.
+ Shoots covered at base with caps of reticulate-pilose sheaths of dead leaves. Sheath without auricle. .... 9. *B. riparia*.
3. Stem joints rather pubescent, rarely glabrous. Lemma covered with rather dense hairs along nerves, and sometimes between them, generally with 1–6 mm long terminal awn .... 4.
+ Stems glabrous or lower joints slightly pubescent. Lemma glabrous along nerves but in lower part at base sometimes scabrous or appressed-puberulent, usually awnless or with relatively short, 1–3 mm long terminal awn .... 5. *B. inermis*.
4. Lemma (even though only in young florets) pubescent throughout back with rather long hairs nearly equal in length .... 5.
+ Lemma rather pilose along marginal and middle nerves, with developed hair bands, glabrous or pubescent between them, but pubescence usually concentrated in lower part of lemma and consisting of short hairs or spinules .... 6.
5. Panicles pyramidal, lax; spikelets on slender flexuous, sometimes nutant branches. Palea with dense soft, more or less curved hairs along keels, becoming more erect and distant in fruit ..... 1. *B. alpina*.
+ Panicles oblong, narrow, compressed. Spikelets on erect, rather thick, obliquely erect branches. Palea with long dense erect cilia along keels negligibly interspersed with short spinules ....
   .... 11. *B. taimyrensis*.
6. Branches of panicles and sometimes even stems under panicles covered with short, usually soft, slightly crispate hairs ....

.... 7.

+ Branches of panicles scabrous due to spinules, sometimes fairly long, transformed into short stiff apically curved hairs. Stems glabrous or slightly scabrous under panicles .... 8.

7. Stem joints as well as stems pubescent with only recurved hairs. Lower branches of panicles short, equal to or not more than double length of spikelets; upper spikelets often sessile .... 7. *B. korotkiji.*

+ Stem joints and stems near joints with hairs usually decurved and very rarely recurved. One of lower branches of panicles a few times (2 or more times) longer than sessile spikelets; upper spikelets on very short branches .... 8. *B. pavlovii.*

8. Leaves flat, yellowish or grayish-green, narrow (2–4, rarely up to 6 mm), relatively long and usually narrowly acuminate. Lemma with fairly broad bands of long hairs along marginal nerves; if hairs shorter lemma usually glabrous between nerves. Panicles oblong or ovate, greenish-brown, dark violet, sometimes speckled violet .... 9.

+ Leaves flat, glaucous green, relatively broad (4–8 mm) and short, acute toward tip. Lemma with very short hairs along marginal nerves or sometimes spinules; with sparse, short appressed hairs or spinules, less often subglabrous between nerves. Panicles pyramidal or ovate, speckled (spikelets greenish-brown at base; dark violet at tip); green only in shaded specimens .... 2. *B. altaica.*

9. Branches of panicle glabrous or rather scabrous. Sheaths and leaf blades (except for lowest ones) glabrous, occasionally with distant long hairs on one or both surfaces. Plant of meadows and tundras .... 10.

+ Branches of panicle sharply scabrous, sometimes subpuberulent. Sheaths and all leaf blades covered with dense, somewhat entangled and shorter hairs on both surfaces. Plant of exposed arid habitats .... 6. *B. karavajevii.*

10. Lemma covered between nerves at very base (usually not above center) with hairs much shorter than on nerves, sometimes pubescence distinct only in upper florets of spikelet .... 11.

+ Lemma glabrous between nerves; occasionally with small glabrous patch at very base; slightly above, near center of lemma, short-scabrous or with scattered spinules .... 10. *B. sibirica.*

11. Panicles narrow-oblong, branches obliquely erect or appressed to main rachis, glabrous or more or less scabrous. Lemma pubescent along marginal nerves almost up to top with long, semidistant or appressed hairs .... 12.

+ Panicles broad-oblong or ovate, branches spreading or ascending, flexuous or nutant, sharply scabrous. Lemma pubescent along marginal nerves slightly above center, pubescence varying from long

appressed hairs to minute spinules .... 3. *B. austrosibirica.*
12. Spikelets grayish-violet. Panicle branches obliquely erect only at anthesis, later appressed to main rachis. Keels of palea with rather long erect cilia. Pubescence between nerves on back of lemma generally long persisting; hairs along midrib not rarely reaching awn base but usually invariably above middle. Awns of lemma 1–5 mm long .... 11. *B. taimyrensis.*
+ Spikelets greenish-violet; panicle branches obliquely erect, not appressed. Keels of palea with soft flexuous hairs. Pubescence between nerves on back of lemma disappearing with age; hairs along midrib usually extend from base to middle. Awns of lemma 1–2 (4) mm long, often absent .... 12. *B. vogulica.*

1. **B. alpina** (Malyschev) Peschkova 1986 in Novosti sist. vyssh. rast. 23: 26. —*Bromus sibiricus* f. *alpinus* Malyschev. —*Bromopsis arctica* var. *alpina* (Malyschev) Peschkova.

Rhizomatons plant, 30–60 cm tall, sometimes forming loose mats. Stems glabrous, pubescence varying from barely visible to moderate only at joints. Leaf sheath more often glabrous, leaf blades with sparse distant hairs on upper surface. Panicles 5–12 cm long, pyramidal, rather nutant, lax, branches very slender, fairly long, rather scabrous, slightly flexuous. Spikelets 1.5–2.5 cm long, speckled, brownish-violet, rachilla with soft hairs. Glumes with diffuse hairs along sides and center of rib, less often glabrous. Lemma more or less uniformly with dense hairs (pubescence along nerves weak) on back, with 2–5 mm long terminal awn. Palea with dense soft curved hairs along keels becoming more erect and distant in fruit.

Rocky tall-grass meadows in high hills, near scrubs, sometimes thin forests; descending in northern regions into forest belt along river valleys. **East. Sib.:** IR—Pr (source of Khomolkho river in Bodaibinsk region, Mama settlement in Mamsk-Chuisk region), BU—Se, Yuzh (Ubur-Khongoldoi river in Tunkinsk alps—class. hab.—and others), ChI—Ka, Shi (Orocha river upper course in Uryumkan river basin), YAK—Vi. —Endemic. Map 59.

2. **B. altaica** Peschkova 1986 in Novosti sist. vyssh. rast. 23: 26.

Plant with short rhizome forming loose mats. Stems 25–100 cm or more tall, with pubescent or sometimes glabrous joints. Leaves flat, rather broad ((2) 4–8 mm), relatively short, crowded in lower half of stem, glabrous or pilose together with lower sheaths. Panicles 7–20 (25) cm long, pyramidal or ovate, with spreading long, often flexuous branches, densely covered with long spinules, sometimes transformed into short hairs. Spikelets 1.8–3 (4.5) cm long, multicolored, greenish-brown and dark violet, green only in shaded specimens. Rachilla pilose or with short spinules. Glumes ovate-lanceolate, glabrous. Lemma with sparse appressed short hairs along marginal nerves, midrib with scattered short spinules or glabrous, with short spinules or hairs between nerves near base, with 0.5–5 mm long terminal awn (awn length variable within the same panicle). Palea with short spinules or pilose along keels. Altay (Aktash settlement in Ulagansk region), $2n = 28$.

Alpine and subalpine meadows of moderate and upper hill belts, river valleys, larch forests and turfed rocky slopes. **West. Sib.**: AL—Go (Tarkhatta river—class. hab.—and others). **Cen. Sib.**: TU (far south: Tsagan-Shibetu mountain range and Sangilen upland). —Kazakhstan Altay, Mongolia. Map 64.

3. **B. austrosibirica** Peschkova 1986 in Novosti sist. vyssh. rast. 23: 28.

Densely foliate plant with long rhizome. Stems single, 30–100 cm or more tall, more or less pubescent or glabrous only at joints. Leaves usually narrow, 2–4 (6) mm broad, acuminate, flat, glabrous or pubescent with sparse long distant hairs on one or, rarely, both surfaces. Panicles oblong, with slender obliquely erect branches, often flexuous or nutant. Branches covered with long spinules, sometimes transformed into short stiff curved hairs. Spikelets 1.5–3 (4.5) cm long, greenish-brown, sometimes with violet tinge. Rachilla usually comparatively puberulent. Glumes ovate-lanceolate, scabrous along keel. Lemma pilose along marginal nerves (length of hairs highly variable); midrib glabrous or subglabrous; short pubescence visible between nerves at base of lemma, sometimes seen only in upper florets in spikelets (youngest ones). Palea with dense, uniformly arranged spinules along keels, sometimes transformed into stiff cilia.

Forests, predominantly larch and pine, moderate and upper hill-forest belt; forest fringes, forest meadows, glades, exposed slopes; ascending along river valleys into subbald-peak belt and descending into hill steppe belt. **West. Sib.**: AL—Go (upper course of Tobozhok and Bukusun rivers). **Cen. Sib.**: KR—Kha, Ve, TU (Marachevka village in Tandinsk region—class. hab.—and others). **East. Sib.**: IR—An, Pr, BU—Se, Yuzh, ChI—Shi (Mal. Sokhondo town, pass in Yablonov mountain range), YAK—Vi. —Endemic. Map 60.

4. **B. benekenii** (Lange) Holub 1973 in Folia Geobot. Phytotax. (Praha) 8 (2): 167. —*Bromus ramosus* auct. p.p. —*Bromus benekenii* (Lange) Trimen. —*Bromopsis ramosa* subsp. *benekenii* (Lange) Tzvelev.

Perennial, tall (up to 150 cm) plant forming lax mats. Stems few, scabrous under panicles; above joints and partly at joints, puberulent with decurved hairs. Sheaths of lower and middle leaves with lanceolate or deltoid auricles at tip; covered with long jointed hairs like leaf blades, sometimes interspersed with short, slightly curved hairs. Leaves flat, broad (0.5–1.2 cm). Panicles with obliquely erect or nutant branches, scabrous; lower branches in 2 to 5 together, highly unequal in length. Spikelets green, 2.5–3.5 cm long, rachilla puberulent. Lemma lanceolate, gradually acuminate, covered on sides with fairly long appressed hairs, with scabrous 5–8 m long terminal awn. Palea distinctly shorter than lemma, with short spinules, usually unequal in length along keels. Anthers 3–4 mm long.

Linden and spruce-fir forests. **West. Sib.**: KE (Kuznets Alatau), AL—Ba. —Europe, Caucasus, Mid. Asia, Mediterranean, West. Asia. Described from Denmark.

5. **B. inermis** (Leysser) Holub 1973 in Folia Geobot. Phytotax. (Praha), 8 (2): 167. —*Bromus inermis* Leysser. —*Zerna inermis* (Leysser) Lindman.

Perennial, with long decumbent rhizome. Stems 30–150 cm tall, usually glabrous, sometimes rather pilose but then joints more modestly pubescent. Sheaths and leaf blades usually glabrous, occasionally pubescent with long squarrose hairs, without auricles, rarely with very short obtuse auricles. Panicles (7) 10–25 cm long, erect, with obliquely erect scabrous branches. Glumes unequal, obtuse and broad, terminally tunicate. Rachilla covered with short spinules or hairs. Lemma glabrous or scabrous on back near base, obtuse, awnless or with short 1–3 mm long terminal awn. Palea with very short, crowded hairs along keels, sometimes transformed into spinules. West. (Chuisk road in Altay) and Central (environs of Kyzyl) Siberia, $2n = 28$.

Meadows, forest glades, fringes, meadow steppes, among shrubs, sand-pebble riverbeds, roadsides, farm hedges, fallow land. **West. Sib.**: TYU—Khm, Tb, KU, OM, TO, NO, KE, AL—Ba, Go. **Cen. Sib.**: KR—Pu, Tn, Kha, Ve, TU. **East. Sib.**: IR—An, Pr, BU—Se, Yuzh, ChI—Ka, Shi, YAK—Vi. —Europe, Caucasus, Mediterranean, Asia, Far East (introduced plant). Described from central Europe (environs of Galle).

6. **B. karavajevii** (Tzvelev) Czer. 1981, Sosudistye rast.: 337. —*B. pumpelliana* subsp. *karavajevii* (Tzvelev) Tzvelev. —*Bromus ircutensis* auct. non Kom.

Plant with long rhizome, sometimes forming lax mats. Stems 6–100 cm or more tall, densely pubescent at and above joints, villous due to decurved hairs, rather scabrous under panicle. Sheaths and leaf blades densely pubescent with distant long hairs, usually slightly decurved. Panicles narrow, but not appressed, erect or slightly nutant, branches short (2–5–7 cm long), obliquely erect, slightly away from main rachis, usually covered with dense short spinules. Spikelets 1.5–2.5 cm long, usually brown-violet. Rachilla hirsute. Glumes glabrous or with short sparse hairs-spinules along nerves. Lemma rather pilose along nerves and sometimes on back at base, with 2–4 mm long terminal awn; rarely, awn very short. Palea with short and stiff cilia along keels.

Meadow steppes, steppified meadows, arid larch forests and their fringes. **Cen. Sib.**: KR—Kha (Ulen' village, Son station), Ve. **East. Sib.**: IR—An, YAK—Vi, Al (Churapcha village—class. hab.—and others). — Endemic. Map 61.

7. **B. korotkiji** (Drobov) Holub 1973 in Folia Geobot. Phytotax. (Praha) 8 (2): 168. —*B. pumpelliana* subsp. *korotkiji* (Drobov) Tzvelev. —*Bromus korotkiji* Drobov. —*B. ircutensis* Kom.—*Zerna korotkiji* (Drobov) Nevski.

Perennial, with long branched rhizome. Stems 50–80 cm tall, together with lower sheaths rather pubescent with recurved subsquarrose hairs, most dense at and under joints. Ligule short, with short cilia along fringe. Panicles 7–15 cm long, compressed, slightly nutant, branches short, 1–2 (up to 5) cm long, pilose, densely appressed to main rachis. Spikelets 1.5–3 cm long, light brown, sometimes slightly violet, rachilla with long and soft hairs. Lemma more or less obtuse, awnless or with short (up to 1.5 mm) awn, crinite on marginal nerves almost up to top; hairs mostly very short. Palea covered with dense soft hairs along keels that transform into cilia in upper part.

Sand deposits along steppe river valleys and Baikal coasts. **East. Sib.**: IR—An (Ol'khon island), BU—Se (Ulan-Burga river, Barguzin tributary—class. hab.—and others), Yuzh (Ikhe-Ogun river in Tunkinsk valley, Posol'sk and Boyarsk settlements), ChI—Ka (Kalakan river, tributary of Vitim river), Shi (Srednii Zakultui ravine in Olovyaninsk region and Tsarik-Narasun area near Onon river). —Endemic. Map 63.

8. **B. pavlovii** (Roshev.) Peschkova, comb. nova. —*Bromus pavlovii* Roshev. 1926 in Sev. Mongoliya 1: 161. —*B. ircutensis* auct. non Kom. — *Bromopsis korotkiji* auct. p.p. —*B. pumpellianus* subsp. *korotkiji* auct. p.p.

Perennial with long decumbent rhizome. Stems 80–100 cm tall, subglabrous, sparsely pubescent with rather short decurved hairs at and below joints. Leaf sheaths densely covered with soft, almost silky, densely appressed decurved hairs. Ligule 1–5 mm long, appressed-pilose on outer surface, sometimes glabrous. Panicles 12–30 cm long, narrowly compressed, slightly nutant, branches pilose, highly unequal, longest 6–12 cm, shortest 1.5–2 cm long, densely appressed to main rachis. Spikelets 2–3.5 cm long, light or slightly colored, rachilla subglabrous in lower part, crinite in upper. Lemma awnless or with short (up to 1 mm) awn; along lateral ribs densely hairy with long and divergent hairs; smooth, glabrous or subglabrous in middle part. Palea covered with short stiff hairs or spinules along keels.

Sand dunes along banks of steppe lakes and rivers. **Cen. Sib.**: TU (Lake Tere-Khol' in Erzinsk region). **East. Sib.**: BU—Yuzh (Kumyn hills in Kyakhtinsk region). Described from NW Mongolia.

9. **B. riparia** (Rehm.) Holub 1973 in Folia Geobot. Phytotax. (Parha) 8 (2): 158. —*Bromus riparius* Rehm.

Rhizome with short decumbent branches. Stems 60–75 cm tall, usually few, base covered with dead, reticulate-piloses sheaths, glabrous under panicle, scabrous or with short decurved hairs in lower half. Sheath without auricles and, like leaf blades, distant-crinite. Leaves narrow, 2–3 mm broad, long and acuminate. Panicles somewhat compressed, with

60

scabrous obliquely erect branches. Spikelets 1.8–3 cm long, green, rachilla spasely pilose. Lemma lanceolate, puberulent along sides or throughout surface, gradually acuminate into scabrous (3) 5–8 mm long awn. Palea distinctly shorter than lemma, with short spinules along keels. Anthers about 5 mm long.

Dry valley meadows and meadow steppes. **West. Sib.**: KU (Chistoe settlement in Shumikhinsk region). —Europe (southeast), Mid. Asia, introduced in Far East. Described from Ukraine.

10. **B. sibirica** (Drobov) Peschkova 1987 in Novosti sist. vyssh. rast. 23: 31. —*Bromus sibiricus* Drobov 1914 in Tr. Bot. muz. Akad. nauk 12: 229, p.p. quod var. *glaber* Drobov. —*Zerna pumpelliana* auct. —*Bromopsis pumpelliana* auct.

Plant with long rhizome 50–100 cm tall, foliate up to midstem and above. Stems glabrous, rather pubescent (sometimes very densely) only at joints and above with decurved hairs. Leaf sheaths and blades glabrous or more or less densely pubescent on one, very rarely both surfaces, with long divergent hairs; leaf blades flat, relatively narrow (2–4 (6) mm), acuminate. Panicles 7–15 (20) cm long, often (though not always) narrow, oblong, with slightly nutant, sometimes flexuous scabrous branches. Spikelets greenish-brown, sometimes tinged violet. Rachilla covered with long spinules or hairs. Glumes slightly obtuse, glabrous. Lemma with hair bands (sometimes reduced to spinules) along marginal nerves and midrib extending up to middle or slightly above; usually glabrous or with very short, faintly visible spinules between nerves, with 1–5 mm long terminal awn. Palea with dense spinules or cilia along keels. Yakutia (Chersk settlement), $2n = 56$.

Plains and lower hill belt, coniferous and mixed forests, meadows and forest glades, steppified meadows, coastal sand and pebble beds, exposed slopes. **Cen. Sib.**: KR—Tn, Kha, Ve, TU. **East. Sib.**: IR—An, Pr, BU—Se, Yuzh, ChI—Ka, Shi, YAK—Vi, Al (bank of Amga river near bridge on Ust'-Maisk road—class. hab.—and others), Yan, Ko.—Mongolia. Map 62.

11. **B. taimyrensis** (Roshev.) Peschkova sp. nova 1986 in Novosti sist. vyssh. rast. 23: 31, nomen nudum. —*Bromus sibiricus* var. *taimyrensis* Reshev, 1930 in Dokl. AN SSSR, 5: 108, nomen nudum. —*Bromopsis pumpelliana* var. *taimyrensis* (Roshev.) Tzvelev, 1976, Zlaki SSSR: 219, nomen nudum. —*B. arctica* auct.: Peschkova, 1986 in Novosti sist. vyssh. rast. 23: 28.

Plante laxe caespitosae breviter rhizomatosae (15) 20–50 (65) cm altae, nodis inferne approximatis. Caules in nodis plus minusve pilosi, sub panicula glabri. Folia plana, in parte caulis inferiora sita, ut vaginae glabra vel supra patenter pilosa, interdum utrinque pilosa. Paniculae 5–10 (15) cm longae, rectae angustae, ramulis post anthesin rachidi generali arcte appressis sparse asperis vel subglabris. Spiculae 1.5–2.5 cm longae, violaceae vel griseolo-violaceae, raro viridulo-fuscae, rachilla longe pilosa. Glumae glabrae vel sparse pilosae. Lemmata secus nervos marginales ac medium usque fere ad apicem longe et dense pilosa, inter nervos plerumque brevius ac sparsius pilosa, interdum glabrascentia, apice in aristam 1–5 mm longam abeuntia. Paleae superiores ad carinam pilis plus minusve longis, ciliis 0.5–1 mm longis vel aculeolis tectae.

Typus: Taimyr Orientalis, in fluxu inferiore fl. Jamu-Nera, 74°50′ lat. bor. et 106° long. orient. (systema lacus Taimyrensis), ad declive arenosum ardum 6 VIII 1928 No. 686, A.I. Tolmatschew (LE).

Affinitas. Species B. *vogulicae* (Soczava) Holub et B. *arcticae* (Shear) Holub affinis, a quibus panicula angusta, ramulis rachidi generali arcte appressis sparse asperis differt.

Distributio. Paeninsula Taimyr, Putorana, Jakutia arctica.

Loosely cespitose plant with short rhizome, 15 (20)–50 (65) cm tall, with joints proximate in lower part. Stems rather pubescent at joints, glabrous under panicle. Leaves flat, crowded in lower half of stem; like sheath, glabrous or distant-pilose on upper surface, sometimes on lower surface as well. Panicles 5–10 (15) cm long, erect, narrow; with scabrous branches closely appressed to main rachis after anthesis. Spikelets 1.5–2.5 cm long, violet and grayish-violet, rarely greenish-brown. Rachilla crinite. Glumes glabrous or diffusely pilose. Lemma with dense long hairs along marginal nerves and midrib; pubescence almost reaching awn tip, hairs between nerves usually very short and sparse, sometimes lacking, with 1–5 mm long terminal awn. Palea with rather long hairs-cilia, or spinules 0.5–1 mm long along keels. Central Siberia, $2n = 42$ (Baselak Lake on Putoran plateau) and 56 (Talnakh settlement).

Sand and pebble bed terraces of rivers and lakes, lichen-larch thin forests, bushy meadows, meadow slopes. **Cen. Sib.**: KR—Ta (Yamu-Nery river—class. hab.—and others), Pu. **East. Sib.**: YAK—Ar, Ol, Yan. — Endemic. Map 56.

B. *arctica* (Shear) Holub erroneously cited by some investigators for Siberia, is not found there; it grows for certain in Kamchatka and is common in Alaska and northwestern Canada.

12. **B. vogulica** (Soczava) Holub 1973 in Folia Geobot. Phytotax. (Praha) 8 (2): 169. —B. *pumpelliana* subsp. *vogulica* (Soczava) Tzvelev.

Loosely cespitose plant with short rhizome. Stems 25–50 (70) cm tall, few, glabrous, pubescent with decurved hairs at and above joints, with proximate internodes in lower part. Sheaths of lower leaves densely pubescent with distant slightly decurved hairs; leaf blades flat, diffusely pubescent on upper, occasionally both surfaces. Panicles with nutant, slightly scabrous branches; during anthesis and thereafter; before anthesis narrow, with branches appressed closely to main rachis. Spikelets 10–20 mm long, dichromatic, speckled, greenish-brown, with slight violet tinge. Rachilla with distant soft hairs. Glumes with isolated hairs along midrib or glabrous. Lemma along marginal nerves with fairly broad and dense band of long hairs almost reaching awn tip, usually glabrous or rather pilose between nerves, rather obtuse, obtuse or with short, up to 2 (4) mm long terminal awn. Palea with long soft hairs along keels.

Arid rubble slopes in bald-peak belt. **West. Sib.**: TYU—Yam, Khm. **Cen. Sib.**: KR—Pu (Khantaisk Lake). —Urals. Described from Urals (Kozhim river—Kos'ya river tributary). Map 65.

12. Bromus L.

1. Awns of dry plants and fruiting plants erect, not laterally declinate, or lacking .... 2.
+ Awns of dry plants and fruiting plants distinctly laterally declinate .... 4.
2. Sheaths of all leaves glabrous or covered with very short hairs in lower leaves. Anthers longer than 1 mm .... 3.
+ Sheaths of all leaves or only lower ones covered with fairly long (along 1 mm) hairs. Anthers about 0.5 mm long .... 3. *B. mollis.*
3. Spikelets with more or less distinct pinkish-violet tinge. Lemma invariably with fairly long (3–8 mm) awns. Anthers 3–4 mm long .... 1. *B. arvensis.*
+ Spikelets pale green. Lemma awnless or with 1–6 mm long awns. Anthers 1.5–2.5 mm long .... 4. *B. secalinus.*
4. Lemma narrowly obovate, with slightly protruding rounded corners on sides and barely visible terminal teeth up to 0.2 mm long. Palea almost half narrower than lemma; usually with more than 15 rigid cilia on each keel .... 2. *B. japonicus.*
+ Lemma broadly obovate, with highly prominent corners on sides and small but very distinct terminal teeth about 0.3 mm long. Palea almost third narrower than lemma, usually with less than 15 rigid cilia on each keel .... 5. *B. squarrosus.*

1. **B. arvensis** L. 1753, Sp. Pl.: 77

Stems 30–80 cm tall, glabrous. Leaves linear, flat, up to 6 mm broad, puberulent together with sheaths. Panicle broadly spreading, more or less violet, with long scabrous branches bearing 2–5 (7) spikelets. Spikelets lanceolate or oblong, somewhat narrowed toward tip, 5- to 10-flowered. Glumes lanceolate, glabrous, upper slightly longer than lower. Lemma about 8 mm long, oblong-lanceolate, rather angular along margins, with 2 short terminal teeth; scabrous erect short awn emerging from base of teeth. Palea nearly as long as lemma but slightly narrower, ciliate along keels. Anthers 3–4 mm long, linear.

As weed among hedges, fallow lands and in farms. **West. Sib.**: KE (Munzha river near Vaseevsk settlement), AL—Ba (Troitskoe village). **Cen. Sib.**: KR—Ve (environs of Krasnoyarsk). —Europe, Caucasus,

Mediterranean, West. Asia, introduced in other regions. Described from southern Europe.

2. **B. japonicus** Thunb 1784 in Murray, Syst. Veg. ed. 14: 119.

Stems 20–60 cm tall, numerous, aggregated into mat, glabrous, pilose in joints. Leaves narrowly linear, 1.5–4 mm broad, pubescent with long fine hairs on upper surface as well as outer surface of sheath. Panicles 5–12 cm long, green, initially spreading, lax, later secund, with slender scabrous nutant branches bearing 1–3 spikelets each. Spikelets lanecolate, 6- to 12-flowered. Glumes oblong-lanceolate, glabrous, upper 2–3 mm longer than lower. Lemma 8–9 mm long, oblong-elliptical, indistinct and roundly angular along margins above center, 2 sharp terminal teeth (teeth about 2 mm long); awn initially erect, later laterally declinate; awn 9–12 mm long in upper florets in spikelets, and, considerably shorter in lower florets. Palea distinctly (by 2 mm) shorter than lemma, with distant long cilia along keels. Anthers 1–1.5 mm long, oblong-oval.

Weed on roadsides and in farms. **Cen. Sib.**: KR—Ve (Ust'-Yansk village and Minusinsk town). —Europe, Causcasus, Mediterranean, Asia, introduced in other regions. Described from Japan.

3. **B. mollis** L. 1762, Sp. Pl. ed. 2: 112.

Stems 20–45 cm tall, covered with short decurved hairs. Leaves relatively short, flat, 1–3 mm broad; together with sheaths, pubescent with long distant hairs. Panicles 2–5 cm long, whitish-green, erect, dense, with short dense hairs, branches obliquely erect with 1 or 2 spikelets. Spikelets oblong, (2-) 6- to 10-flowered. Glumes somewhat unequal, broadly oblong or lanceolate, pubescent with long and short hairs. Lemma 8–9 mm long, oblong-elliptical, not angular along edges, pubescent with various size hairs, with 2 short sharp terninal teeth; awn erect, as long or slightly shorter than lemma. Palea distinctly (by 1 mm) shorter and half width of lemma, with interrupted cilia along keels. Anthers 0.5–0.7 mm long.

Weed, introduced. **Cen. Sib.**: KR—Ta (Noril'sk town), Kha (Charkovo settlement in Ust'-Abakansk region). —Europe, Caucasus, Mediterranean, Asia Minor, introduced in other regions. Described from southern Europe.

4. **B. secalinus** L. 1753, Sp. Pl.: 76.

Stems up to 1 m tall, glabrous, with short, rather sparse hairs only at joints. Leaves narrow (3–5 (8) mm), glabrous or slightly puberulent on upper surface, sheath glabrous. Panicles 10–16 cm long, oblong, pale green or yellowish, lax, with long, scabrous obliquely erect branches with 1 or 2 spikelets. Spikelets ovate or oblong-ovate, 3- to 10-flowered, not superposed, with distinct awn. Glumes ovate, glabrous, somewhat unequal, upper somewhat longer than lower. Lemma 7–8 mm long, ovate, glabrous, with highly involuted edges; 2 short, obtuse teeth at tip, awn short (1–2 mm) or lacking. Palea as long as lemma, with long cilia along keels. Anthers about 2 mm long.

Weed in farms, on waste and fallow lands. **West. Sib.**: TYU—Tb, TO (near Tomsk, Pozdneevo village), NO (near Kargat settlement, Biisk steppe), KE (Kondoma and Tumanovo villages). **Cen. Sib.**: KR—Ta (Valek settlement), Kha (Tashtyp village), Ve. **East. Sib.**: BU—Yuzh (Mishikha station). —Europe, Caucasus, Mediterranean, West. Asia, introduced in other regions. Described from Northern Europe.

5. **B. squarrosus** L. 1753 Sp. Pl.: 76.

Stems 10–50 cm tall, slender, glabrous, with isolated hairs only at joints. Leaves 1–3 mm broad, flat or longitudinally folded, together with sheaths pubescent with fine, long distant hairs. Panicles 3–12 cm long, lax, secund, pale green, with long, slender, scabrous, nutant branches, usually with single spikelet. Spikelets ovate-oblong, strongly laterally compressed, 5- to 12-flowered. Glumes oblong-lanceolate, glabrous, occasionally pilose; upper considerably longer than lower. Lemma 8–9 mm long, broadly oblong-rhombic, highly angular along margins, with 2 very short (0.3 mm long) but distinct terminal teeth; awn up to 10 mm long, initially erect, later strongly laterally recurved. Palea almost 1/3 narrower than lemma, with distant cilia along keels, usually less than 15 on each keel. Anthers about 1 mm long.

Flat steppes. **West. Sib.**: AL—Ba (Klyuchi village). **Cen. Sib.**: KR—Pu (Dudinka town). —Europe, Caucasus, Mediterranean, Mid. and West. Asia, West. China (Junggar), the Himalayas; introduced in other regions. Described from "France, Switzerland and Siberia".

13. Avena L.

1. **A. fatua** L. 1753 Sp. Pl.: 80. —*A. septentrionalis* Malzev.

Plant 50–120 cm tall. Stems glabrous or pubescent at joints. Leaf blade flat, 5–10 mm broad, ciliate, very rarely pilose along margin at base. Panicles lax, 15–30 cm long, spreading. All florets in spikelet with joints (falling singly as fruit ripens). Glumes 20–25 mm long, acuminate. Segments of rachilla glabrous under lower floret and hirsute above. Lemma 15–20 mm long, glabrous or pubescent with long (about 30 mm), geniculate awn convoluted in lower part.

Weed, in farms, on roadsides. **West. Sib.**: TYU—Tb, KU, OM, TO, NO, KE, Al—Ba, Go. **Cen. Sib.**: KR—Kha, Ve, TU. **East Sib.**: IR—An. —Almost all extratropical countries of both hemispheres. Described from Europe.

14. Avenula (Dumort.) Dumort.

1. Palea covered with spinules along keels. Hairs in upper part of rachilla segments up to 3 mm long .... 2.

+ Palea glabrous along keels. Rachilla covered with 3–7 mm long hairs .... 4. *A. pubescens.*
2. Sheath of cauline leaves closed for less than third length from base. Hairs in upper part of rachilla segments 0.8–1.3 mm long .... 3.
+ Sheath of cauline leaves closed for half or more from base. Hairs in upper part of segments of rachilla 2–2.5 mm long. Stems thick, with short decumbent shoots, hardly forming mats. Leaf blade 5–12 mm broad, flat .... 1. *A. dahurica.*
3. Panicles 4–9 cm long, dense. Spikelets brown, with golden tinge High-altitude plant .... 2. *A. hookeri* s. str.
+ Panicles 5–15 cm, multispiculate, lax. Spikelets pale green or yellowish. Plant of moderate and low altitudes ....
.... 3. *A. hookeri* subsp. *schelliana.*

1. **A. dahurica** (Kom.) Holub 1976 in Folia Geobot. Phytotax. (Praha) 11: 295. —*Avenastrum dahuricum* (Kom.) Roshev.—*Helictotrichon dahuricum* (Kom.) Kitag.

Tall (up to 100 cm) plant with thick stems (2–3 mm thick) and decumbent rhizome, usually not forming mats. Leaves broad (5–12 mm), flat, short, sharply narrowed at tip, with thick whitish fringe along margin. Leaf ligules about 5 mm long. Leaf sheath closed half or more from base. Panicles few-flowered, usually, with speckled, 12–20 mm long spikelets. Lemma 11–14 mm long, with 1.5–2 mm long hairs at base and geniculate awn on back.

Meadows, forest glades, among shrubs, thin forests. **East. Sib.:** ChI—Shi (Shilka river—class. hab.—and others), YAK—AL (M. Nimnyr village), Yan.—Far East, Manchuria, Korea. Map 72.

2. **A. hookeri** (Scribner) Holub s. str. 1976 in Folia Geobot. Phytotax. (Praha) 11: 298. —*Avenastrum asiaticum* (Roshev.) Grossh. —*Helictotrichon asiaticum* (Roshev.) Grossh. —*H. hookeri* (Scribner) Henr. —*Avena versicolor* auct. non Vill.

Plants 15–50 cm tall, forming small loose mats. Leaves flat, 3–5 mm broad, scabrous along margin, with whitish cartilaginous veins. Ligules 4–5 mm long, truncate at tip. Panicles 3–6 cm long, compressed, not branched. Spikelets 10–18 mm long, brown or gold. Lemma covered at base with about 1 mm long hairs and geniculate awn on back. Altay (Aktash settlement), $2n = 14$.

Alpine meadows, tundras, rocks in high-altitude belt. **West. Sib.:** AL—Go. **Cen. Sib.:** KR—Kha, Ve, TU. **East. Sib.:** IR—An (Borgotoi river). —West. China (Junggar), North America, Mongolia. Described from Rocky Mountains (North America). Map 68.

3. **A. hookeri** (Scribner) Holub subsp. **schelliana** (Hackel) Lomonosova comb. nova. —*Avena schelliana* Hackel 1892 in Tr. Peterb. Bot. sada, 12: 419. —*Avenastrum schellianum* (Hackel) Roshev. —*Helictotrichon schellianum* (Hackel) Kitag. —*Avenula schelliana* (Hackel) Holub.

Stems 30–80 cm tall, numerous, forming small loose mats. Leaf blades 2–5 mm broad, flat or more or less longitudinally folded, with whitish cartilaginous fringe. Panicles 5–15 cm long, multispiculate, branched. Spikelets greenish-silver, sometimes with brown or violet tinge. Lemma 9–11 mm long, covered at base with 0.5–0.8 mm long hairs. Awns 13–15 mm long, geniculate.

Steppes, rocky slopes, rocks. **West. Sib.**: TYU—Tb (between Kalachinsk and Lagushkin villages, Larikha village), KU, OM (Yasnaya Polyana and Krasnoyarsk villages), NO, KE (Shibanovo and Kulebakino villages), AL. **Cen. Sib.**: KR—Kha, Ve, Tu. **East. Sib.**: IR, BU, ChI—Shi, YaK—Vi, Al, Yan. —Europe, Mid. Asia, Mongolia, Far East, Nor. China. Described from Amur. Map 71.

4. **A. pubescens** (Hudson) Dumort. 1868 in Bull. Soc. Bot. Belg. 7: 68. —*Avena pubescens* Hudson. —*Avenastrum pubescens* (Hudson) Opiz. —*Helictotrichon pubescens* (Hudson) Dumort.

Plants 40–100 cm tall, with short, decumbent shoots. Stems single or forming loose mats. Leaf sheaths pilose or glabrous. Leaf blade flat, 3–12 mm broad, pubescent or glabrous. Panicles spreading, loose, with long slender scabrous branches. Spikelets 11–16 mm long, greenish-silver, with crinite awn. Lemma with curved awn on back, emerging near middle. Palea glabrous along keels. Altay (Mëny village), $2n$ =14.

Meadows, forests, among shrubs. **West. Sib.**: NO, TO, KE, AL—Ba, Go. **Cen. Sib.**: KR—Kha, Ve, TU. **East. Sib.**: IR—An, Pr, BU—Yuzh. —Europe, Mid. Asia, West. China (Junggar). Described from Great Britain. Map 70.

15. Helictotrichon Besser

1. Leaf blade cross section without compact ring of subepidermal mechanical tissue .... 2.
+ Leaf blade cross section with compact ring of subepidermal mechanical tissue .... 3.
2. Leaf blade glabrous, smooth, lustrous outside (underside) ....
.... 4. *H. mongolicum* s. str.
+ Leaf blade more or less densely pilose outside ....
.... 5. *H. mongolicum* subsp. *sajanense*.
3. Sheath of cauline leaves glabrous or with up to 0.2 mm long hairs. Leaf blade glabrous or covered with diffuse short bristles outside
.... 4.
+ Sheath of cauline leaves covered with 0.6–1.2 mm long distant hairs. Leaf blade often also more or less pilose outside ..... 3. *H. krylovii*.
4. Leaf ligule pilose on back and along margin. Leaf blade 0.7–1.0 (1.2) mm diam, without bristles outside .... 5.
+ Leaf ligule elongate, glabrous or with isolated short hairs along margin. Leaf blade 0.5–0.7 mm diam, usually covered with diffused

short bristles outside .... 2. *H. desertorum.*
5. Plants with decumbent subsurface shoots, forming loose mats ....
.... 6. *H. sangilense.*
+ Plants without decumbent subsurface shoots, forming dense mats
.... 1. *H. altaicum.*

1. **H. altaicum** Tzvelev 1968 in Rast. Tsentr. Azii 4: 101. —*H. desertorum* auct. p.p. non Less.

Compactly cespitose plant 20–70 cm tall, with remnants of dead leaf sheaths at base. Leaf blades longitudinally folded, setaceous, 0.7–1.0 (1.2) mm broad, glabrous outside, without bristles and spinules. Radical leaves 2 or 3 times shorter than stem, rarely as long. Leaf sheath glabrous. Leaf ligules shortened, ciliate at tip, puberulent or scabrous outside. Panicles 4–10 cm long. Spikelets greenish-yellow, sometimes with faint violet tinge. Lemma 8–10 mm long, with 2–3 mm long hairs at base and geniculate awn on back. Altay (Kosh-Agach village, Terekhtinsk mountain range), $2n = 14$.

Steppes, rocky slopes. **West. Sib.:** AL—Ba (south), Go. **Cen. Sib.:** KR—Kha, Ve, TU. **East. Sib.:** IR—An, Pr, BU—Yuzh. —Mongolia. Described from Mongolia. Map 66.

2. **H. desertorum** (Less.) Nevski 1937 in Sov. bot. 4: 41. —*Avena desertorum* Less. —*Avenastrum desertorum* (Less.) Podp.

Large, cespitose plant 30-70 cm tall. Radical leaves numerous, setaceous, 0.5–0.7 mm diam, long, reaching inflorescence, covered outside with diffuse bristles. Ligule elongate, usually fimbriate-lacerate teminally, glabrous or with isolated spinules along margin. Panicles 6–8 cm long. Spikelets 10–12 mm long.

Steppes, rocky slopes. **West. Sib.:** KU, OM (Yasnaya Polyana village between Abrashkina, Andreevsk, Kalachinsk and Lagushkina villages), NO, KE, AL—Ba. **Cen. Sib.:** KR—Ve (Noshino village, Lebyazh'e Lake). —Europe, Urals, Mid. Asia. Described from South. Urals. Map 67.

3. **H. krylovii** (Pavlov) Henr. 1940 in Blumea 3: 431. —*Avenastrum krylovii* Pavlov.

Plant 20–40 cm tall, forming small compact mats. Radical leaves setaceous, longitudinally folded, 1–2 mm diam, short, not more than half of stem, pubescent with long hairs or glabrous with long hairs prominent along margin. Leaf sheath densely pubescent with long, distant hairs. Leaf ligule up to 0.2 mm long, densely pubescent along margin. Panicles 4–6 cm long, rather lax, oblong. Spikelets speckled. Lemma with geniculate awn.

Steppes, arid meadows, thin forests. **East. Sib.:** YAK—Vi (environs of Yakutsk—class. hab.—and others), Yan.—Chukchi. Map 73.

4. **H. mongolicum** (Roshev.) Henr. s. str. 1940 in Blumea 3: 431. — *Avenastrum mongolicum* (Roshev.) Roshev.

Plant 10–30 (50) cm tall, forming small compact mats. Leaf blade longitudinally folded, rarely flat, 1.5–2 (2.5) mm broad, glabrous and lustrous outside, with some lacerated strands of sclerenchyma in cross section. Leaf sheath glabrous or rather pubescent. Panicles 3–6 cm long, lax. Spikelets brownish-violet, speckled. Lemma (5) 7–9 mm long, with 1–1.5 mm long hairs at base; geniculate awn on back.

Alpine grasslands, rocky slopes and rocks in high-altitude belt. **West. Sib.**: AL—Go. **Cen. Sib.**: TU. **East. Sib.**: IR—An, BU—Se (source of Pravyi Ulyun river, Rytyi cape, Argoda river), Yuzh (Munku-Sardyk hill—class. hab.—and others), ChI—Shi (Mal. Sokhondo bald peak). —Mongolia, Tibet, Mid. Asia (Saur mountain range). Map 69.

5. **H. mongolicum** (Roshev.) Henr. subsp. **sajanense** Lomonosova subsp. nova.

Plantae caespitosae 20–35 cm altae. Laminae foliorum complicatae vel planae ad 2 mm latae unacum vaginis plus minusve dense pilosae, inflorescentces affingentes. Paniculae laxa variegatae, spiculis ca 10 mm longis.

Typus: Sajan orientalis, alpes Kitojenses, in fluxu superiore fl. Kitoj, regio subalpina in lariceto herboso raro, ad saxum 3 VII 1958, L. Malyschev.

Affinitas. A *H. mongolico* s. str. laminis longis pilosis differt.

Cespitose plant 20–35 cm tall. Leaf blade longitudinally folded or flat, up to 2 mm broad, together with sheath rather densely pilose, reaching inflorescence. Panicles lax, speckled, with spikelets about 10 mm long.

Tundras, rocks, high-altitude belt. **Cen. Sib.**: TU (Karga river upper course). **East. Sib.**: BU—Se (upper course of Kita river—class. hab.—and Zabit). —Endemic.

6. **H. sangilense** Krasnob. 1977 in Bot. zhurn. 62 (6): 854.

Perennial plant 70–80 cm tall, with decumbent subsurface shoots forming lax mats. Leaf blade longitudinally folded, setaceous, 0.8–1 (1.2) mm diam, glabrous outside, pubescent within. Ligule of cauline leaves about 1 mm long, pubescent. Panicles 6–7 cm long. Spikelets 12–13 mm long, 3- or 4-flowered. Lemma 9–10 mm long, with awns up to 2 cm long, emerging near middle.

Steppes, rocky slopes in upper mountain belt. **Cen. Sib.**: TU (upper course of Naryn river—class. hab.—and others). —Endemic.

16. Trisetum Pers.

1. Stems and branches of panicles glabrous, smooth or scabrous due to diffuse and very short spinules .... 2.
+ Stems rather densely pubescent with distant short hairs under panicle as well as panicle branches .... 5.
2. Callus glabrous or with very short hairs not reaching base of lemma; latter usually golden-brown, rather hard. Rachilla with long

(1–8 mm) hairs. Fairly large plant with short decumbent subsurface shoots and loose, although relatively narrow, panicle .... 3.
+ Callus with short hairs, slightly covering base of lemma; latter usually violet-brown or green, herbaceous-membranous. Rachilla with short, up to 1 (rarely more) mm long hairs. Rather low, loosely cespitose plant with fairly dense spicate panicle .... 4.
3. Panicles loose, oblong, 10–25 cm long. Plant 50–100 cm tall ....
.... 5. *T. sibiricum* s. str.
+ Panicles dense, oval, 3–8 (10) cm long. Plant up to 40 cm tall ....
.... 6. *T. sibiricum* subsp. *litorale*.
4. Leaves 0.5–2 (3) mm broad, scabrous or puberulent, often with rigid short cilia along margin, longest near point of blade deflection. Lower glumes usually covering awn base of adjoining florets. Lemma with very short spinules, dull, not lustrous. Palea distinctly shorter than lemma .... 1. *T. agrostideum*.
+ Leaves 2–4 mm broad, covered with long, fine, sparse hairs on both surfaces, occasionally on one or only along margin. Lower glumes usually terminating at base of awn, rarely somewhat higher. Lemma smooth, slightly lustrous. Palea nearly as long as lemma .... 2. *T. altaicum*.
5. Spicate panicles compact, not interrupted, lobed only sometimes in lower part, obtuse or rounded, grayish or brownish-violet at tip .... 6.
+ Spicate panicles relatively lax, interrupted often in lower part but sometimes even throughout length, narrowed or slightly acuminate at tip, pale green or slightly violet .... 3. *T. molle*.
6. Glumes unequal, upper distinctly longer and broader than lower; tips barely reach awn base of adjoining florets. Spicate panicles narrowly cylindrical, oblong or oblong-ovate, usually rather uniform in breadth .... 7. *T. spicatum*.
+ Glumes nearly equal and almost wholly covering adjoining florets. Spicate panicles broader, oval or oblong-ovate, sometimes broadened at tip .... 4. *T. mongolicum*.
1. **T. agrostideum** (Laest.) Fries 1842, Nov. Fl. Suec. Mant. 3: 180. —
T. subalpestre (Hartman) L. Neuman.
Stems 15–35 (50) cm tall, usually several, glabrous, forming small mats together with short and narrow radical leaves. Leaves 0.5–2 mm broad, flat or convolute, grayish-green, glabrous, or with very short hairs on upper surface; often with short cilia along margin, longest at point deflection of leaf blades. Ligules up to 1 (rarely up to 1.5) mm long, lacerated at tip, sometimes with cilia. Panicles 2–6 (10) cm long, somewhat compressed, narrow, 6–10 mm broad, but not compact, branches appressed to main rachis, smooth, sometimes slightly scabrous due to short spinules. Spikelets 5–6 mm long, violet, rarely green. Glumes

70

unequal; lower usually acuminate and longer than point of deflection of awn of adjoining floret. Lemma rather scabrous due to very short spinules, dull; awn emerging somewhat above middle, twice indistinctly geniculate. Palea distinctly (by 1–2 mm) shorter than lemma. Callus base puberulent. Anthers 0.4–0.8 mm long. East. Siberia (NE Baikal coast; Bogatyr' Lake on Putoran plateau; Utinsk pass on Kolyma upland), $2n = 28$.

Arctic and high-altitude tundra, rocky slopes and debris, arable grasslands, river and lake sand-pebble beds, rubble hill tops, forests and scrubs below bald peaks and their clearances; descends along river valleys into midportion of forest belt. **Cen. Sib.**: KR—Pu, Tn (Baikit settlement), Ve (Kizir river near Kinzelyuk river estuary), TU (upper course of Kara-Khem river in Tapsa river basin on Akademik Orbuchev mountain range and Ulet-Khol' Lake, Kaa-Khemsk region). **East. Sib.**: IR—An, Pr, BU—Se, Yuzh, ChI—Ka, Shi (Sokhondo and Mal. Sokhondo hills), YAK—Ar, Vi (Mogdy village in Mirninsk region and Peledui village in Lensk region), Al (Teplyi Klyuch settlement, Mal. Nimnyr, Evota pass), Yan (Segyan-Kyuël' settlement, Khunkhada river in Tompo river basin). —Scandinavia, Far East. Described from Sweden (Lapland). Map 74.

2. **T. altaicum** Roshev. 1922 in Bot. mat. (Leningrad) 3: 85.

Forming tiny mats. Stems 15–50 cm tall, glabrous, smooth. Leaves 2–4 mm broad, yellowish-green on both surfaces, rarely only on underside or with long fine distant hairs on margin. Ligules 1.5–3 mm long, crenate or lacerate. Panicles 2–10 cm long, 1–2 cm broad, rather loose, branches glabrous, smooth or slightly scabrous, obliquely erect. Spikelets 4–6 mm long, 2- or 3-flowered, violet, sometimes green. Glumes unequal, upper usually distinctly (frequently almost half) shorter than lower, sharply narrowed at tip; tips of lower glumes usually barely reaching or crossing site of awn deflection of adjoining floret. Lemma glabrous, usually smooth and slightly lustrous; awn arising somewhat above middle or in upper third, twice indistinctly geniculate, slightly contorted, fairly long (almost twice lemma length). Palea nearly as long as lemma. Callus with very short hairs at base. Anthers 0.8–1.5 mm long. Altay (Aktash settlement), $2n = 14$.

Alpine belt (1,500–2,400 m above sea level), alpine and subalpine meadows, sand-pebble beds along river valleys, thickets of subalpine shrubs and thin forests, wet rocks and turfed rocky slopes; descends into forest belt along rivers. **West. Sib.**: KE, AL—Go. **Cen. Sib.**: KR—Pu, Kha, Ve, TU. **East. Sib.**: IR—An, BU—Se, Yuzh, ChI—Shi (Sokhondo hill, source of Zun-Agutsakan, Kirkun and Bal'dzhi rivers in Onon basin). —Mid. Asia, West. China (Junggar), Mongolia (Hentei). Described from Altay. Map 75.

3. **T. molle** (Michaux) Kunth 1829, Rev. Gram. 1: 101. —*T. spicatum* subsp. *molle* (Michaux) Hultén. —*T. spicatum* f. *elatior* auct. non Krylov.

Loosely cespitose plant 10–40 (50) cm tall. Stems pubescent with soft usually decurved hairs under inflorescence or throughout length. Leaves 1–3 mm broad, scabrous or puberulent, with long cilia along margin; not rarely, together with sheaths, with long and soft hairs, sometimes subglabrous. Panicles 3–8 cm long, about 1 (rarely more) cm broad, rather green, sometimes slightly violet, elongate, rather lax, interrupted in lower part or throughout length, acuminate, branches with soft hairs. Glumes slightly unequal, upper longer and broader than lower, both distinctly shorter than adjoining florets. Lemma covered on back with exremely small hairs or very short spinules. Awn geniculate once or, occasionally twice, contorted. Callus base puberulent. Anthers 0.3–0.8 (up to 1) mm long. Krasnoyarsk region (Khakoma Lake on Putoran plateau), $2n = 42$.

Forest and subbald-peak belts on sand-pebble beds along river valleys and near water reservoirs, in dwarf arctic birch and dwarf-tree thickets, thin larch forests, near ice crusts, dumps, on rocky-rubble slopes; found at low levels (200–300 m above sea level) in high latitudes. **Cen. Sib.:** KR—Pu, Tn. **East. Sib.:** IR—Pr (Khomolkho river in Bodaibinsk region), BU—Se, ChI—Ka, YAK—Ar, Ol (Doldyn river near D'yakhaa river estuary in Oleneksk region), Al, Yan, Ko. —Far East, Japan, North America. Described from Canada. Map 90.

Siberian specimens do not match fully with type and obviously belong to 2 distinct geographic races. The northern race, extending from Kolyma to Putoran and from the Arctic to Baikal, is characterized by very dense green panicles interrupted only in lower part; glumes only sligthly unequal and usually longer, narrowly acuminate but not covering flowers completely. Sheath densely covered with distant hairs, ligules longer, slightly pilose outside, stems almost throughout length pubescent with long decurved hairs. The southern race found southeast of Stanovoi upland is characterized by slender interrupted and very narrow, usually violet panicles; glumes strongly unequal, strongly acuminate, distinctly shorter than adjoining florets. Sheath often glabrous or with diffuse divergent hairs, ligules very short, glabrous outside; stems pilose under inflorescence, sometimes glabrous in lower part.

4. **T. mongolicum** (Hultén) Peschkova 1979 in Fl. Tsentr. Sib. 1: 97. —*T. spicatum* subsp. *mongolicum* Hultén. —*T. spicatum* auct. p.p.

Rather low, (5) 10–25 (40) cm, cespitose plant. Stems densely pubescent with short hairs under panicles and branches of inflorescence. Leaves 1–4 mm broad, flat, occasionally convolute, grayish or yellowish-green, sometimes violet, scabrous or with short (velutinous) hairs usually on both or one surface; very rarely leaves pubescent with long, distant hairs (predominantly in largest specimens). Ligules 0.5–1.5 mm long, crenate or lacerate. Panicles 1–6 cm long, 6–18 (25) mm broad, dense, spicate, often violet, ovate or oblong, obtuse at tip; sometimes, panicles in lower part somewhat lobed, in well-developed specimens—lax: var. *elatior* (Krylov) Peschkova, comb. nova (*T. spicatum* f. *elatior* Krylov, Fl.

Altay, 1914: 1612). Spikelets 2- or 3-flowered, rachilla puberulent. Glumes nearly equal, lower slightly narrower and shorter than upper, almost completely covering adjoining florets. Lemma glabrous; sometimes scabrous on back due to barely visible spinules. Awn geniculate once, rarely twice, not contorted, obliquely erect or recurved almost at right angle. Callus base puberulent. Anthers 0.5–0.8 (1) mm long. Altay (Aktash village in Kuraisk mountain range), $2n = 28$.

Alpine belt at altitudes of 1,600–2,800 m above sea level in rubble, dryad and lichen tundras, on rocky, frequently carbonate slopes and talus, alpine grasslands, in high-mountain scrubs and subalpine thin forests; descending along river valley pebble beds into lower belts upto 800–1,000 m altitude. **West. Sib.**: AL—Go. **Cen. Sib.**: KR—Kha, Ve, TU (Peste-Karasuk river on southern slope of Tannu-Ol—class. hab.—and others). **East. Sib.**: IR—An, BU—Se, Yuzh, ChI—Ka (Serdnii Sakukan river on Kodar mountain range). —Mid. Asia, West. China, Tibet, the Himalayas. Map 76.

5. **T. sibiricum** Pupr. s. str. 1845, Beitrg. Planzenk. Reich. 2: 65.

Plant with short rhizome up to 1 m tall. Stems single or more, glabrous, smooth. Leaves 2–10 mm broad, flat, relatively short, glabrous, not rarely covered on upper surface with long, distant hairs, sometimes puberulent on lower surface; lower sheaths usually scabrous due to short dense spinules or hairs, sometimes long, sparsely hairy. Ligules 1–2 mm long, obtuse, crenate. Panicles 7–20 cm long, 3–6 (8) cm broad, narrowly oblong, spreading poorly, branches up to 4 cm long, scabrous or subglabrous. Spikelets 6–8 mm long, brownish-yellow, sometimes violet; rachilla with fairly long (1–1.8 mm) hairs. Glumes unequal, usually green along midrib, occasionally brownish-violet; lower almost half shorter and narrower than upper, broadly white scarious along margin. Lemma compact, brownish or cinnamon-golden, scabrous on back due to very short spinules; awn emerging from upper third of back, horizontally recurved or even decurved. Palea equal to or slightly shorter than lemma. Callus glabrous or with very short hairs, barely reaching lemma base. Anthers 1.8–2.5 mm long. Altay (Aktash settlement), $2n = 14$.

Mixed and deciduous forests, scrubs, forest glades and fringes, wet meadows and swamps; ascending along river valleys high into mountains. **West. Sib.**: TYU—Khm, Tb, KU, OM, TO, NO, KE, AL—Ba, Go. **Cen. Sib.**: KR—Tn, Kha, Ve, TU. **East. Sib.**: IR—An, Pr, BU—Se, Yuzh, ChI—Shi, YAK—Ol, Vi, Al, Yan, Ko. —Europe, Mid. Asia, Far East, Nor. China, Mongolia, North America (Alaska). Described from Malozemel'sk tundra (Belaya river).

6. **T. sibiricum** subsp. **litorale** Rupr. ex Roshev. 1922 in Izv. Glavn. bot. sada RSFSR, 21: 90.

Stems 15–60 cm tall, aggregated together with numerous radical leaves into loose mats. Panicles 3–7 cm long, 2–3 cm broad, ovate or oblong during anthesis, subspicate in fruit, branches up to 2 cm long. Glumes violet or brown along back. Krasnoyarsk region (Syndasko river in Taimyr), $2n = 14$.

Swamps, scrubs; tundra and subbald-peak belts along river and lake sand beds. **Cen. Sib.**: KR—Ta, Pu. **East. Sib.**: YAK—Ar, Ol, Yan, AI (Verkh. Amga settlement in Aldan region). —Arctic Europe, Urals, nor. Far East, North Amrica (Arctic Alaska). Described from Kanin peninsula.

7. **T. spicatum** (L.) K. Richter 1890, Pl. Eur. 1: 89.

Cespitose plant 8–25 (40) cm tall. Stems, predominantly in upper part, pubescent with long, divergent or decurved dense hairs. Leaves narrow, 1–2.5 (3) mm broad, flat or convolute, usually puberulent or subglabrous, with long hairs along margin, occasionally crinite. Ligules short, up to 1 mm long, crenate and short-ciliate. Panicles 1.5–5 cm long, 5–15 mm broad, brownish-violet, narrowly cylindrical, sometimes oblong-ovate, dense, lobed or interrupted below. Spikelets most often 2-flowered. Glumes unequal; lower perceptibly shorter and narrower than upper, not covering adjoining flowers. Lemma glabrous; like glumes, green along midrib, violet around it, with brownish-cinnamon scarious margin. Awn almost invariably geniculate twice and highly contorted. Callus base puberulent. Anthers 0.6–1 mm long. Krasnoyarsk region (Baselak Lake on Putoran plateau), $2n = 28$.

Bald peak belt of high latitudes and tundra zone, along pebble beds of hill rivers, sand banks of lakes, on rocks and talus, in meltwater puddles; rubble wet sites. **West. Sib.**: TYU—Yam, Khm. **Cen. Sib.**: KR—Ta, Pu. **East. Sib.**: YAK—Ar. —Arctic Europe, nor. and cent. Urals, Far East. Scandinavia, North America. Described from northern Sweden (Lapland). Map. 77.

× Trisetokoeleria Tzvelev

1. × **T. gorodkowii** (Roshev.) Tzvelev (*Koeleria asiatica* subsp. *asiatica* × *Trisetum sibiricum* subsp. *litorale*). Known from type collection. **West. Sib.**: TYU—Yam (Gydansk tundra, Gyda-Yam gulf basin, Esena-To Lake, 11 Vll 1927, B. Gorodkov).

2. × **T. taimyrica** Tzvelev (*Koeleria asiatica* subsp. *asiatica* × *Trisetum agrostideum*). Collected from 2 sites. **Cen. Sib.**: KR—Ta (Novaya river—left tributary of Khatanga, above Bogatyr' river estuary, 6 VII 1910, No. 146, N. Vargina; Syndasko river, in sedge tundra, 22 VII 1979, N. Vodop'yanova, R. Krogulevich and others).

17. Koeleria Pers.—Koeleria

1. Bases of stems and vegetative (sterile) shoots not covered with reticulately split old sheaths and not with bulbous thickening. Leaves glabrous, scabrous or pilose .... 2.

74

+ Bases of stems and vegetative shoots covered with reticulately split or filiform old sheaths and with bulbous thickening.Leaves densely puberulent on both surfaces .... 11. *K. glauca.·*
2. Plants usually forming mats .... 3.
+ Plants with long, branched rhizomes, not forming mats ....
.... 14. *K. skrjabinii*
3. Mats very loose, rhizome slender, decumbent .... 4.
+ Mats compact or loose, rhizome not decumbent .... 8.
4. Glumes obtuse or obtuse-acuminate .... 5.
+ Glumes short-, long- or awnlike-acuminate .... 6.
5. Leaf blades glabrous, sheath glabrous, occasionally diffusely pilose, glumes glabrous, lemma obtuse .... 9. *K. delavignei* s. str.
+ Leaf blades more or less pilose, sheath densely pilose, glumes pilose or scabrous, lemma acute ...... 10. *K. delavignei* subsp. *veresczaginii.*
6. Panicles oblong, violet or grayish-violet; spikelets usually pilose, occasionally glabrous .... 7.
+ Panicles pyramidal, dark violet, spikelets usually glabrous ....
.... 3. *K. atroviolacea.*
7. Mats not large, stems 30–40 cm tall, leaf blades usually squarrose-pilose outside, panicles grayish-violet .... 13. *K. ledebourii.*
+ Mats large, stems 10–30 cm tall, leaf blades glabrous or diffusely appressed-pilose outside, panicles violet .... 2. *K. asiatica.*
8. Vegetative shoots 2 or 3 together and covered by common cap of sheath remnants or single; mats very compact, old sheaths undivided, leaf blades flat, occasionally convolute, glabrous or scabrous due to short obtuse spinules or pilose; very stiff .... 9.
+ Vegetative shoots single, mats compact or rather loose, old sheaths laciniate or more or less undivided, leaf blades usually convolute, pilose, sometimes scabrous, with occasional hairs, rigid .... 13.
9. Vegetative shoots usually 2 or 3 together and covered by common envelope of old sheaths .... 10.
+ Vegetative shoots single, with 3 or 4 green leaves .... 11.
10. Leaf blade glabrous or scabrous due to short spinules, panicles loose, stems puberulent throughout length .... 12. *K. karavajevii.*
+ Leaf blade for most part densely long-pilose, panicles very compact, stems under panicle pilose for 1–3 cm .... 1. *K. altaica.*
11. Leaf blade convolute or flat, (1) 1.5–2 mm broad, lemma densely pilose throughout surface .... 12.
+ Leaf blade mostly flat, 2–3 (4) mm broad, lemma glabrous or diffusely pilose only at base .... 15. *K. thonii.*
12. Spikelets 5.5–6.5 mm long, leaf blade 2 mm broad, lemma acuminate, squarrose-pilose .... 4. *K. chakassica.*
+ Spikelets 4–4.5 mm long, leaf blade 1–1.5 mm broad, lemma acute, densely appressed-pilose .... 16. *K. tzvelevii.*

13. Mats compact, spikelets 3–5 mm long, lemma acute .... 14.
+ Mats rather loose, spikelets 5.5–7 mm long, lemma long- or awnlike-acuminate .... 15.
14. Stems (20) 30–40 (70) cm tall, sheath remnants more or less undivided, leaf blade of vegetative shoots 10–18 cm long, convolute, rarely flat, pilose; spikelets 4–5 mm long .... 5. *K. cristata* s. str.
+ Stems 15–35 cm high, sheath remnants laciniate, leaf blade of vegetative shoots 3–7 cm long, convolute, scabrous, sometimes with sparse hairs; spikelets 3–4 mm long ....
.... 7. *K. cristata* subsp. *mongolica*.
15. Leaf sheath semiappressed-pilose, not inflated, mats rather loose, spikelets (5) 5.5–6.4 mm long, upper glume barely shorter than adjoining lemma .... 6. *K. cristata* subsp. *hirsutiflora*.
+ Leaf sheath squarrose-pilose, slightly inflated, mat loose, spikelets 6–7 mm long, upper glume considerably shorter than adjoining lemma .... 8. *K. cristata* subsp. *seminuda*.

1. **K. altaica** (Domin) Krylov 1928, Fl. Zap. Sibiri 2: 261, p.p., excl. var. *glabrifolia*. —*K. eriostachya* subsp. *caucasica* var. *altaica* Domin.

Plant forming very compact mats. Stems 10–30 (35) cm tall, pilose, violet under panicle for 1–3 cm and at joints, with numerous undivided brown sheath remnants at base. Vegetative shoots 2 or 3 together, sometimes single, but with (3) 4 green leaves. Leaf blade of vegetative shoots 3–7 cm long, 1.5–2 mm broad, convolute or semiconvolute, rigid, glaucous green, crinite outside, scabrous within due to numerous short spinules or pilose; with cartilaginous teeth along margin. Sheath of cauline leaves densely pilose in lower part, scabrous in upper, slightly inflated, leaf blades 1.5–3 (4) cm long, longitudinally convolute, rarely flat, pilose outside, pilose or scabrous within. Panicles 2–4 (6) cm long, 6–12 mm broad, compact, sometimes interrupted, oblong, violet. Spiklelets 5–5.5 mm long, 2 (3)-flowered; glumes acuminate, violet, softly appressed pilose or pilose only along keel; lower 3.5–4 mm long, lanceolate, upper 4 mm long, broadly lanceolate, distinctly (by one-third) shorter than adjoining lemma. Lemma 4.5–5 mm long, acuminate or with short awn, appressed-pilose throughout length, violet. Anthers 1.8–2 mm long. Altay (Kosh-Agach village), $2n = 28$.

Hill steppe and forest steppe belts along slopes, limestone cliffs, rubble talus, grass-herbaceous steppes. **West. Sib.**: AL—Go. **Cen. Sib.**: TU. —West. China, Mongolia. Described from Altay. Map 82.

Variation is noticed in pubescence of glumes (from densely pilose to glabrous), panicle shape (from oblong to cylindrical) and compactness. *K. altaica* differs from other species in densely hirsute glaucous-green leaves of vegetative shoots, violet panicles, and crinite lemma and palea.

2. **K. asiatica** Domin 1905 in Bull. Herb. Boiss., ser. 2 (5): 947.

Plant forming loose mats. Stems 7–30 (40) cm tall, usually densely pilose throughout length, violet or rather greenish, strong, not thickened at base, covered by numerous, undivided straw-brown sheath remnants. Leaf blade of vegetative shoots 3–10 cm long, 1.5–2 mm broad, flat or convolute, not rigid, bright green, glabrous outside, occasionally appressed-pilose; glabrous or scabrous along ribs within, with short teeth or occasional hairs along margin. Sheaths of lower cauline leaves covered with numerous squarrose or decurved hairs; of upper ones inflated, violet, densely pilose; leaf blade 1.5–4.5 cm long, 2 (3) mm broad, convolute or flat, glabrous or pilose outside; diffusely pilose or subglabrous along veins within, with tuft of short hairs at base. Panicles 1.5–3 (4) cm long, 8–10 mm broad, compact, oval-oblong, usually dark violet. Spikelets 4–6 mm long, 2- or 3-flowered, glumes acuminate, pilose or glabrous, with short hairs along keel; lower 3 mm long, lanceolate, upper 4–4.2 mm long, broadly lanceolae, distinctly shorter than adjoining lemma. Lemma 4–5 mm long, acuminate awnlike, crinite, sometimes subglabrous, considerably longer than upper one. Anthers 1.2–2 mm long. Putoran plateau, $2n = 14$.

Arctic dryad and moss tundras, sandy spots and slopes, fallow sections, coastal slopes, pebble beds and hilly rocky tundras, rubble and weakly turfed slopes. **West. Sib.**: TYU—Yam. **Cen. Sib.**: KR—Ta (Taimyr peninsula—class. hab.—and others), Pu. **East. Sib.**: YAK—Ar, Yan. — Urals (north), Far East (north), North America (northwest). Map 79.

3. **K. atroviolacea** Domin 1907 in Biblioth. Bot. (Stuttgart) 65: 252. — *K. geniculata* Domin. —*K. asiatica* subsp. *atroviolacea* (Domin) Tzvelev.

Plant forming very loose mats. Stems 10–30 cm tall, pilose throughout length, dark, strong, not thickened at base, enveloped in more or less few undivided grayish-brown sheath remnants. Vegetative shoots single. Leaf blades of vegetative shoots 5–12 cm long, 1.5–2 mm broad, flat or longitudinally folded, not rigid, green; with prominent midrib outside, glabrous or scabrous on both surfaces, occasionally diffusely pilose. Sheaths of cauline leaves puberulent, disinctly inflated, leaf blade 2–5 cm long, 1–2 (3) mm broad, flat, glabrous or puberulent on both surfaces. Panicles 2–5 cm long, 7–10 mm broad, rather loose, pyramidal, occasionally oblong, sometimes interrupted at base, dark violet or speckled. Spikelets 5.5–6 mm long, 2- or 3-flowered. Glumes acuminate, violet with reddish fringe, lustrous, glabrous, occasionally with short cilia or weakly pilose along keel; lower 3.5–4 mm long, narrowly lanceolate, upper 4.5–5 mm long, broadly lanceolate, narrowed toward base, slightly (by 0.5–1 mm) shorter than adjoining lemma. Lemma about 5 mm long, acuminate, glabrous or diffusely pilose in lower half. Anthers 2 mm long. East. Sayan (Tunkinsk mountain range), $2n = 14$.

Alpine grasslands in high hills, sedge-dryad meadows, scrubs, moss-lichen, herbaceous-lichen, moss, cobresia tundra and rocks, debris, and slopes in upper forest belt. **West. Sib.**: AL—Go. **Cen. Sib.**: KR—Kha (Oznachennoe village—class. hab.—and others), Ve (Kurtushibinsk

mountain range, Zolotaya river upper course), TU. **East. Sib.**: BU—
Yuzh.—Mongolia. Map 80.

4. **K. chakassica** Reverd. 1964 in Fl. Krasnoyarsk. kraya 2: 62.—*K.
cristata* subsp. *chakassica* (Reverd.) Tzvelev.

Plant forming small, compact mats. Stems (30) 40–50 cm tall, pubescent, pinkish, lustrous, stout for 1 cm under panicle, not thickened at base, covered with more or less few undivided brown sheath remnants. Leaf blade of vegetative shoots 5–10 cm long, 2 mm broad, flat or longitudinally convolute, rigid, glaucous, glabrous like sheath. Sheath of cauline leaves glabrous, lustrous, leaf blade usually convolute, glaucous, glabrous. Panicles 5–7 cm long, 0.8–1 cm broad, compact or rather loose, oblong, with pink or violet tinge. Spikelets 5.5–6.5 mm long, glumes acuminate, glabrous, with short spinules only along keel; lower 4–4.5 mm long, upper 5 mm, nearly same in length as adjoining lemma. Lemma 5–5.5 mm long, acuminate, squarrose-pilose throughout surface.

Rocky steppes, marble debris, granite gravel. **Cen. Sib.**: KR—Kha (between Bol. Vorota ulus and Bateni village on Yenisey river—class. hab.—and others), Ve (Krasnoturansk region: Lebyazh'e Lake; Novoselovsk region: Chernaya Koma village), TU. —Endemic. Map 83.

5. **K. cristata** (L.) Pers. s. str. 1805, Syn. Pl. 1: 97, quoad nom. —*K. gracilis* Pers., nom. illeg. —*K. krylovii* Reverd.

Plant forming compact mats. Stems (20) 30–40 (70) cm tall, with short pubescence, pinkish or green, lustrous, slender under panicle for 0.5–1 cm and under joints; enveloped in more or less few undivided grayish-brown sheath remnants at base. Leaf blades of vegetative shoots 10–18 (20) cm long, 0.5–1.5 mm broad, convolute, occasionally flat, rigid, glaucous green, puberulent on both surfaces, with tufts of long hairs at base. Sheaths of lower cauline leaves puberulent or scabrous, leaf blades 2–6 cm long, 0.5–1.5 mm broad, usually convolute, appressed-pilose outside, sometimes glabrous; scabrous or pilose along ribs within. Panicles 5–10 (12) cm long, 4–8 mm broad, compact, cylindrical, gradually narrowed toward both ends. Spikelets 4–5 mm long, 2- or 3-flowered, glumes acute, glabrous, with crest of short translucent spinules along keel; lower 2.5–3 mm long, with single nerve, upper 3–4 mm long, lanceolate, with 3 nerves nearly same length as adjoining lemma, barely longer or shorter. Lemma 3–4.5 mm long, acute, glabrous or scabrous, occasionally diffusely pilose at base.

Herbaceous, herbaceous-grass and meadow steppes, in steppified, floodplain and solonetz meadows, floodplain pebble beds, fallow land, fringes of pine forests, roadsides. **West. Sib.**: TYU—Tb, KU, OM, TO, NO, KE, AL—Ba, Go. **Cen. Sib.**: KR—Kha, Ve, TU. **East. Sib.**: IR—An, Pr, BU—Se, Yuzh, ChI—Ka, Shi, YAK—Vi (environs of Yakutsk town; Sergelyakh and Chuchuur-Mura villages), AL (Tommotsk region, "Pyatiletka" state farm). —Eurasia and North America. Described from Great Britain. Map 88.

Highly polymorphic species. Hybridizes with other species. Specimens from south-eastern Altay occupy intermediate position between subsp. *cristata* and subsp. *transiliensis* (Reverd. ex Gamajun.) Tzvelev.

6. **K. cristata** (L.) Pers. subsp. **hirsutiflora** (Domin) Vlassova, comb. nova. —*K. gracilis* Pers. var. *hirsutiflora* Domin 1907 in Biblioth. Bot. (Stuttgart) 65: 216.

Plant forming rather compact mats. Stems 50–60 (70) cm tall, pilose, pinkish, stout under panicle for 1–5 (6) cm and at joints; covered by more or less few undivided light brown sheath remnants at base. Leaf blades of vegetative shoots 7–19 cm long, 1.5–2.5 mm broad, flat, occasionally convolute, green, hirsute on both surfaces, sheath pilose. Sheath of lower cauline leaves semiappressed-pilose, leaf blades 4–6 cm long, 1.5–2 mm broad, convolute, pilose on both surfaces. Panicles 5–8 cm long, 8–15 mm broad, loose or somewhat compact, oblong-cylindrical, greenish-pink. Spikelets 5.5–6.4 mm long, 2- or 3-flowered, glumes acuminate, scabrous, pilose or with short spinules along keel; lower 3.5–4 mm long, upper 5.5–6 mm, hardly shorter than adjoining lemma. Lemma 5 mm long, acuminate, puberulent in lower half. Anthers 2 mm long.

Rocky and rubble slopes in steppes and forest steppes. **East. Sib.**: IR—An (environs of Balagansk—class. hab.—and others), Pr, BU—Se, Yuzh (Zun-Murino village). —Endemic. Map 86.

Specimens from Baikal Lake coasts differ in very short leaves and compact violet panicles.

V.V. Reverdatto introduced *K. krylovii* based on species *K. gracilis* Pers. of group *Hirsutiflora* Domin. This species combines 3 variants: var. *pilifera* (glumes glabrous, lemma pilose), var. *gordjaginii* (glumes and lemma pilose) and var. *hirsutiflora* (glumes pilose, lemma glabrous). Further, V.V. Reverdatto assumed that plants of group *Hirsutiflora* are localized only in Siberia and hence definitely distinguishable from European *K. gracilis* morphologically (pubescence of glumes and lemma) as well as geographically and should be recognized as an independent species. But specimens of *K. gracilis* with more or less pilose lemma are found even in the European part of the USSR while plants with glabrous lemma are known from Siberia. Thus, geographical and morphological boundaries of the new species remained quite indistinct. Many investigators regarded *K. krylovii* as related to *K. altaica*. This was partly because type *K. krylovii* from Kuznets Alatau (LE) has more or less pilose spikelets, greenish-violet panicles and green, more or less pilose leaves of vegetative shoots, i.e., the very same characteristics which P.N. Krylov[2] cited for *K. altaica* (Domin) Krylov. He pointed out that *K. altaica* grows in Altay high hills in alpine meadows and tundras but descends even lower into the forest belt. P.N. Krylov gave a fairly broad treatment of *K. altaica* (stems thickened at base, leaves green or glaucescent, flat or convolute, smooth or pilose). V.V. Reverdatto[3] regarded plants with pilose spikelets from Altay hill tundra belt as *K. altaica*. N.N. Tzvelev excluded these plants from *K. altaica* and regarded them as a variant of *K. ledebourii*: *K. asiatica* Domin subsp. *ledebourii* (Domin) Tzvelev var. *glabrifolia* (Kryl.) Tzvelev; he regarded plants from the hill-steppe belt as *K. altaica*. The position of *K. krylovii* is further complicated

[2] P.N. Krylov. 1928. Flora Zapadnoi Sibiri [Flora of Western Siberia]. Tom. Otd-nie rus. Botan. o-va. Tomsk, 2nd ed., pp. 139–376.
[3] V.V. Reverdatto. 1964. Flora Krasnoyarskogo kraya [Flora of Krasnoyarsk Region]. Tomsk University, Tomsk, 2nd issue, 139 pp.

by the fact that specimens of this species from Khakasia described in the publication do not differ distinctly from specimens of *K. cristata* for which V.V. Reverdatto adduced 18 variants (including those with pilose leaves and lemma). Thus, we do not accept the name *K. krylovii* as an independent species since its size and boundaries are not completely clear and time and again have given rise to confusion. Insofar as var. *hirsutiflora* is concerned (Domin included it in group *Hirsutiflora*), specimens similar to type (LE) are often found in southern Central Siberian plateau and hence it is possible to consider them a distinct subspecies close to *K. cristata* (L.) Pers. s. str.

7. **K. cristata** subsp. **mongolica** (Domin) Tzvelev 1971 in Novosti sist. vyssh. rast. 7: 71. —*K. tokiensis* Domin subsp. mongolica Domin. — *K. macrantha* (Ledeb.) Schultes.

Plant forming small, rather compact mats. Stems 15–35 cm tall, pubescent, greenish under panicle for 0.5 cm and under joints; covered by numerous distinctly laciniate sheath remnants at base. Leaf blades of vegetative shoots 3–7 (12) cm long, about 1.5 mm broad, convolute, rigid, greenish glaucous, scabrous on both surfaces, sometimes with sparse hairs; sheath scabrous but not pilose. Sheath of cauline leaves glabrous or scabrous; leaf blade 2–3 cm long, 1.5 mm broad, weakly scabrous on both surfaces. Panicles 2–6 (7) cm long, 6–10 mm broad, oblong, greenish, lustrous. Spikelets 3–4 mm long, 2-flowered, glumes acute, glabrous; lower 2.5–3 mm long, upper 2.8–3.2 mm, slightly shorter than adjoining lemma. Lemma (2.5) 3–3.2 mm long, acute, glabrous, occasionally diffusely pilose at base. Anthers 2–2.2 mm long.

In fine rubble, sand and desert-steppe slopes in steppes. **East. Sib.:** BU—Yuzh, ChI—Shi. —Mongolia, NE China. Described from Mongolia. Map 87.

8. **K. cristata** subsp. **seminuda** (Trautv.) Tzvelev 1971 in Novosti sist. vyssh. rast. 7: 71. —*K. seminuda* (Trautv.) Gontsch. —*K. sibirica* (Domin) Gontsch.

Plant forming rather loose mats. Stems 40–50 cm tall, sparsely pilose throughout length, greenish-violet, stout; not thickened at base, covered by more or less undivided light cinnamon sheath remnants. Leaf blade of vegetative shoots 10–15 cm long, 1.5–2 mm broad, flat or convolute, rigid, usually densely pilose on both surfaces, sometime weakly scabrous. Sheaths of all cauline leaves squarrose-pilose, slightly inflated; leaf blade 4–8 cm long, 1–2 (3) mm broad, flat or convolute, pilose or scabrous on both surfaces. Panicles (4) 6–9 cm long, 10–12 mm broad, loose, interrupted at base, oblong, pinkish-violet. Spikelets 6–7 mm long, 2- to 4-flowered, glumes acuminate, with broad violet fringe, glabrous or puberulent along keel; lower 4–5 mm long, upper 5–6 mm, considerably shorter than adjoining lemma. Lemma 5–6 mm long, acuminate awnlike, sometimes with short awn, appressed- or squarrose-pilose in lower half or throughout surface. Anthers 1.9–2.5 mm long.

Ascending up to subbald-peak belt in forest zone, grassy meadows in riverine floodplains, forest glades, steppe slopes, river sand, pebble beds, carbonate slopes, sedge swamps. **Cen. Sib.**: KR—Tn. **East. Sib.**: IR—Pr, BU—Se, YAK—Ar (Lena river, Ayakit settlement—class. hab.), Ol (Siktyakh settlement), Vi, AL, Yan. —Endemic. Map 89.

9. **K. delavignei** Czern. ex Domin s. str. 1907 in Biblioth. Bot. (Stuttgart) 65: 247. —*K. barabensis* (Domin) Gontsch. —*K. incerta* Domin.

Plant forming very loose mats with short subsurface shoots. Stems 25–50 (70) cm tall, pilose, green, slender for 0.5–1 cm under panicle, not thickened at base, covered with undivided light brown sheath remnants. Leaf blade of vegetative shoots 10–30 cm high, 1–2 mm broad, convolute, sometimes flat, not rigid, pale green, usually glabrous, sometimes sparsely hairy. Sheath of cauline leaves glabrous or scabrous, leaf blade 4–10 cm long, 1 (3) mm broad, usually with tuft of long hairs at base, convolute or flat, glabrous or scabrous. Panicles 6–12 cm long, 5–10 mm broad, lax, oblong-cylindrical, light-colored or greenish-violet. Spikelets 3.5–5 mm long, 2- or 3-flowered; glumes obtuse, greenish, glabrous, with short cilia along keel; lower 3 mm long, narrowly lanceolate, upper 4 mm long, broadly lanceolate, barely shorter than adjoining lemma. Lemma 3–4 mm long, obtuse or obtuse-acuminate, glabrous or scabrous, greenish or violet-pink with golden fringe. Anthers 1.2–1.8 mm long.

Floodplains, often solonetz grass-herbaceous, wild rye-alkali grass, sedge-herbaceous, dry valley and forest meadows, forest resorts and birch stands. **West. Sib.**: TYU—Tb, KU, OM, TO (environs of Tomsk town), No, KE, AL—Ba, Go. **Cen. Sib.**: KR—Kha, Ve, TU (Tselinnyi settlement). —European USSR. Described from Ukraine. Map 81.

10. **K. delavignei** subsp. **veresczaginii** Tzvelev ex Vlassova 1987 in Bot. zhurn. 72 (12): 1667.

Plant with very loose mat and short subsurface shoots. Stems 60–90 cm tall, with short pubescence for 1–5 cm under panicle, sheath remnants at base of shoots few, brown, undivided. Leaf blades of vegetative shoots (15) 20–30 cm long, 2–2.5 mm broad, soft, flat or convolute, rather pilose on both surfaces, occasionally glabrous. Sheath of cauline leaves usually densely pilose, leaf blades 4–12 cm long, 3 mm broad, with tuft of long hairs at base, usually flat, pilose on both surfaces. Panicles 10–12 cm long, 1–1.5 cm broad, oblong-cylindrical, loose, greenish-violet. Spikelets 6 mm long, glumes obtuse-acuminate, scabrous or pilose; lower 3.5–4 mm long, upper 5–5.2 mm long. Lemma short- or obtuse-acuminate, 5 mm long, scabrous or diffusely pilose.

Hill-forest belt and high hills on slopes, limestone rocks, glades, subalpine meadows. **West. Sib.**: AL—Go (Uznezya—class. hab.—and others). —Endemic.

11. **K. glauca** (Sprengel) DC. 1813, Catal. Pl. Horti Monspel.: 116.

Plant forming rather large mats. Stems 20–60 cm tall, pilose, greenish-yellow, stout for more than 2 cm under panicle, with bulbous thickening at base, covered by split dark brown sheath remnants. Vegetative shoots 2 or 3 together and covered by common envelope of dead leaf sheaths. Leaf blades of vegetative shoots 8–16 cm long, 2–3 mm broad, usually flat, rigid, greenish glaucous, covered on both surfaces by short hairs, appearing velutinous. Lower sheath of cauline leaves scabrous or puberulent, not inflated; leaf blades 2–7 cm long, 0.5–1.5 mm broad, convolute or flat, scabrous. Panicles (2) 5–10 cm long, 5–15 mm broad, compact or somewhat lax, cylindrical. Spikelets 4–5 mm long, 2- or 3-flowered, glumes obtuse, glaucescent, glabrous; lower 3 mm long, lanceolate, upper 4–5 mm long, broadly lanceolate, shorter, as long or longer than adjoining lemma. Lemma 3–3.5 mm long, obtuse, puberulent. Anthers 2–2.4 mm long.

Sand in pine forests, sand of river terraces, dunes around lakes. **West. Sib.**: TYU—Tb, KU, NO (Akademgorodok; Kuibyshev town), AL—Ba. **Cen. Sib.**: KR—Tn (Yenisey river 20 km above Podkamennoi Tunguska river estuary), Kha (Chernogorsk), Ve. **East. Sib.**: IR—An, Pr (Kirensk region, Beloborodovo village; Zhigalovskii region Yakimovka village), BU—Yuzh. —Europe. Described from Europe. Map 78.

12. **K. karavajevii** Govor. 1971 in Novosti sist. vyssh. rast. 8: 22.

Plant forming compact mats. Stems 12–30 cm tall, sparsely hairy throughout length, thickened at base, covered by numerous undivided imbricately folded sheath remnants. Vegetative shoots usually 2 or 3 together and enveloped in sheaths of dead leaves. Leaf blade of vegetative shoots 3–6 (7) cm long, 1 mm broad, convolute, rigid, glaucesent, scabrous or subglabrous outside, rather glabrous or short-pubescent within. Sheaths of cauline leaves densely puberulent, leaf blades 2–4 cm long, 1 mm broad, convolute; glabrous or weakly scabrous outside, short-pubescent within. Panicles 2.5–3.5 cm long, 1 cm broad, lax, oval with pinkish-violet tinge. Spikelets 3.5–5.2 mm long, 2-flowered, glumes scabrous, acuminate; lower 3 mm long, upper 4 mm long, considerably shorter than adjoining lemma. Lemma 4 mm long, acuminate, weakly pilose or glabrous. Anthers 1.5–2.5 mm long.

On sand dunes. **East. Sib.**: YAK—Ol (Arangastakh Lake—class. hab.: Khorongki river terrace). —Endemic.

13. **K. ledebourii** Domin 1907 in Biblioth. Bot. (Stuttgart) 65: 164. — *K. asiatica* subsp. *ledebourii* (Domin) Tzvelev.

Plant forming loose mats. Stems 30–40 cm tall, pilose almost throughout length, violet, stout, not thickened at base, covered by numerous undivided light brown sheath remnants. Vegetative shoots single. Leaf blades of vegetative shoots (10) 15–20 cm long, 1.5–2 mm broad, convolute, occasionally flat, rigid, green, usually squarrose-pilose outside; scabrous, occasionally glabrous within. Sheaths of cauline leaves weakly pilose or glabrous, slightly inflated, leaf blades 5–10 cm long, flat or convolute; pilose outside, glabrous within. Panicles 3–5 (10) cm long, (7) 10–14 mm broad; compact or interrupted, oblong, pale violet at base.

Spikelets 6–6.5 mm long, 2 (3)-flowered, glumes acuminate awnlike, from densely pilose to subglabrous; lower 4 mm long, narrowly lanceolate, upper 5 mm long, broadly lanceolate, shorter than adjoining lemma. Lemma 5.5–6 mm long, acuminate awnlike, densely pilose or scabrous. Anthers 1.9–2 mm long.

Alpine meadows, kobresia and lichen tundras in high hills. **West. Sib.**: AL—Ba, Go. —Urals (Iremel' hill), Mid. Asia (Tarbagatai, Junggar Alatau, Tien Shan). Described from western Altay. Map 84.

14. **K. skrjabinii** Karav. et Tzvelev 1971 in Novosti sist. vyssh. rast. 8: 23.

Plant not forming mats, with long decumbent subsurface shoots. Stems 25–60 cm tall, puberulent almost throughout length and at joints, not thickened at base, covered with numerous undivided brown sheath remnants. Leaf blades of vegetative shoots 10–15 cm long, 2–2.5 mm broad, flat or convolute, rigid, glaucous green; glabrous or scabrous outside, scabrous within. Sheath of cauline leaves puberulent or scabrous, leaf blades 3–5 cm long, 1.5–3.5 mm broad, convolute, glaucescent, very rigid; glabrous outside, scabrous within. Panicles 3.5–9 (10) cm long, 8–10 mm broad, cylindrical, lax. Spikelets 6–7.5 mm long, 2- to 4-flowered; glumes acuminate, glabrous, scabrous along keel; lower 4–5 mm long, upper 5–6.5 mm long. Lemma 5.4–6.5 mm long, acuminate, glabrous, scabrous or, at base, diffusely pilose. Anthers 2.5–3.5 mm long.

Weathered sands. **East. Sib.**: YAK—Vi (Nidzheli Lake—class. hab.—and others).—Endemic.

15. **K. thonii** Domin 1907 in Biblioth. Bot. (Stuttgart) 65: 139.

Plant forming somewhat compact mats. stems 30–60 cm tall, pubescent, violet-pink, stout for 1–6 cm under panicle and under joints; covered by numerous undivided brown sheath remnants at base. Leaf blades of vegetative shoots 10–15 cm long, 2–3 (4) mm broad, usually flat, occasionally longitudinally folded, rigid, glaucous, glabrous on both surfaces, or scabrous within. Sheath of cauline leaves glabrous, weakly scabrous only in upper part, leaf blade 4–8 cm long, 2–3.5 mm broad, convolute, glabrous or scabrous. Panicles 8–12 (14) cm long, 10–15 mm broad, narrowly pyramidal, lustrous. Spikelets 6–8 mm long, 3- or 4-flowered, glumes acuminate, glabrous, very rarely scabrous; lower 4 mm long, narrowly lanceolate, upper 5 mm long, broadly lanceolate, distinctly shorter than adjoining lemma. Lemma 5 mm long, acuminate awnlike, glabrous or pilose at base. Anthers 2.2–2.5 mm long.

Hill-steppe and forest belts on open sand and sand dunes, pine groves on sand, river valleys. **Cen. Sib.**: KR—Kha (Chernogorsk settlement), Ve (Minusinka river source—class. hab.—and others). **East. Sib.**: IR—An (Ust'-Ilimsk town; Nizhneilimsk settlement; Angara river above Ilim river estuary). —Endemic. Map 85

16. **K. tzvelevii** Vlassova 1987 in Bot. zhurn. 72 (12): 1668.

Plant forming compact, rather small mats 2–3 cm diam. Stems 40–50 cm tall, glabrous, pubescent for 0.5–1 cm under panicle, sheath remnants at base of shoots several, brown, laciniate or more or less undivided. Leaf blades of vegetative shoots 8–10 (15) cm long, 1–1.5 mm broad, rigid, usually flat, glabrous, scabrous only on upper surface due to short spinules. Sheaths of cauline leaves scabrous, leaf blades 2–3 cm long, convolute or semiconvolute, glabrous below, scabrous above. Panicles 6–7 cm long, 1 cm broad, oblong, lax. Spikelets 4–4.5 mm long, glumes acute, glabrous, with short spinules along keel; lower 3–3.5 mm long, upper 4 mm long. Lemma acute, 3.5–4 mm long, densely pilose, greenish-violet.

Clay-sand banks of rivers and lakes. **East. Sib.**: ChI—Shi (Kailastui village—class. hab.—and others). —Endemic in Transbaikal.

18. Deschampsia Beauv.

1. Awns of lemma erect, shorter than or equaling spikelet, sometimes 1–1.5 mm longer than latter .... 2.
+ Awns of lemma geniculate, long, 2–3 mm longer than spikelet ... . 10. *D. turczaninowii.*
2. Spikelets large, 4.5–6 mm long, not many, loosely arranged on branches .... 3.
+ Spikelets much smaller, 3–4 mm long, many, up to 150 in panicle .... 6.
3. Panicles usually loose or more or less spreading, with slender, sometimes flexuous branches. Spikelets pinkish or violet-green, 5–6 mm long .... 4.
+ Panicles very compact, often spicate, with greatly shortened, straight upturned branches. Spikelets dark violet, 4.5–5 mm long .... 3. *D. brevifolia.*
4. Plants glaucous-green, forming compact mats without decumbent subsurface shoots; radical leaves several, fairly rigid, scabrous. Stems with thick envelope of broad light brown scarious sheaths of dead leaves at base 5
+ Plants green, loosely cespitose, with long decumbent subsurface shoots; radical leaves few, soft, usually glabrous. Stems without envelope of dead leaf sheaths at base. Branches slender and flexuous. Spikelets only at branch ends .... 7. *D. obensis.*
5. Panicles pyramidal, with some rather thick, ascending branches. Spikelets 2- or 3-flowered, pink- or pale-violet. Lower glumes more or less broadly lanceolate, obtuse. Lateral nerves of lemma poorly visible or absent; awn weak, arising close to base and not exceeding tip of lemma .... 5. *D. glauca*
+ Panicles broadly spreading, with slender horizontally declinate branches. Spikelets 2 (3)-flowered, green or golden-green. Lowei

84

glumes narrowly lanceolate, acuminate. Lateral nerves of lemma sharply projecting; awns strong, arising close to base and exceeding spikelets .... 11. *D. vodopjanoviae.*

6. Leaves and panicle branches glabrous. Lemma with sharply projecting lateral nerves .... 7.

+ Leaves and panicle branches poorly or highly scabrous. Lemma with poorly distinct lateral nerves .... 8.

7. Rather low, 15–20 (30) cm tall plant forming somewhat compact mats. Radical leaves short, reaching 1/5 stem length, numerous, declinate swordlike, longitudinally folded .... 2. *D. borealis.*

+ Tall, up to 70 cm plant not forming mats. Radical leaves very long, reaching 1/3 stem length, few, flat, erect, rarely longitudinally folded .... 9. *D. sukatschewii.*

8. Spikelets dark violet, with golden fringe. Lemma with distinct lateral nerves. Awns scarcely surpassing spikelets. Hill plant .... 9.

+ Spikelets light green, sometimes pale violet. Lemma with poorly distinct lateral nerves; awns weak, sometimes exceeding spikelets. Plant of plains .... 10.

9. Panicles lax, spreading, branches slender, long, horizontally declinate. Spikelets 3.5–4 mm long, loosely arranged. Stems 30–40 cm tall, aggregated into compact mat, without envelope of dead leaf sheaths at base. Radical leaves short, numerous, slightly scabrous along veins .... 1. *D. altaica.*

+ Panicles compressed, often spicate, with spikelets flabellately arranged. Spikelets 4–4.5 mm long. Stems 15–20 (25) cm tall, aggregated into somewhat loose mat, with envelope of light brown dead leaf sheaths at base. Radical leaves long, flat or longitudinally convolute, rather few, glabrous or with few spinules along veins .... 6. *D. koelerioides.*

10. Stems stout, up to 70 cm tall. Leaf blades of radical and cauline leaves long, highly scabrous, especially along marginal veins. Panicles broadly spreading, 10–20 cm long, with large number (up to 150) of spikelets; branches long, rather horizontally declinate, strongly scabrous. Glumes lanceolate, acuminate. Awns of lemma rather stout, arising close to base, scarcely surpassing spikelets .... 4. *D. cespitosa.*

+ Stems 30–50 cm tall, relatively slender and weak in lower part. Leaf blades of radical and cauline leaves covered with sparse and slender spinules. Panicles 7–10 cm long, with ascending, glabrous branches, sometimes with sparse spinules. Glumes lanceolate, obtuse. Awns of lemma weak, attached in upper third; not exceeding spikelets .... 8. *D. submutica.*

1. **D. altaica** (Schischkin) Nikiforova comb. nova. —*D. cespitosa* var. *altaica* Schischkin 1927 in Krylov, Fl. Zap. Sib. 2: 231.

Stems 20–30 (40) cm tall, with compact envelope of narrow brownish-cinnamon dead leaf sheaths at base forming compact mat. Radical leaves numerous, short, green, covered on upper surface by poorly distinct slender spinules, flat or longitudinally folded. Panicles violet, lax, spreading, 5–8 cm long; branches slender, long, branched, slightly flexuous, horizontally declinate or gently ascending, terminating in spikelets. Spikelets 2 (3)-flowered, 4–4.5 mm long. Glumes narrowly lanceolate, keeled, equal: lower narrower than upper, dark violet, with broad bright golden fringe. Lemma viloet, lanceolate, with slightly distinct lateral nerves. Awns strong, erect, surpassing spikelets by 0.5–1 mm, attached in lower fourth. Anthers 1–1.2 mm long.

Alpine meadows, hill tundras, pebble riverbeds. **West. Sib.**: AL—Ba (Tigeretskii mountain range—class. hab.—and others), Go. **Cen. Sib.**: KR—Kha, Ve (Kulymys mountain range; Agul river), TU. **East. Sib.**: BU—Yuzh. —Endemic. Map 98.

2. **D. borealis** (Trautv.) Roshev. 1934 in Fl. SSSR 2: 247. *D. arctica* var. *borealis* (Trautv.) Krylov. —*D. cespitosa* subsp. *borealis* (Trautv.) A. et D. Löve.

Stems 10–20 (30) cm tall, aggregated into small, somewhat compact mat, with envelope of narrow brown dead leaf sheaths at base. Radical leaves numerous, narrow, tender, soft, glabrous, longitudinally convolute, ensiform and slightly deflexed from stem, with inflated sheaths. Cauline leaves reaching 1/2 length of stem. Panicles lax, 3–7 cm long, with slender, long, slightly flexuous, often horizontally declinate lower branches. Spikelets 3–3.5 mm long, 2-flowered, golden-violet. Glumes strongly unequal: lower narrowly lanceolate, acuminate, considerably narrower and shorter than upper. Lemma and palea with truncate and irregularly dentate margin: lemma twice broader than palea, with 2 sharply distinct lateral nerves. Awns stout, erect, arising from lower third. Anthers 1.2–1.7 mm long.

Pebble and sandy riverbeds, rubble tundra. **West. Sib.**: TYU—Yam. **Cen. Sib.**: KR—Ta (Taimyr peninsula—class. hab.—and others), Pu. **East. Sib.**: YAK—Ar. —Europe (north), Far East, North America. Map 92.

3. **D. brevifolia** R. Br. 1824 in Parry, Jour. Voy. N.W. Pass (Suppl. App.): 291. —*D. arctica* (Sprengel) Merr. —*D. cespitosa* subsp. *brevifolia* (R. Br.) Tzvelev.

Stems 20–30 cm tall, with compact envelope of light brown, broad, scarious dead leaf sheaths at base. Radical leaves numerous, usually flat, glabrous, sometimes longitudinally folded, scabrous, soft, half length of stem. Panicles compressed, 6–15 cm long, with dark violet, very short, rather thick, glabrous obliquely erect branches. Spikelets 5–5.5 mm long, 2 (3)-flowered. Glumes violet, sharply lanceolate, keeled, rather similar, nearly as long as lemma. Latter lanceolate, with truncate irregularly dentate margin; lower with poorly distinct lateral nerves. Awns stout, erect, as long or scarcely longer than spikelets. Putoran plateau (Khakoma Lake), $2n = 52$.

Tundra, rocky, clayey, sandy exposures, near water reservoirs. **Cen. Sib.**: KR—Ta, Pu. **East. Sib.**: YAK—Ar. —Europe (north), Far East, North America. Described from Canada. Map 93.

4. **D. cespitosa** (L.) Beauv. 1812 in Ess. Agrost.: 91, 149, 160.

Stems 30–100 cm tall, forming very compact mat together with numerous radical leaves. Leaves longitudinally folded, rigid, long, grayish-green; sharply scabrous along veins, especially marginal veins. Panicles large, 10–25 cm long, spreading, with highly scabrous subhorizontally inclined branches; spikelets quite densely arranged on branches. Spikelets 2 (3)-flowered, small, 3–3.5 mm long. Glumes yellowish-green, green and pale violet, with golden lustrous margin, lanceolate, unequal, lower narrowly lanceolate; upper broadly lanceolate. Lemma and palea light violet, equal in length, obtuse, with dentate margin; lemma with less distinct lateral nerves and slender awn not exceeding spikelets. Anthers 1.2–1.5 mm long. Altay (Cherga village), $2n = 26$.

Meadows, forest glades, banks of rivers and lakes. **West. Sib.**: TYU—Tb, KU, OM, TO, NO, KE, AL—Ba, Go. **Cen. Sib.**: KR—Tn (Kuz'movka village), Kha, Ve, TU. **East. Sib.**: IR—An, Pr, BU—Yuzh, YAK—Vi (Sangar settlement). —Europe, Caucasus, Mid. Asia, Far East, Mediterranean, West. Asia, West. China, Mongolia, North America. Described from Europe. Map 94.

5. **D. glauca** Hartman 1820 in Handb. Scand. Fl.: 448. —*D. cespitosa* subsp. *glauca* (Hartman) Hartman.

Stems 20–30 cm tall, with numerous broad, flat, lustrous, light brown dead leaf sheaths at base, as a result of which mats readily disintegrate. Radical leaves flat or longitudinally folded, with glaucous tinge, weakly scabrous. Panicles more or less spreading, with long, glabrous, violet branches, slightly horizontally inclined or recurved. Spikelets 5.5–6 mm long, 2- or 3-flowered, loosely arranged in panicle: upper spikelet exceeding length of glumes. Latter broadly lanceolate, keeled, acuminate, violet- or lilac-golden. Lemma and palea pale green or pink, obtuse; lemma with poorly distinct lateral nerves. Awns slender, shorter than spikelet. Anthers 1.7–2 mm long.

Tundra, rocky and pebble floodplains of brooks, swamps. **West. Sib.**: TYU—Yam. **Cen. Sib.**: KR—Ta, Pu. **East. Sib.**: YAK—Ar. —Nor. Europe, Far East, North America. Described from Sweden. Map 95.

6. **D. koelerioides** Regel 1868 in Bull. Soc. Nat. Moscow 41: 299. —*D. cespitosa* subsp. *koelerioides* (Regel) Tzvelev.

Stems 15–20 (30) cm tall, with envelope of few light brown dead leaf sheaths at base. Radical leaves flat or longitudinally folded, weakly scabrous along margin, reaching fourth of stem length. Panicles compressed, subspicate, 3–6 cm long. Branches short, rather thick, directed upward, glabrous or slightly scabrous. Spikelets 4–4.5 mm long, flabellately arranged at ends of branches. Glumes nearly equal in size, acuminate, keeled, yellowish or dark violet, with golden scarious margin. Lemma and palea 3.5–4 mm long, with scarious, lacerated margin: lemma with poorly distinct lateral nerves. Awns arising near base stout, not contorted, erect or slightly curved, as long or barely longer than spikelet. Anthers 2–2.5 mm long.

Grasslands along river banks in alpine belt. **West. Sib.**: AL—Ba, Go. **Cen. Sib.**: KR, TU, **East. Sib.**: BU—Yuzh, ChI—Shi (Sokhondo bald peak). —Mid. and West. Asia, Mongolia. Described from Tien Shan. Map 96.

7. **D. obensis** Roshev. 1932 in Izv. Bot. sada SSSR 30: 771. —*D. cespitosa* subsp. *obensis* (Roshev.) Tzvelev.

Stems 20–40 cm tall, slender, without envelope of dead leaf sheaths at base, forming small loose mats, with long subsurface rhizome. Radical leaves reaching inflorescence, glabrous or with occasional spinule along marginal veins. Panicles large, spreading, equal to half plant length, with smaller number of spikelets arranged singly at ends of long slender glabrous flexuous branches. Spikelets 2-flowered, 5–6 mm long. Glumes equal in size, narrowly lanceolate, light green or pinkish, with light golden fringe. Lemma and palea pale green, with truncate dentate margins; lemma with 2 darker colored lateral nerves. Awns weak, arising from lower third, usually not surpassing spikelet. Anthers 2–2.2 mm long.

Shifting and weakly turfed sand of river valleys. **West. Sib.**: TYU—Yam (Nakhodka bay—class. hab.—and others). **Cen. Sib.**: KR—Ta, Pu. **East. Sib.**: Yak—Ar. —Europe (north), Far East. Map 97.

8. **D. submutica** (Trautv.) Nikiforova stat. nova. —*Aira cespitosa* var. *submutica* Trautv. 1877 in Tr. Peterb. bot. sada 5: 141. —*D. sukatschevii* subsp. *submutica* (Trautv.) Tzvelev.

Stems 30–50 cm tall, relatively slender and weak in lower part, with envelope of brown dead leaf sheaths at base. Radical leaves long, numerous, flat or longitudinally convolute, slightly scabrous along veins. Panicles 7–10 cm long, somewhat spreading, secund, pale green with few spikelets. Branches slender, long, glabrous, with occasional spinules and loosely arranged spikelets. Latter 2- or 3-flowered, 4–4.5 (5) mm long, upper floret surpassing glumes and exserted from spikelet. Glumes light green or violet-green, lanceolate, obtuse; lower shorter and narrower than upper. Lemma lanceolate, with 2 poorly distinct lateral nerves; awns weak, slender, attached in upper third, shorter than spikelet. Anthers about 1.5 mm long.

Sandy and pebbled river banks, willow groves. **Cen. Sib.**: KR—Pu, Tn (Noginsk settlement), Ve (Nazimovskoe and Mikhalevo villages). **East. Sib.**: YAK—Ar, Ol (Olenek river—class. hab.—and others), Vi, Al, Yan, Ko. —Far East.

The question of genetic relationships in the species is debatable. We are inclined to regard it as a northern ecogeographical race, related to species *D. cespitosa*, from which it differs in weak stem, flat, tender, long leaves covered with tiny spinules, with few but much larger spikelets and less spreading panicle. At the same time, the absence of compact mat and presence of weakly scabrous, sometimes glabrous leaves brings this species closer to *D. sukatschewii*.

9. **D. sukatschewii** (Popl.) Roshev. 1934 in Fl. SSSR 2: 246. —*D. cespitosa* subsp. *orientalis* auct. non Hultén.

Stems 50–70 cm tall, erect, not forming mats, without envelope of dead leaf sheaths. Radical leaves few, soft, flat, tender, long, green, invariably glabrous and smooth. Panicles large, spreading, apically often nutant, pale green, 1/4 to 1/3 length of stem. Branches slender, long, glabrous, with large number (up to 150) of spikelets. Spikelets small, 2.5–3.2 (3.5) mm long, 2-flowered, light or reddish green. Glumes sharply lanceolate, strongly unequal: lower shorter and narrower than upper by almost half. Lemma lanceolate, with sharply distinct lateral nerves and stout awn surpassing spikelet. Anthers 1.5–1.7 mm long. Buryatia (northeastern coast of Lake Baikal), $2n = 26$.

Meadows, sandy and pebbled river banks. **Cen. Sib.**: TU. **East. Sib.**: IR An, Pr, Bu—Se, Yuzh, ChI—Ka, Shi (Novyi Olov village—class. hab. —and others). —Far East, Mongolia, East. Asia. Map 91.

10. **D. turczaninowii** Litv. 1922. in Spisok rast. Gerb. russk. fl. 8: 158. —*D. cespitosa* subsp. *turczaninowii* (Litv.) Tzvelev.

Stems 30–60 cm tall, numerous, densely foliate, aggregated into very compact mat, with envelope of glossy, broad, light brown dead leaf sheaths at base. Radical leaves long, flat, sometimes longitudinally convolute, scabrous, reaching panicles. Latter spreading, 9–15 cm long, with some shortened, slightly scabrous or almost smooth recurved branches. Spikelets 2- or 3-flowered, large, 4.5–6.5 (7) mm long, violet with golden tinge. Glumes lanceolate, acuminate, with sharply distinct midrib. Lemma acuminate, with distinct lateral nerves, with about 1.5 mm long hairs at base. Awns thick, geniculate, surpassing spikelet by 2–4 mm, arising from lower fourth. Anthers 1.5–1.7 mm long.

Sand and pebble beds on Baikal bank. **East. Sib.**: IR—An, Pr (Zayarsk settlement), BU—Se, Yuzh. —Endemic. Map 99.

11. **D. vodopjanoviae** Nikiforova 1987 in Bot. zhurn. 72 (12): 1666.

Mat compact. Stems 30–40 cm tall, erect, with envelope of broad, light brown dead leaf sheaths at base. Radical leaves numerous, longitudinally convolute, stiff, with occasional slender spinules along veins. Panicles 10–15 cm long, lax, with slender, long, nearly smooth branches gently recurved. Spikelets 2- or 3-flowered, 3.5–4 (4.5) mm long, loosely arranged at ends of long, slightly flexuous branches. Glumes lanceolate, acuminate, nearly equal, violet-green, with broad light golden fringe. Lemma obtuse, with 2 sharply distinct lateral nerves. Awns erect, stout, attached in lower fourth scarcely longer than spikelet. Anthers 1.7–2 mm long.

Rubble slopes, pebble beds and sand shoals of hill rivers in subbald-peak belt. **Cen. Sib.**: KR—Pu (Khaya-Kyuël' Lake—class. hab. —and others). —Endemic. Map 100.

Hybrids of *D. sukatschewii* and *D. glauca* resemble *D. sukatschewii* in shape of panicles, size of spikelets, shape of glumes, and presence of lateral nerves in lemma, and *D. glauca* in presence of compact envelope of dead leaf sheaths at stem base.

Hybrids differ from *D. sukatschewii* in compact mat and envelope of sheaths of flat, light brown, dead radical leaves; from *D. glauca* in smaller spikelets, 3.5–4.5 (not 5–6) mm, panicle spreading farther, and lateral nerves of lemma well defined.

19. Lerchenfeldia Schur

1. **L. flexuosa** (L.) Schur subsp. **montana** (L.) Tzvelev 1970 in Novosti sist. vyssh. rast. 7: 44. —*Deschampsia flexuosa* auct. non Nees.

Rhizome short, decumbent. Stems erect or ascending, 30–70 cm tall. Leaves longitudinally convolute, setaceaous, outer surface glabrous and smooth, inner (upper surface) densely covered with papillae. Panicles lax, compressed after anthesis, branches very slender, flexuous, smooth. Spikelets 5–6 mm long, with pinkish-violet tinge. Glumes scarious with single nerve, longer than lemma. Lemma with geniculate awn arising from base on back and exserted 1.5–3 mm from spikelet.

High mountain meadows, tundras, and thin forests. **West. Sib.**: TYU —Yam (Nyda river), Khm (upper course of Man' river). **Cen. Sib.**: KR— Ve (Eniseisk town). —Europe, Caucasus, Asia Minor, North America. Described from Europe.

In the type subspecies, spikelets 3.6–5 mm long, not colored, panicle branches densely covered with short spinules. Found in coniferous and mixed forests, and forest glades of Europe, the Far East as well as Japan, China, and North America.

20. Milium L.

1. **M. effusum** L. 1753, Sp. Pl.: 64

Rhizome decumbent, with short subsurface shoots. Stems single, erect, glabrous, smooth, 70–150 cm tall. Leaves up to 1.8 cm broad, smooth, scabrous only along margin. Ligules up to 8 mm long, oblong, obtuse, scarious. Panicles broad, 15–35 cm long, lax, with long nutant branches, scabrous due to short spinules. Spikelets 1–flowered, about 3 mm long, green or slightly violet. Glumes ovate, membranous, convex, with 3 nerves, slightly longer than lemma. Lemma awnless, hard, cartilaginous, glossy, ripe brown.

Wet coniferous and deciduous forests. **West. Sib.**: TYU—Yam, Khm, Tb, KU, OM, TO, NO, KE, AL—Ba, Go. **Cen. Sib.**: KR—Pu (environs of Igarka town), Tn, Kha, Ve, TU. **East. Sib.**: IR—An, Pr, BU—Se, Yuzh, YAK—Vi (environs of Olekminsk). —Eurasia, North America. Described from Europe. Map 102.

21. Calamagrostis Adanson

1.   Rachilla rudiment very short, glabrous or absent. Glumes very long and acuminate, subulate .... 2.
+   Rachilla rudiment consists of hairs appressed to palea .... 5.
2.   Lemma with 3 nerves. Callus hairs usually 1.5–2 times longer than lemma. Palea 1.5–2.5 times shorter than lemma .... 3.
+   Lemma with 5 nerves. Callus hairs nearly as long or slightly longer than lemma. Palea 1.5 times shorter than lemma .... 5. *C. canescens.*
3.   Awn arising near middle on back of lemma .... 4.
+   Awn arising directly from acuminate or obtuse, sometimes bidentate tip of lemma .... 20. *C. pseudophragmites.*
4.   Spikelets 5–7 mm long. Upper glume 0.8–1 mm shorter than lower .... 8. *C. epigeios.*
+   Spikelets 7–9 (10) mm long. Upper glume 1–1.5 mm shorter than lower .... 15. *C. macrolepis.*
5.   Panicle branches scabrous, very rarely smooth or almost so, but then panicles very dense, with short branches .... 6.
+   Panicles small, spreading during anthesis, 3–8 cm long, with a few long smooth spicate branches, 1–3 at each joint .... 7. *C. deschampsioides.*
6.   Callus hairs not exceeding 2/3 length of lemma .... 7.
+   Callus hairs as long (or almost so) as lemma .... 21.
7.   Awns well developed, coiled in lower part, distinctly geniculate, usually far exserted from spikelet .... 8.
+   Awns less developed, erect or barely curved, not exserted from spikelet .... 13.
8.   Plant forming compact mats, without decumbent subsurface shoots. Surface shoots with numerous dead leaf sheaths at base. Panicles usually spicate .... 9.
+   Plant forming loose mats, with short decumbent subsurface shoots. Surface shoots without dead leaf sheaths at base but with scaly leaves without or with short blade .... 11.
9.   Leaf blade scabrous on upper surface due to spinules or glabrous .... 10.
+   Leaf blade densely puberulent on upper surface .... 21 *C. purpurascens.*
10.   Glumes subulately acuminate. Callus hairs short, considerably shorter than half length of lemma. Awns arising from lower fourth, often close to base .... 2. *C. arctica.*
+   Glumes acuminate but not subulate. Callus hairs reaching almost half length of lemma. Awns arising above lower fourth, usually at level of one-third .... 10. *C. kalarica.*
11.   Leaf blade rather scabrous or with occasional long hairs on

upper surface, narrower (3–7 mm) and long. Panicles rather lax. Spikelets 4–5.5 (6) mm long .... 12.

+ Leaf blade entirely glabrous and smooth on upper surface, flat and fairly broad (3–10 mm), slightly shortened. Panicles compact. Spikelets 6–9.5 mm long .... 11. *C. korotkyi*.

12. Hairs at base of floret 1/5 to 1/4 length of lemma .... 3. *C. arundinacea*.

+ Hairs at base of floret 1/3 length of lemma ...... 1. *C. andrejewii*.

13 (7). Sheath blade joint rather pilose .... 14.

+ Sheath blade joint glabrous .... 16.

14. Palea distinctly shorter than lemma. Callus hairs 1/2 to 2/3 length of lemma. Panicles quite lax. Sheath blade joint usually less pilose (up to subglabrous), stems with more interrupted glabrous joints .... 15.

+ Palea as long as lemma, callus hairs 1/3 to 1/2 as long as lemma. Panicles fairly compact. Stem joints puberulent .... 17. *C. obtusata*.

15. Leaf blades 2.5–10 mm broad, grayish-green on upper surface .... 18. *C. pavlovii*.

+ Leaf blades 2–5 mm broad, green on both surfaces .... 6. *C. chalybaea*.

16. Rachilla rudiment long (1.5–2 mm), densely pilose which, together with hairs covering it, is almost as long as floret ..... 17.

+ Rachilla rudiment very short (0.6–1.5mm), with shorter and less numerous hairs not reaching floret tips, usually totally glabrous in lower part .... 18.

17. Panicles very compact, spicate, with highly shortened branches, usually rather pinkish-violet. Glumes acuminate .... 14. *C. macilenta*.

+ Panicles more lax, with very long branches, greenish with pinkish-violet tinge. Glumes barely acute, somewhat obtuse .... 24. *C. salina*.

18 Glumes acute, often scabrous throughout surface, leptodermatous. Plants of marshes, marshy meadows, tundras and bald peaks .... 19.

+ Glumes subulate or fairly gradually acuminate, usually scabrous, coriaceous-membranous only along midrib. Plants of Arctic or southern Siberian hills .... 20.

19. Glumes 4–6 mm long, callus hairs as long as lemma, occasionally slightly shorter. Awn somewhat curved in lower part, later straight .... 13. *C. lapponica*.

+ Glumes (2.5) 3–3.5 mm long, callus hairs 2/3 to 3/4 length of lemma. Awn erect .... 16. *C. neglecta*.

20. Glumes subulate-acuminate, narrowly lanceolate, 5–6 mm long

.... 23. *C. sajanensis.*

+ Glumes fairly acuminate, 3–4 mm long, broadly membranous, golden, lustrous along margin .... 9. *C. holmii.*

21 (6). Stems 40–150 cm tall, relatively thick. Leaf blades 3–8 mm broad, strongly declinate from stem. Glumes 3.5–5 mm long, acuminate .... 22.

+ Stems 25–80 cm tall, slender. Leaf blades narrow, 1.5–3.5 mm broad, with weakly convoluted margin, stiff and usually appressed to stem or slightly declinate. Glumes 2.5–4 mm long, acute .... 25. *C. tenuis.*

22. Panicles usually rather lax, widely spreading, spikelets 3–6 mm long. Leaves usually with grayish tinge, 4–8 (10) mm broad .... 23.

+ Panicles fairly dense, with closely aggregated up to 4.5 mm long spikelets. Leaves bright green, 3–6 (7) mm broad. Awns arising from middle of lemma or slightly above ....
.... 22. *C. purpurea.*

23. Sheath blade joint glabrous, stems glabrous below joints ... 24.

+ Sheath blade joint puberulent. Stems rigid, appressed-pilose below joints .... 4. *C. barbata.*

24. Awns arising close to middle of lemma and well developed .... 12. *C. langsdorfii.*

+ Awns arising in upper third of lemma and usually greatly reduced .... 19. *C. phragmitoides.*

1. **C. andrejewii** Litv. 1911 in Spisok rast. Gerb. russk. fl. 7: 157. — *C. arundinacea* (L.) Roth × *C. obtusata* Trin.

Rhizome short, decumbent. Stems 70–100 cm tall . Leaves flat, scabrous or with occasional long hairs on upper surface. Panicles up to 20 cm long, narrow, compressed, with sharply scabrous branches. Spikelets about 4 mm long, lanceolate, green or violet. Lemma about 3.5 mm long, broadly lanceolate, with geniculate awn arising slightly above base and exserted for 2–3 mm on side of spikelet. Callus hairs 1/3 length of lemma.

Deciduous, dark coniferous and aspen forests. **East. Sib.:** IR—An (Ishideï and Moka settlements), Pr (watershed of Lena and Tutury rivers). —European USSR. Described from Komy ASSR (near village Mordino).

2. **C. arctica** Vasey 1893 in Bull. U.S. Dept. Agr., Div. Bot. 13 (2): tab. 55.

Stems 20–40 cm tall, aggregated in small numbers into mat, with gray sheath remnants of dead leaves at base. Leaves 1.5–5 mm broad, flat or longitudinally convolute, glabrous, smooth. Panicles 2–9 cm long, 1–1.5 cm broad, spicate, with shortened, rather scabrous branches. Spikelets 4–6 mm long, pinkish-violet. Lemma with geniculate awn arising from lower fourth, sometimes close to base. Callus hairs 1/3 length of lemma. Rachilla rudiment 1–2 mm long, with up to 3 mm long hairs.

Rocky slopes in hill tundras. **East. Sib.**: BU—Se (Ukuolkit river), ChI—Ka. —Chuksh area Alaska, Aleutian islands. Described from St. Paul Island in the Bering Sea. Map 104.

3. *C.* **arundinacea** (L.) Roth 1789, Tent. Fl. Germ. 2 (1): 89.

Stems few in number, up to 1–1.5 m tall, erect, forming loose mats. Leaves 4–7 mm broad, rigid, scabrous or with occasional long hairs on upper surface. Panicles up to 25 cm long, erect, narrowed toward both ends, moderately spreading during anthesis, deflorated ones compressed. Spikelets 4–6 mm long, oblong-lanceolate, green or pinkish. Lemma elongate-lanceolate, crenate at tip, with long (5–9 mm) geniculate awn arising from lower third on back and distinctly exserted from spikelet; its upper geniculation usually longer than lower, highly twisted. Palea slightly shorter but considerably narrower than lemma. Callus hairs short, 4–5 times shorter than floret. Rudiment of rachilla about 1 mm long, with 2–3 mm long hairs. Gornyi Altay (Ust'-Sema settlement), $2n = 28$.

Arid forests, among shrubs, on rocky slopes of forest belt. **West. Sib.**: TYU—Khm, Tb, KU, OM, TO, NO, KE, AL—Ba, Go. **Cen. Sib.**: KR—Kha, Ve. **East. Sib.**: IR—An, ChI—Shi. —Europe, Caucasus, Mediterranean. Described from Europe. Map 101.

4. **C. barbata** V. Vassil. 1963 in Feddes Repert. 68 (3): 216. —*C. purpurea* subsp. *barbata* (V. Vassil) Tzvelev.

Rhizome decumbent. Stems 60–120 (180) cm tall, sometimes weakly branched in middle and lower joints, frequently scabrous under inflorescence, rigidly appressed pubescence at or under joints. Sheath glabrous, but somewhat scabrous due to very short spinules. Leaves 4–8 (10) mm broad, flat, glabrous and scabrous, grayish. Sheath blade joints with tufts of hairs. Panicles 9–15 cm long, lax, spreading, with densely scabrous branches. Spikelets 4.5–6 mm long, glumes scabrous throughout surface, sometimes with diffuse long piliform spinules near keel. Lemma 2.6–3.2 mm long, with hairs as long as lemma on callus, with slender erect awn on back arising close to middle and hardly surpassing its tip.

Wet meadows and sedge marshes. **East. Sib.**: ChI—Shi. —Far East, Japan, China. Described from Okhotsk coast (Tauisk-Ol'sk region, near Kava estuary). Map 105.

5. **C. canescens** (Web.) Roth 1789, Tent. Fl. Germ. 2 (1): 93. —*C. lanceolata* Roth.

Stems 70–130 cm tall, branched. Leaves 2.5–5 mm broad, narrowly linear, flat, slightly scabrous, glabrous or covered on upper surface with long and slender hairs. Panicles not compact, nutant, 10–12 cm long, 2–3.5 cm broad. Spikelets 4–5 mm long, violet, glumes with very short spinules only along keels. Lemma broadly lanceolate, with very poorly developed, often nearly completely reduced awn arising in upper third. Callus hairs slightly longer than lemma. Rachilla rudiment absent or very short (up to 0.5 mm long), glabrous, occasionally with relatively few hairs.

Floodplain and marshy meadows, wet birch fellings. **West. Sib.**: TYU—Tb, KU, OM, TO, NO, KE, AL—Ba. **Cen. Sib.**: KR—Kha, Ve. — Europe, Caucasus. Described from FRG. Map 106.

6. **C. chalybaea** (Laest.) Fries 1843 in Hartm., Handb. Skand. Fl. ed. 4: 26. —*C. obtusata* Trin. × *C. canescens* (Web.) Roth.

Rhizome short, decumbent. Stems 80–120 cm tall, glabrous, unbranched. Leaves 2–5 mm broad, linear, sharply scabrous, green on upper surface, without glaucescent shade. Panicles 10–20 cm long, fusiform. Spikelets 3.5–4 mm long, glumes lanceolate, almost equal. Lemma acuminate, with geniculate awn arising barely above base and scarcely reaching its tip. Callus hairs exceeding half length of lemma or as long. Rachilla rudiment 1–1.5 mm long, with long hairs, surpassing lemma.

Thin forests, forest glades. **West. Sib.**: TYU—Tb, OM. —European USSR, Scandinavia. Described from Sweden. Map 107.

7. **C. deschampsioides** Trin. 1836, Sp. Gram. Icon. et Descr.: tab. 354.

Stems 20–30 cm tall, slender, ascending, often rooting at nodes forming rather loose mats. Leaves narrowly linear, longitudinally convolute. Panicles 3–8 cm long, with few spikelets, spreading, oblong or pyramidal. Spikelets 4–4.5 mm long, brownish-lilac, lanceolate. Lemma bidentate at tip, with curved awn arising near middle. Callus hairs half length of lemma. Rudiment of rachilla 1.5 mm long, with hairs as long as callus hairs.

Turfed terraces of brooks and river estuaries in hypnaceous-peat moss tundras. **Cen. Sib.**: KR—Ta (Cyndasko river, Krestovskii Island in Yenisey river estuary), TU (Mugur-Aksy settlement). **East. Sib.**: YAK—Ar. —Arctic zone of European USSR, Far East, Japan, China, Alaska, Canada. Described from Kamchatka peninsula. Map 103.

8. **C. epigeios** (L.) Roth 1788. Tent. Fl. Germ. 2 (1): 34.

Plants 80–160 cm tall, with long rhizome and numerous dead leaf sheaths at base. Stems with 2 (3) interrupted joints, erect, strong. Leaves usually broad (up to 10 mm), flat, or narrow, convolute, gray-green, stiff, scabrous. Panicles up to 30 cm long, 1.5–6 cm broad, dense, with short branches covered with minute spinules, deflexed from inflorescence rachis in fruit and thereafter. Spikelets 5–7 mm long, narrowly lanceolate, greenish or violet. Glumes nearly equal, tapered long-subulate at tip, cristate-scabrous along keel. Lemma 1.5–2 times shorter than glumes, almost completely membranous, callus hairs almost twice longer than lemma, with erect slender awn on back; awn arising from middle. Rachilla rudiment lacking.

Meadows, thin forests, among shrubs, on sand and pebble beds. **West. Sib.**: all regions. **Cen. Sib.**: all regions. **East. Sib.**: IR—An, Pr, BU—Se, Yuzh, ChI—Ka, Shi, Yak—Ar, Ol, Vi, Al. —Europe, Caucasus, Mid. Asia, Far East, Iran, the Himalayas, Central Asia, Mongolia, Japan, China. Described from Europe.

9. **C. holmii** Lange 1885 in Holm, Nov. Zemlia's Veg.: 16. —*C. bungeana* V. Petro. —*C. steinbergii* Roshev. —*C. evenkiensis* Reverd.

Stems 10–40 cm tall, aggregated into compact mats, erect or ascending at base, with 2 or 3 joints. Leaves 0.5–0.7 mm broad, narrowly linear, longitudinally folded, smooth. Panicles 2–5 cm long, compressed. Glumes 3–4 mm long, gradually acuminate, brownish, with golden margin. Lemma broadly membranous in upper part; smooth, somewhat scabrous only in upper part; awn erect, arising from lower third and barely surpassing tip of lemma. Callus hairs rather few, equal to or less than half length of lemma. Rachilla rudiment 0.6–1 mm. Yakutia (Shandrin river), $2n = 28$.

Along slopes of marine and river terraces, banks of brooks. **West. Sib.**: TYU—Yam. **Cen. Sib.**: KR—Ta, Pu. **East. Sib.**: YAK—Ar, Ol, Yan, Ko. —Europe (eastern Bol'shezemel'sk tundra), Far East (north), Nor. America. Described from Novaya Zemlya. Map 112.

10. **C. kalarica** Tzvelev 1964 in Arkt. fl. SSSR 2: 74.

Plant 20–40 cm tall, forming compact mats; shoots tangled with dead leaf sheaths at base. Leaves 2–5 mm broad, flat or loosely folded longitudinally, glabrous and smooth on both surfaces. Panicles 3–9 cm long, 1–1.5 cm broad, spicate, with highly shortened branches, smooth in lower part, slightly scabrous in upper. Spikelets 4–6 mm long, pinkish-violet, lanceolate, gradually acuminate. Lemma covered with very small spinules, long geniculate awn arising from lower third of lemma. Callus hairs 1/3 to 1/2 length of lemma. Rachilla rudiment 1–2 mm long, pilose.

Tundras in alpine belt. **East. Sib.**: BU—Se, ChI—Ka (class. hab.— water divide of Kuanda and Kalar rivers). —Endemic. Map 108.

11. **C. korotkyi** Litv. 1918 in Spisok rast. gerb. russk. fl. 55: No. 2750. —*C. turczaninowii* Litv. —*C. korotkyi* subsp. *turczaninowii* (Litv.) Tzvelev.

Stems 40–90 cm tall, erect, poorly foliated, forming mats together with numerous radical leaves. Leaves up to 10 mm broad, flat, stiff, without projecting veins, glabrous and smooth on upper surface, glaucescent; rather scabrous beneath. Panicles 7–15 cm long, usually spicate, compressed, narrow, with shortened scabrous branches. Spikelets 6-9 mm long, green or somewhat violet, narrowly lanceolate, acute. Lemma 4–5.5 mm long, with long (7–11 mm), geniculate awn twisted in lower part and arising from lower quarter. Callus hairs 2 or 3 times shorter than lemma. Rachilla rudiment 1–1.5 (2) mm long, with 3–4 mm long hairs. Buryatia (Yuzhno-Muisk mountain range), $2n = 42$.

Rocky and rubble slopes, rocks in dry thin forests. **East. Sib.**: IR—Pr, BU—Se, Yuzh, ChI—Ka, Shi (environs of Eravnisk Lakes—class. hab.—and others). —Mongolia, Japan, China, Far East. Map 109.

12. **C. langsdorfii** (Link) Trin. 1824, Gram. Unifl.: 225. —*C. purpurea* subsp. *langsdorfii* (Link) Tzvelev. —*C. fusca* Kom. —*C. canadensis* subsp. *langsdorfii* (Link) Hultén.

Rhizome decumbent. Stems up to 150 cm tall, with 4–6 interrupted joints; smooth and glabrous under inflorescence and joints; usually branched at lower and middle joints. Leaves (3) 4–10 mm broad, flat, usually strongly declinate from stem; upper surface grayish, glabrous and scabrous, sometimes with scattered long hairs. Ligules of upper leaves 5–10 (12) mm long. Panicles 10–20 (25) cm long, spreading, with scabrous branches bearing rather few spikelets. Latter (3) 3.5–6 mm long, grayish-green or dull reddish. Glumes oblong-lanceolate, covered with numerous, relatively long spinules. Lemma somewhat shorter than glumes, callus hairs as long as (or nearly so) lemma; awn slender, 2–3 mm long arising on back close to middle or slightly below and not exceeding lemma. Rachilla rudiment 0.2–0.7 mm long. Putoran plateau, $2n = 28$ (Khakoma Lake), $2n = 42$ (Talnakh settlement).

Wet and moist valley meadows, among shrubs in forest glades and burned-out forests, banks of water reservoirs. **West Sib.**, **Cen. Sib.**, **East. Sib.**—all regions. —Europe, West. China, Mongolia, East. Asia, North America. Described from Aleutian Islands. Map 110.

13. **C. lapponica** (Wahlenb.) Hartm. 1820, Handb. Scand. Fl: 46.

Stems 30–100 cm tall, slender, forming rather loose mats. Leaves 1–3 mm broad, narrowly linear, acuminate, longitudinally convolute, almost invariably scabrous on lower surface, diffusely scabrous or with sparse long hairs on upper surface. Panicles 7–15 cm long, 0.8–2 cm broad, compact, narrow. Spikelets 4–6 mm long, brownish-violet, lanceolate. Lemma 3.5–4 mm long, with curved awn arising from lower third or middle on back and somewhat surpassing it. Callus hairs as long or slightly shorter than lemma. Rachilla rudiment about 1 mm long, with up to 2–3 mm long hairs.

Tundra, sand-pebble riverbeds, thin forests, among shrubs. **West Sib.**: TYU—Yam, Khm, AL—Go. **Cen. Sib.**: KR—Ta, Pu, Tn (Verkh. Derognya river), Ve, Tu. **East. Sib.**: IR—An, Pr, Bu—Se, Yuzh, ChI—Ka, Shi, YAK—Ar, Ol, Yan. —Nor. Europe, Nor. Mongolia, Far East, Korean peninsula, Alaska, Canada, Greenland. Described from Sweden.

14. **C. macilenta** (Griseb.) Litv. 1921 in Bot. mat. (Leningrad), 2: 119.

Stems 15–50 cm tall, aggregated into mats, erect or sometimes ascending, covered with dead leaf sheaths at base. Leaves 1–3 mm broad, convolute, less often flat, scabrous. Panicles 3–7 cm long, 0.6–1.5 cm broad, compressed, narrow, subspicate, dense, with short sharply scabrous branches. Spikelets 4–5 mm long, lanceolate, violet. Lemma 3–4 mm long; awn geniculately twisted in lower part, arising from lower quarter on back and slightly exserted from spikelet. Callus hairs short, 2 or 3 times shorter than lemma. Rachilla rudiment 1.5–2 mm long, with long dense hairs as long as spikelet. Altay (Kosh-Agach village) and Tuva (Ak-Chira village), $2n = 28$.

Solonetz meadows, sand-pebble beds, floodplain forests. **West. Sib.:** AL—Go (Charysh river—class. hab.—and others). **Cen. Sib.:** KR—Kha, TU. **East. Sib.:** IR—An, BU—Se, Yuzh, ChI—Shi. —Mongolia. Map 113.

15. **C. macrolepis** Litv. 1921 in Bot. mat. (Leningrad) 2: 125. —*C. epigeios* subsp. *macrolepis* (Litv.) Tzvelev.

Plant with long rhizome. Stems 1–1.5 m tall, thick, stout, foliated almost up to top. Leaves 3–10 mm broad, stiff, grayish-green, usually flat but with edges rolled up on drying, glabrous or with short spinescent hairs, scabrous. Panicles 15–20 cm long, 2–3.5 cm broad, compressed, fairly dense, lobed. Spikelets 7–10 mm long, green or violet in upper part, with unequal glumes (upper 1–1.5 mm shorter than lower). Lemma 3–4 mm long, with erect or slightly curved awn arising close to middle on back. Callus hairs 1.5 times longer than lemma. Rachilla rudiment lacking.

River sand, pebble beds, solonchak meadows. **West. Sib.:** OM, NO, AL—Ba. **Cen. Sib.:** KR—Ve. **East. Sib.:** BU—Yuzh, ChI—Shi. —Southern European USSR, Caucasus, Mid. Asia, Iran, Mongolia, the Himalayas, Japan, China. Described from Tajikistan (Shugnan, Anderob settlement on Pyandzhe). Map 114.

16. **C. neglecta** (Ehrh.) Gaertner, Meyer et Scherber 1799, Fl. Wett. 1 : 94.

Rhizome long, with decumbent shoots. Stems 50–100 (150) cm high, single, slender, reddish at base. Leaves 1.5–3 mm broad, narrow, longitudinally convolute, setaceous; only some leaves flat and scabrous. Sheath blade joint glabrous. Panicle 6–15 cm long, 0.5–2 cm broad, narrow, subspicate. Spikelets 2–3.5 mm long, brownish-violet, rarely green, oblong-ovate. Lemma 2–2.7 mm long, with erect awn arising from lower third or middle on back and usually not exceeding lemma tip. Callus hairs 2/3 to 3/4 shorter than lemma. Rachilla rudiment 0.3–0.6 mm long, with up to 1–1.5 mm long hairs.

Marshes, wet meadows, sand banks of rivers and lakes. **West. Sib., Cen. Sib., East. Sib.**—all districts. —Eurasia, North America, Canada, Greenland. Described from Sweden. Map 111.

17. **C. obtusata** Trin. 1824, Gram. Unifl. 225.

Rhizome long, with decumbent shoots. Stems tall (up to 130 cm), aggregated into compact mats. Leaves 3–8 mm broad, flat, glabrous or with occasional slender hairs in upper part. Panicles narrow, more lax at anthesis, up to 25 cm long, 1–5 cm broad. Spikelets 3–4 mm long, oblong-ovate, acute, green or violet. Lemma somewhat shorter than glumes, with short awn curved in upper part and arising from lower fourth on back and laterally exserted from spikelet. Callus hairs more or less few, not surpassing half length of lemma. Rachilla rudiment 0.5–0.7 mm long, with up to 2 mm long hairs.

Dark coniferous and mixed forests, forest meadows. **West. Sib.:** TYU—Tb (Tobol'sk vicinity—class. hab.—and others), KU, OM, TO, NO, KE, AL—Ba, Go. **Cen. Sib.:** KR—Pu, Tn, Ve, TU. **East Sib.:** IR—An, Pr, BU—Se, Yuzh, ChI—Shi, YAK—Vi, Al. —Eastern Europe, Far East, Mongolia. Map 115.

18. **C. pavlovii** Roshev. 1932 in Izv. Bot. sada AN SSSR, 30: 295. — *C. obtusata* Trin. × *C. purpurea* (Trin.) Trin. —*C. varia* auct. non Host—*C. krylovii* Reverd.

Rhizome decumbent, Stems up to 170 cm tall, foliated. Leaves 4–10 mm broad, flat, grayish-green, scabrous on upper surface. Panicles 15–20 cm long, up to 5 cm diam, compressed. Spikelets 3.5–5 mm long, violet, rarely greenish, lanceolate. Lemma about 3.5 mm long, barely longer than palea. Awn arising from lower 1/4 to 1/3 on back, somewhat geniculate, extending beyond back very slightly. Callus hairs 2/3 to 3/4 length of lemma. Rachilla rudiment about 1 mm long, with 2–3 mm long hairs.

Forests and forest meadows, ascending to alpine belt. **West. Sib.:** TO, KE, AL—Go. **Cen. Sib.:** KR—Kha, Ve, TU. **East. Sib.:** IR—An, Pr, BU—Se, Yuzh. —Mid. Asia, Mongolia. Described from Mid. Asia (Junggar Alatau). Map 119.

19. **C. phragmitoides** Hartman 1832, Handb. Skand. Fl. ed. 3: 20. — *C. flexuosa* Rupr. —*C. elata* Blytt. —*C. purpurea* subsp. *phragmitoides* (Hartman) Tzvelev.

Stems 80–170 cm tall, erect. Leaves 3–7 mm broad, flat, scabrous. Ligules 3–10 mm long, scarious and laciniate at tip. Panicles 12–20 cm long, fairly spreading at anthesis. Spikelets 4.5–7 mm long, violet. Glumes lanceolate, acuminate, almost similar, lower usually slightly longer than upper. Lemma 3–4 mm long, oblong-ovate, bidentate at tip, with erect short awn arising from upper third and slightly projecting above lemma teeth. Palea 1.5 times shorter than lemma. Callus hairs slightly longer or nearly equaling lemma. Rachilla rudiment 1/2–2/3 mm, with hairs as long as callus hairs.

Wet meadows, swamp fringes, river and lake banks. **West. Sib.:** TYU—Tb, OM, TO, NO, KE, AL—Ba. **Cen. Sib.:** KR—Tn (Kuz'movka settlement), Ve (environs of Krasnoyarsk and Minusinsk). —Nor. and Cen. Europe, Caucasus. Described from Sweden. Map 120.

20. **C. pseudophragmites** (Haller fil.) Koeler 1802, Descr. Gram.: 106.

Stems up to 90 cm high, stout, erect. Rhizome long, decumbent. Leaves 3–6 mm broad, stiff, often longitudinally folded, scabrous due to short spinules. Panicles 8–20 cm long, 2–4 cm broad, ovate or oblong, lax, spreading, with long, scabrous branches. Spikelets violet. Glumes linear-subulate, acuminate, unequal; lower 5–8 mm long, 2–3 mm longer than upper. Lemma 3–3.5 mm long, bidentate at tip. Awn erect, 1.5–2.5 mm long, arising from tip of lemma. Callus hairs as long as spikelets and 1.5–2 times longer than lemma. Rachilla rudiment lacking.

Rocky slopes, banks of water reservoirs, coastal sand and pebble beds, among shrubs. **West. Sib.**: TYU—Khm, Tb, KU, OM, TO, NO, AL—Ba, Go. **Cen. Sib.**: KR—Ve, TU. **East. Sib.**: IR—An, BU—Se, Yuzh, YAK—Al, Yan, Ko. —Europe, Far East, Central Asia, Korean Peninsula, Japan. Described from Switzerland. Map 116.

21. **C. purpurascens** R. Br. 1823 in Richardson, Bot. App. Franklin Gaerney: 731. —*C. wiluica* Litv. ex V. Petrov.

Stems 20–50 cm tall, enveloped by dead leaf sheaths at base, together with radical leaves aggregated into dense, compact mats. Leaves 2–5 mm broad, narrowly linear, flat or convolute, densely puberulent on upper surface. Panicles 3–10 cm long, about 1 cm broad, spicate, narrow, rather scarce in lower part. Spikelets 5–6.5 mm long, pinkish-violet. Lemma barely shorter than glumes, with long, geniculate awn twisted in lower part, arising near base or in lower fourth on back and far exserted from spikelet. Callus hairs short, barely reaching fourth of lemma in length. Rachilla rudiment about 2 mm long, uniformly pubescent with hairs. Yakutia (Chersk settlement), $2n = 42$.

Hilly rubble and rocky tundras, on rocks, in sparse larch upper forest zone. **Cen. Sib.**: KR—Ta, Pu. **East. Sib.**: BU—Se, YAK—Ar, Ol, Vi, Al, Yan, Ko. —Alaska, Canada, Greenland. Described from Canada. Map 117.

22. **C. purpurea** (Trin.) Trin. 1824, Gram. Unifl.: 219. —*C. langsdorfii* auct. —*C. poplawskae* Roshev.

Rhizome decumbent. Stems 40–90 cm high, forming loose mats, smooth under inflorescence and joints. Leaves (2) 3–6 mm broad, flat, bright green, usually glabrous and scabrous. Ligules of upper leaves 6–8 mm long. Panicles 7–10 (15) cm long, poorly spreading, branches ascending, scabrous with densely arranged spikelets. Spikelets 2.5–4.5 (5) mm long, reddish-violet. Glumes scabrous throughout surface due to short spinules (occasionally elongated hairlike). Lemma 2.3–3.6 mm long, callus hairs as long as lemma, awn slender, arising above middle on back and often surpassing lemma tip by 1–1.5 mm. Rachilla rudiment 0.2–0.7 mm long, with about 3 mm long hairs. Buryatia (Kotera river), $2n = 28$.

Wet meadows and among shrubs along hill slopes, sand-pebble riverbeds. **West. Sib.**,: TYU—Tb, KU, OM, TO, NO, KE, AL—Ba, Go. **Cen. Sib.**: KR—Pu, Tn, Kha, Ve, TU. **East. Sib.**: IR—An, Pr, BU—Se, Yuzh, ChI—Shi. —Europe, West. China, Mongolia, East. —Asia. Described from Baikal region. Map 118.

23. **C. sajanensis** Malyschev 1961 in Bot. mat. (Leningrad) 21: 452. —*C. epigeios* (L.) Roth × *C. macilenta* Litv.

Stems 30–60 cm tall, forming small compact mats together with radical leaves. Leaves up to 3 mm broad, narrow linear, stiff, glaucous green, scabrous. Panicles 5–8 cm long, 1–2 cm broad, compressed, oblong, with short, sharply scabrous branches. Spikelets 5–6 mm long, narrowly lanceolate, violet, turning brown later. Lemma 3–3.6 mm long, with erect about 2 mm long awn arising from middle of back and surpassing lemma by 0.3–1 mm. Callus hairs barely shorter than lemma, numerous. Rachilla rudiment 0.7–1 mm long, with 2.5–3 mm long hairs.

Riverbed sand. **East. Sib.**: BU—Yuzh (Mondy village—class. hab.—and others). —Mongolia.

24. **C. salina** Tzvelev 1965 in Novosti sist. vyssh. rast.: 27.

Stems 30–90 cm tall, forming loose mats. Leaves 1–3.5 mm broad, flat or loosely folded longitudinally, grayish-green, scabrous. Panicles 4–12 (15) cm long, 0.4–2 cm broad, rather lax. Spikelets 3.5–4.5 mm long, lanceolate, obtuse or acute, green or with faint violet tinge. Lemma with short erect or suberect awn, arising from lower third on back. Callus hairs a third or half as long as lower glumes. Rachilla rudiment 1.5–2.5 mm long, with numerous long hairs.

Saline meadows. **East. Sib.**: BU—Yuzh (Zun-Bie area along Zagustai river), ChI—Shi (Abagaitui village—class. hab.—and others). —Mongolia, Central Asia. Map 121.

25. **C. tenuis** V. Vassil. 1940 in Bot. mat. (Leningrad) 8: 66. —*C. angustifolia* subsp. *tenuis* (V. Vassil.) Tzvelev.

Stems 25–80 cm tall, slender, glabrous and smooth, forming mats. Leaves 1–3.5 mm broad, flat or longitudinally folded, without tufts of hairs on sheath blade joints. Panicles 7–13 cm long, somewhat compressed, with sharply scabrous branches. Spikelets small, 2.5–3 (4) mm long, narrowly lanceolate, pinkish-violet. Lemma 3 mm long, with erect awn arising from middle on back and slightly surpassing lemma. Callus hairs almost as long as floret. Rachilla rudiment about 1 mm long, with up to 3 mm long hairs.

Floodplain meadows, sand-pebble beds, banks of rivers and lakes. **East. Sib.**: BU—Se, Yuzh, ChI—Ka, Shi. —Far East. Described from Anadyr river basin. Map 122.

22. Apera Adanson

1. **A. spica-venti** (L.) Beauv. 1812, Ess, Agrost.: 151.

Annual, 50–120 cm high with slender fibrous roots. Leaves up to 5 mm broad, scabrous on both surfaces. Panicles up to 30 cm long, spreading, with sharply scabrous branches. Spikelets about 2.5 mm long, single-flowered, green or violet along edges. Glumes lanceolate, acute, unequal; upper longer and broader than lower. Lemma elliptical, with short spinules in upper half awn erect, long (2–4 times surpassing spikelet), arising slightly below tip. Callus with very short hairs or glabrous. Rachilla rudiment 0.5 mm long, glabrous.

Cultivated and fallow land, plantations, sandy sites. **West. Sib.**: TYU—Tb, KU, OM, TO, KE (Kuznets basin). **Cen. Sib.**: KR—Kha, Ve, TU. **East. Sib.**: IR—An, Pr, BU—Yuzh, YAK—Vi (environs of Olekminsk). —Europe, Caucasus, Mid. Asia and Asia Minor, Far East. Described from Europe. Map 123.

23. Agrostis L.

1. Palea 1.5–3 times shorter than lemma ....2.
+   Palea absent or very short, up to 0.5 mm long .... 12.
2. Plant with decumbent subsurface shoots .... 3.
+   Plant without decumbent subsurface shoots but with procumbent surface shoots .... 8.
3. Ligule of upper cauline leaf 2–6 mm long; of lower leaves as long or longer than broad. Leaf blades more or less scabrous on both surfaces. Clavate thickenings under spikelets covered with spinules .... 4.
+   Ligule of upper cauline leaf 0.5–1.5 mm long; of lower leaves shorter than broad. Leaf blades subglabrous or with diffuse spinules on upper surface. Clavate thickenings under spikelets invariably glabrous and smooth .... 15. *A. tenuis*.
4. Panicles narrowly lanceolate after anthesis, with branches closely appressed to main rachis. Leaves grayish or glaucescent green, short, acute, radical leaves numerous, upright. Procumbent surface shoots present with subsurface shoots .... 5.
+   Panicles oblong-ovate or pyramidal after anthesis, with declinate branches. Leaves dark or light green, acuminate, radical leaves usually few. Surface shoots lacking .... 6.
5. Lemma awnless .... 12. *A. sibirica*.
+   Lemma with slender straight awn arising from upper third on back .... 3. *A. bodaibensis*.
6. Callus of lemma subglabrous, with very short indistinct, barely visible hairs (shorter than 0.2 mm). Palea almost half length of lemma. Panicles dark violet, with brown tinge, branches long, slender, branched above lower third .... 7.
+   Callus of lemma with short, distinctly visible hairs (0.3–0.5 mm long). Palea usually 2/3 length of lemma. Panicles green or grayish-violet, branches relatively thick and short, branched almost from base .... 7. *A. gigantea*.
7. Panicles 12–20 (25) cm long, branches weak, often flexuous and slightly nutant, lowermost usually 7–12 (15) cm long. Glumes (1–8) 2–2.3 mm long, narrowly lanceolate, slightly

lustrous 1/4 to 1/3 longer than lemma .... 6. *A. divaricatissima.*

+ Panicles, 7–12 (rarely more) cm long, branches very stout, straight, obliquely erect or laterally turned, lowermost rarely longer than 7 cm. Glumes 1.5–1.8 (2.0) mm long, lanceolate, rather dingy, 1/6 to 1/5 longer than lemma .... 10. *A. mongolica.*

8 (2). Stems (25) 30–80 cm tall. Nearly all stoloniform shoots terminating in panicle. Latter 4–15 cm long, green, whitish or grayish-violet, with highly scabrous branches .... 9.

+ Stems 15–25 (40) cm tall. Stoloniform shoots largely vegetative. Panicles 2–5 (10) cm long, reddish-violet, with sub-glabrous or slightly scabrous branches .... 11.

9. Spikelets 1.5–2.5 mm long, green or grayish-violet, gradually acuminate. Lower glumes usually with dense spinules all along keel .... 10.

+ Spikelets 1.2–1.6 mm long, minute, numerous, whitish-green or slightly violet, acute or obtuse. Lower glumes glabrous or with few spinules in upper half along keel .... 1. *A. albida.*

10. Lemma awnless .... 13. *A. stolonifera.*

+ Lemma with long awn exserted from spikelet .... 18. *A.* × *ussuriensis.*

11. Panicles with subhorizontally declinate short branches with few spikelets, more or less scabrous. Anthers 0.6–1 mm long .... 8. *A. jacutica.*

+ Panicles highly compressed with subglabrous branches bearing many spikelets densely appressed to main rachis. Anthers 1–1.4 mm long .... 14. *A. straminea.*

12 (1). Lemma awnless. Anthers 0.3–0.6 mm long. Panicles highly elongate, 10–30 cm long, nearly as long as stem or not shorter than half, broadly spreading .... 13.

+ Lemma with straight or geniculate awn arising close to middle (slightly above or below it) on back. Anthers 0.5–1.5 mm long. Panicles 3–8 times shorter than stem, compressed or spreading but not broad .... 14.

13. Glumes about 2 mm long, acute. Lemma slightly (less than 1/4) shorter than lower glume. Radical leaves fewer, mostly flat, narrow, long .... 5. *A. clavata.*

+ Glumes 2–2.5 (3) mm long, strongly acuminate. Lemma 1.5–2 times shorter than lower glume. Radical leaves many, setaceous, very short .... 11. *A. scabra.*

14. Anthers 0.6–0.7 mm long .... 15.

+ Anthers 0.8–1.5 mm long .... 16.

15. Panicle branches faintly scabrous or subglabrous. Lemma with geniculate awn arising close to middle on back or slightly

below it. Callus sides with distinct hairs 1/6 to 1/5 length of lemma .... 4. *A. borealis.*

+ Panicle branches highly scabrous due to many spinules. Lemma with straight or geniculate awn arising above middle on back. Callus sides with faint tufts of hairs shorter than 1/6 length of lemma .... 2. *A. anadyrensis.*

16. Panicle branches glabrous or subglabrous, rarely with diffuse spinules in upper part. Lower glume glabrous throughout back; with sparse short spinules along keel (usually in upper half, rarely throughout length .... 17.

+ Panicle branches rather scabrous throughout length. Lemma scabrous due to short spinules on back near tip, with dense and acute spinules along keel .... 19. *A. vinealis.*

17. Glumes distinctly unequal. Lemma 0.3–0.7 mm shorter than lower glume, upper half of latter covered with short spinules all along keel .... 18.

+ Glumes nearly equal in size. Lemma almost equal or not more than 0.2 mm shorter than lower glume; latter glabrous, rarely with few spinules near tip, keel hardly developed. Spikelets on distant, subglabrous branches .... 9. *A. kudoi.*

18. Panicles usually spreading at anthesis and thereafter. Radical leaves grayish-green, convolute, rarely flat, very narrow, 0.5–1 (1.5) mm broad, more or less scabrous .... 16. *A. trinii.*

+ Panicles compressed before and after anthesis. Radical leaves yellowish-green, flat, usually 1–2 (4) mm broad, subglabrous or diffusely scabrous .... 17. *A. tuvinica.*

1. *A. albida* Trin. 1845 in Mém. Acad. Sci. Pétersb. (Sci. Phys. Math.) ser. 6: 344. —*A. stolonifera* subsp. *albida* Tzvelev.

Rhizome with slender fibrous roots. Several stems and vegetative shoots forming loose mat. Stems 40–60 cm tall, straight, sometimes geniculate at base. Leaves long (up to 15–20 cm), narrow (1–3 mm), gradually narrowed into slender acuminate tip, narrower than cauline leaves on vegetative shoots, dark or grayish-green, sharply scabrous along margin. Ligules of upper cauline leaves 2 (3) mm long. Panicles 4–15 cm long, 1–3 cm broad, usually compressed after anthesis, branches finely scabrous; clavate thickening at tips poorly developed. Spikelets about 1.5 mm long, numerous, densely sessile, greenish or slightly violet. Glumes nearly equal, acuminate or obtuse, glabrous along keel or lower ones only in upper half, and more often with few spinules close to tip. Lemma barely shorter than glumes, callus with almost invisible hairs; upper ones half of lower. Anthers 0.6–0.9 mm long.

Saline meadows and marshes, near salt lakes. **West. Sib.**: TYU—Tb, KU, OM, NO, AL—Ba. —South European USSR and south. Urals. Described from Lower Volga. Map 127.

2. **A. anadyrensis** Soczava 1934 in Fl. SSSR 2: 746.

Loosely cespitose plant surviving for few years with dense fibrous, filiform roots without subsurface shoots. Stems 20–70 cm tall, fairly stout, slender. Radical leaves developing late, convolute or flat, narrow, long and narrowly acuminate. Cauline leaves 1.5–3.5 mm broad, often flat, rather scabrous. Ligules of upper cauline leaves 2–3 (4) mm long. Panicles elongate, 6–20 cm long, 2–8 cm broad, somewhat spreading at anthesis and thereafter, with long, slender rather sharply scabrous, sometimes rather capitate, obliquely erect branches. Spikelets somewhat violet, rarely green, crowded in upper half of panicle branches. Glumes 2–3 mm long, unequal, lower slightly longer than upper, covered with acute spinules along keel. Lemma shorter than glumes, awn geniculate, rarely straight, arising usually near middle or in upper half on back. Anthers 0.5–0.7 mm long.

Lower mountain belt along river valleys on pebble beds, wet meadows, poplar-chosenia and larch forests, scrubs, roadsides. **East. Sib.**: IR—An (St. Murino on southeast coast of Baikal, probably introduced), BU—Se (Kotel' nikovsk cape), YAK—Ar, Al, Yan, Ko. —Far East, southern Alaska. Described from Magadansk region (Belaya river in Anadyr' basin). Map 124.

3. **A. bodaibensis** Peschkova sp. nova (*A. sibirica* V. Petrov × *A. trinii* Turcz.).

Planta habitu *A. sibiricae* similis a qua lemmate dorso aristis rectis tenuibus e triente superiore oriundis praedito necnon foliis angustioribus plerumque convolutis differt.

Ab *A. trinii* paniculis angustis plus minusve compressis et palea superiore lemmate duplo vel triplo breviore bene distinguitur. Antherae 0.8–1.2 mm longae.

Typus: prov. Irkutensis, pag. Chomolcho, 800 m.s.m., 174 km ab opp. Bodaibo secundum tractum per fodinam Kropotkin, pratum subpaludosum ad viam No. 1630 27 VII 1978, M. Ivanova, L. Beloussova (NS).

Plant similar in habit to *A. sibirica*, differing from it in slender straight awn arising from upper third on back of lemma and very narrow, largely convolute leaves; differs from *A. trinii* in narrow and rather compressed panicles and presence of palea 1/3 or 1/2 length of lemma. Anthers 0.8–1.2 mm long.

River valleys in wet regions. **East. Sib.**: IR—An (Kezhma settlement on Angara river), Pr (Khomolkho settlement in Bodaibinsk region—class. hab.), YAK—Vi (Peledui settlement in Lensk region). —Endemic.

4. **A. borealis** Hartman 1838, Handb. Skand. Fl. ed. 3: 17.

Cespitose perennial. Stems 15–30 cm tall, erect, glabrous. Radical leaves flat, partly longitudinally folded, glabrous, sometimes scabrous only near tip. Cauline leaves 0.5–2.5 mm broad, narrowly linear, acuminate, flat. Ligules of upper cauline leaves 1–2.5 (4) mm long, rather obtuse, crenate or laciniate. Panicles 4–7 cm long, 1–3 cm broad, lax, with subglabrous declinate branches. Spikelets dark purple, glumes 2.5–3.5 mm long, slightly unequal, ovate-lanceolate, acute, glabrous, with acute spinules along keel. Lemma 1/4 to 1/3 shorter than glumes, awn 3–4 mm long, usually arising below middle on back. Anthers 0.5–0.7 mm long.

Hilly and arctic tundras, arable grasslands, sand-pebble riverbeds. **West. Sib.:** TYU—Yam, Khm. —Arctic Europe, Scandinavia, NE North America. Described from Sweden (Lapland). Map 129.

5. **A. clavata** Trin. 1821 in Sprengel, Neue Entdeck. 2: 55.

Plants surviving for few years with slender fibrous roots. Stems (15) 20–75 cm tall, numerous, aggregated into mat. Leaves flat, 1–5 mm broad, acute, bright green, more or less scabrous; radical leaves usually few. Ligules of upper cauline leaves 2–3 mm long. Panicles 6–20 (30) cm long, spreading, nearly half as high as plant, with long slender sharply scabrous branches. Spikelets about 2 mm long, green or pale violet. Glumes nearly equal, slightly scabrous along keel. Lemma slightly shorter than glume, oblong-ovate, awnless; palea very small (about 0.2 mm long) or altogether lacking. Anthers shortly elliptical or oval, 0.3–0.6 mm long, yellow or violet. Baikal region (northern Baikal), $2n = 42$.

Dark as well as light coniferous and mixed forests, forest glades, wet meadows and swamps, riverbed scrubs; very often along roads and tracks; taiga species adapted to acidic soils. **West. Sib.:** TYU—Yam, Khm, OM, NO, TO, KE, AL—Go. **Cen. Sib.:** KR—Pu, Tn, Kha, Ve, TU. **East. Sib.:** IR—An, Pr, BU—Se, Yuzh, ChI—Ka, Shi, YAK—Vi, Yan, Ko.—NE Europe, Caucasus, Far East, Mongolia, Nor. China, Japan, North America. Described from Kamchatka.

Specimens from Putoran plateau, especially from the belt below bald peaks, differ from type in very small size of plants. Stems are solitary; panicles narrow, compressed, violet; panicle branches weakly scabrous or subglabrous; anthers 0.3–0.4 mm long, often dark violet. They can be classified as var. *putoranica* Peschkova, var. nov. —Planta humilis; caules solitarios; panicula angusta, compressa, violacea; ramuli paniculae scabridiuscula vel subglabra; antherae 0.3–0.4 mm lg., saepe atroviolaceae.

Typus: Putorana, lacus Ende, pars orientalis, regio subalpina, in locis humilis argillosis inter prato psychromesophytico, in valle inundationibus, 17 VIII 1970, No. 465 J. Petroczenko.

6. **A. divaricatissima** Mez 1922 in Feddes Repert. 18: 4. —*A. mongolica* auct. non Roshev.

Large (40–100 cm tall) plant with loose mats; subsurface shoots not always present. Stems usually many, straight. Radical leaves, like cauline leaves, poorly developed, 1.5–5 mm broad, narrowly acuminate, sharply and densely scabrous on both surfaces, grayish-green. Ligules of upper cauline leaves 3–5 mm long. Panicles 12–20 (25) cm long; pyramidal, broadly spreading at anthesis and thereafter; 10–15 cm broad, with very

106

slender long (up to 15 cm long) sharply scabrous branches, not branched in lower third. Spikelets (1.8) 2–2.3 mm long, deep violet, slightly brownish. Glumes narrowly lanceolate, gradually acuminate toward tip; lower with few spinules in upper half along keel. Lemma 1/4 to 1/3 shorter than glumes, obtuse, sometimes truncate; occasionally with short (1 mm long) awn, palea nearly half length of lemma. Anthers about 1 mm long.

Wet, often saline meadows along river valleys and close to lakes. **East. Sib**: BU—Se, Yuzh, ChI—Shi. —East. Mongolia, Manchuria, and Korean peninsula. Described from Korean peninsula. Map 128.

7. **A. gigantea** Roth 1788, Fl. Germ. 1: 31. —*A. alba* auct. non L.

Rhizome with subsurface shoots. Stems (15) 30–100 cm tall, straight, solitary or few. Leaves 1–8 mm broad, flat, rather scabrous along fringes and veins. Ligules 2–6 mm long, laciniate at tip. Panicles (5) 7–20 cm long, spreading at anthesis, somewhat compressed before and after anthesis, with sharply scabrous relatively short, obliquely erect branches. Spikelets 1.5–2.5 (3) mm long, greenish or violet. Glumes nearly equal, scabrous along keel. Lemma 1/5 to 1/4 shorter than glumes, usually with very short awnlike extension of midrib or short straight easily shedding terminal awn; with tufts of 0.2–0.5 mm long hairs at base. Palea generally 1/3 shorter than lemma. Anthers 1–1.5 mm long. East. Sayan (Tunkinsk mountain range), $2n = 28$.

Floodplain and dry valley meadows, forest glades, fringes, marshes, forests and scrubs, sand-pebble riverbeds. **West. Sib, Cen. Sib.**: all regions. **East. Sib.**: IR—An, Pr, BU—Se, Yuzh, ChI—Ka, Shi, YAK—Ol, Vi, Al, Yan. —Europe, Mediterranean, West. Asia, West. China (Junggar, Kashgar), the Himalayas, Mongolia. Described from FRG (environs of Bremen).

Highly polymorphic species; plant height, leaf breadth, panicle form and size, length of hairs on callus vary.

8. **A. jacutica** Schicshkin 1934 in Fl. SSSR 2: 747, 179. —*A. stolonizans* auct. fl. Jacut.

Rhizome without subsurface shoots, with dense fibrous, filiform shoots. Vegetative (sterile) surface shoots usually several, often procumbent, geniculate and rooting at lower nodes. Flowering stems 15–35 cm high, ascending or geniculate at base, erect thereafter. Leaves 2–5 cm long, 0.5–1.5 (2) mm broad, gradually acuminate, slightly scabrous on both surfaces, grayish-green, sometimes with reddish tinge; those on flowering stems usually broader and shorter than on vegetative. Ligules of upper cauline leaves up to 2 mm long, those of lower distinctly shorter. Panicles 4–8 cm long, 2–3 cm broad, with few spikelets, rather spreading at anthesis and thereafter; with subglabrous or slightly scabrous, obliquely erect branches, sometimes subhorizontal; clavate thickening at tips of branches slightly developed. Glumes 1.2–2.2 mm long, dark violet with purple hue, rather acute, glabrous or lower with few spinules on back. Lemma distinctly shorter than glumes, obtuse, midrib some-times extended in form of very short awnlike spine; palea 2 or 3 times shorter than lemma. Anthers 0.6–1 mm long, wholly yellow or with violet bands.

Lake banks, river meanders, marshes, wet meadows, sand and peb-ble bed sections of floodplains. **Cen. Sib.**: KR—Pu. **East. Sib.**: IR—Pr (Bol. Patom settlement in Lensk region, Erbogachen in Khatanga region and Sogdiodon in Mamsk-Chui region), Bu—Se (environs of Lower Angara), YAK—Ar (Chersk settlement), Vi (upper part of Sinei river basin, Bagadzha-Yryakh brook, Byrylakhsk station—class. hab.), Al (Khandyga settlement in Tomponsk region, Ynykchan in Ust'-Momsk region), Yan (Tomtor settlement, lower course of Sartang river), Ko (Zyryanka and Lobuya settlements). —Endemic. Map 126.

9. **A. kudoi** Honda 1931 in Miyabe and Kudo, Fl. Hokk. Saghal. 2: 135. —*A. trinii* auct. non Turcz. —*A. vinealis* subsp. *kudoi* (Honda) Tzvelev.

Finely caespitose plant 20–40 cm tall. Stems usually few or several, slender, straight. Radical leaves many, narrow, 0.5–1.5 (2.5) mm broad, narrowly acuminate, flat or convolute (subfiliform), with lower (outer) surface glabrous and smooth, upper (inner) surface scabrous or puberulent. Upper cauline leaf usually flat, slightly recurved, scabrous on upper surface, glabrous beneath; with 1–2 mm long ligules. Panicles 3–10 cm long and 2–5 cm broad; pyramidal, spreading, with relatively few spikelets at anthesis and thereafter; branches glabrous, smooth or slightly scabrous in upper part. Spikelets 2.2–2.6 (3) mm long, brown-ish-violet, sometimes green, lustrous. Glumes nearly equal, lower with barely visible keel; spinules few near tip or lacking. Lemma slightly (by 0.2–0.3 mm) shorter than glumes; awn geniculate, arising close to lower third on back. Anthers 1–1.5 mm long. Putoran plateau (Khakoma Lake), $2n = 28$.

Sand-pebble beds near reservoirs in high altitudes, thickets of river-bed shrubs, valley meadows, rock debris. **Cen. Sib.**: KR—Pu. **East. Sib.**: IR—Pr (Oron Lake), BU—Se, ChI—Ka, YAK—Al, Yan. —Far East, Alaska. Described from Sakhalin. Map 131.

10. **A. mongolica** Roshev. 1926 in Sev. Mongoliya 1: 162.

Compactly caespitose plant with short rhizome and rather fibrous roots. Subsurface shoots not always developed. Stems 30–75 cm tall, several, straight or geniculate at base and sometimes, together with leaf sheaths, violet. Leaves 1–3 mm broad, narrowly acuminate; cauline leaves short, 3–5 cm long; radical leaves long (up to 15 cm), few, densely scab-rous on both surfaces. Ligules of upper cauline leaves 2.5–3.5 mm long. Panicles 7–12 (20) cm long, spreading at anthesis, somewhat compressed later, oblong (up to 10 cm broad), with obliquely erect slender, sharply scabrous or hirsute branches up to 7 (9) cm long, branching above lower third. Spikelets 1.5–1.8 (2) mm long, dark violet. Glumes lanceolate,

108

sharply narrowed at tip; lower with few spinules along keel, rarely without spinules. Lemma 1/5 to 1/4 shorter than glumes; obtuse, awnless or with short straight awn; palea half as long as lemma. Anthers about 1 mm long.

Wet and marshy meadows, saline steppe river valleys and near salt lakes. **Cen. Sib.**: KR—Kha, Ve, TU (Tere-Khol' Lake, Khandagaity settlement). **East. Sib.**: IR—Pr (northwestern coast of Baikal, Ol'khonsk region), BU—Se (Nizh. Kuitun area in Barguzinsk valley; Irkano Lake in Verkh. Angara river valley), Yuzh. —Nor. Mongolia. Described from Mongolia (Uber-Dzhargalante river). Map 139.

11. **A. scabra** Willd. 1797, Sp. Pl. 1: 370.

Finely cespitose plant 15–45 cm tall with numerous slender roots, without decumbent subsurface shoots. Radical leaves several, 3–7 cm long, 0.3–0.6 (1) mm broad, short, setaceous more or less scabrous or subglabrous; cauline leaves more often flat, rarely longitudinally folded, 1–2 mm broad, sharply scabrous along margin and veins. Ligules of upper cauline leaves 2–4 mm long. Panicles 10–30 cm long, 1/3 to 2/3 as long as plant; broadly spreading at anthesis and thereafter; branches very slender, piliform, sharply scabrous, subpilose. Spikelets reddish-violet, concentrated in upper 1/4 or 1/3 of branches. Glumes 2–2.5 (3) mm long, unequal, acuminate; lower with sharp and dense spinules along keel. Lemma awnless, 1.5–2 times shorter than lower glume. Anthers 0.4–0.6 mm long.

Banks of rivers and small reservoirs, sometimes on dumps. **East. Sib.**: YAK—Al (Ynykchan settlement in Ust'-Momsk region), Yan (Segen-Kyuël' settlement in Kobyaisk region), Ko (Zyryanka settlement in Verkhnekolymsk region). —Far East, Japan, Korean peninsula, North America. Described from North America.

12. **A. sibirica** V. Petrov 1930, Fl. Yakut. 1: 175.

Loosely cespitose plant with fibrous dense filiform roots, usually with decumbent subsurface shoots. Vegetative shoots usually numerous, short (not more than 7 cm high). Flowering stems 25–40 cm tall, slender, geniculate in lower nodes, upper internodes very long, significantly exceeding all lower ones taken together. Leaves 2–5 cm long, 2–3 mm broad, flat, acute, grayish or glaucescent green at tip, sharply scabrous on both surfaces, stiff, erect. Ligules of upper leaves 2–3 mm long, of lower ones shorter (1–2 mm). Panicles up to 7 cm long, about 1 cm broad; narrow, linear-lanceolate, compact, many-spiked, greenish or slightly violet even at anthesis; branches usually weakly scabrous, sometimes subglabrous, more rarely sharply scabrous, closely appressed to main rachis. Glumes 1.5–2 mm long, greenish, brown or violet, with few spinules both along keel and near tip. Lemma slightly shorter than glumes, obtuse; callus glabrous or with very short, hardly visible hairs (less than 0.2 mm long). Palea half length of lemma. Anthers 0.7–1 mm long.

Marshy, frequently saline meadows, banks of rivers and lakes, sand-pebble riverbeds, along forest roads, shrubs in floodplains. **West. Sib.**: TO (Kireevsk settlement), KE (Tutal'sk village). **Cen. Sib.**: KR—Tn (Vanavara settlement on Podkamennaya Tunguska river), Kha, Ve, TU. **East. Sib.**: IR—An, Pr, BU—Se (Nizhneangarsk town), Yuzh, YAK—Vi (Olekminsk town—class. hab.—Lensk town and Chapaevo settlement). —Endemic. Map 130.

13. **A. stolonifera** L. 1753, Sp. Pl.: 62. —*A. stolonizans* Besser ex Schultes et Schultes fil. —*A. alba* var. *prorepens* (Koch) Ascherson. —*A. alba* var. *coarctata* (Ehrh.) Blytt.

Loosely cespitose plant, the many rooting surface shoots and stems 30–80 cm tall, ascending or geniculate at base, forming mats. Leaves 3–10 cm long, 1–2 (4) mm broad, narrowly linear, acuminate, scabrous along margin and veins. Ligules of upper cauline leaves up to 2 mm long, of lower usually shorter. Panicles 5–12 cm long, compressed after anthesis, narrow, 0.5–1.5 (2) cm broad, dense, with short sharply scabrous obliquely erect branches. Spikelets 1.5–2.5 mm long, greenish or with violet hue. Glumes gradually acuminate; Lower wholly sharply scabrous along keel or only in upper half. Lemma slightly shorter than glumes, rounded at tip, palea half length of lemma, callus hairs very short (less than 0.2 mm). Anthers about 1 mm long.

Marshy meadows, sand-pebble riverbeds, banks of lakes and meanders, along forest roads. **West. Sib.**: all regions. **Cen. Sib.**: KR—Tn, Kha, Ve, TU. **East. Sib.**: IR—An, Pr, BU—Tuzh (Khoitogol village in Eastern Sayan, Selenga river delta, Bichura settlement), ChI-Ka (Dogopchan settlement in Muisk-Kuandinsk basin), YAK—Vi (Peledui settlement on Lena river). —Europe, Mid. and West. Asia, Mongolia, NE China; introduced in Far East. Described from Europe. Map 132.

14. **A. straminea** Hartman 1819, Gen. Gram. Scand.: 4. —*A. maritima* auct. non Lam.

Rhizome slender, with long rooting shoots. Stems 7–40 cm tall, straight or ascending, many. Leaves 2–8 cm long, 1–2.5 mm broad, flat or longitudinally folded, gradually acuminate, glabrous. Ligules of upper cauline leaves 1–3 mm long, oblong, obtuse. Panicles 2–10 cm long, 0.5–1.5 cm broad, oblong-linear, sometimes spicate, highly compressed, with short glabrous branches, densely appressed to main rachis. Glumes 2.5–3 mm long, dark violet with purple hue, elliptical-lanceolate, acute, nearly equal; glabrous along keel. Lemma distinctly shorter than glumes, obtuse; midrib sometimes extended into very short, readily breaking awn; palea 2–3 times shorter than lemma. Anthers about 1.5 mm long.

Marshy meadows and shoals on sea coast, estuaries of large rivers. **West. Sib.**: TYU—Yam (Kara river near Bol. Vanuita river estuary; Salekhard town; Vakh river between Mal. Palin and Soromin villages). —Northern Europe, Scandinavia. Described from Sweden (Halland province).

15. **A. tenuis** Sibth. 1794, Fl. Oxon.: 36. —*A. vulgaris* With. —*A. capillaris* auct. non L.

Rhizome short, with rather short subsurface shoots. Stems 30–80 cm tall, usually several, aggregated into loose mats. Leaves 1–4 mm broad, flat, very rarely convolute, acuminate, glabrous or with diffuse spinules toward tip; scabrous along margin. Ligules of lower leaves not longer than 1 mm, sometimes almost invisible; of upper leaves 1–1.5 (2) mm long; rounded or with short cusp. Panicles 2–15 cm long, 2–6 cm broad; compressed before anthesis; spreading at anthesis and thereafter, with slender, usually weakly scabrous branches; clavate thickenings under spikelet invariably glabrous. Glumes 1.5–2 mm long, brownish-violet, rarely greenish; lower ones scabrous in upper part along keel. Lemma somewhat shorter than glumes, obtuse; palea half length of lemma. Anthers about 1 mm long.

In floodplain and forest meadows, sand banks of rivers and lakes, grassy aspen and fir forests; introduced in dumps along highways and railroads. **West. Sib.:** TYU—Khm, Tb, OM, TO, NO, KE, AL—Ba, Go. **Cen. Sib.:** KR—Ve. **East. Sib.:** IR—An, BU—Yuzh (southeastern Baikal coast). —Europe, Caucasus, Mid. Asia (Junggar Alatau), West. Asia; introduced in Far East. Described from Great Britain (environs of Oxford). Map 133.

16. **A. trinii** Turcz. 1856 in Bull. Soc. Nat. Moscou 29 (1): 18, in adnot. *A. canina* auct. non L. —*A. vinealis* subsp. *trinii* (Turcz.) Tzvelev.

Rhizome short, with rooting subsurface shoots. Stems (15) 20–70 (90) cm tall, slender, straight, usually few, forming loose mats together with vegetative shoots. Radical leaves very narrow, 0.5–1.5 (2) mm broad, convolute, rarely flat, grayish-green, usually very short, 3–7 (rarely longer than 10) cm, rather scabrous on lower (outer) surface; sharply scabrous or puberulent on upper (inner) surface, sometimes glabrous and smooth. Upper cauline leaves usually convolute, densely scabrous on both surfaces; ligules 0.5–2 (3) mm long. Panicles 7–14 cm long, 2–6 cm broad; spreading, lax at anthesis and thereafter, with branches glabrous or slightly scabrous in upper part. Spikelets 1.6–2.3 mm long, grayish-violet. Glumes unequal; lower somewhat longer than upper, glabrous, smooth with few short spinules only along keel in upper part. Lemma 0.3–0.7 mm shorter than lower glume, with geniculate awn arising in lower third or close to middle on back. Anthers 0.7–1.2 mm long.

Steppe and lower part of forest belt—in steppes, dry valley, steppified and solonetz meadows, forest glades and fringes, sand banks of rivers and lakes. **Cen. Sib.:** KR—Tn (Vanavara village, Kukuingda river), Kha, Ve, TU (southern foothills of West. Sayan). **East. Sib.:** IR—An (Baikal Lake—class. hab. —and others), Pr, BU—Se, Yuzh, ChI—Shi, YAK—Vi, Al. —Far East, Nor. Mongolia and NE. China. Map 134.

17. **A. tuvinica** Peschkova sp. nova.

Rhizoma breve, surculis subterraneis haud longis, interdum nullis. Culma 15–55 cm alti, plerumque numerosi, unacum foliis radicalibus copiosis caespitulos haud magnos tormantes. Folia plana angusta, 1–2 (3.5) mm lata, longe et tenuiter acutata, pallide vel flavido-viridia, secus nervos subtus praecipue plus minusve scabra, haud raro sublevia, margine acuta scabra, culmeum superius ligula 1–2 mm longa praeditum. Panicula 5–13 cm longa, 1–3 cm lata, ante et post anthesin plus minusve compressa, ramulis oblique ascendentibus glabris vel superne scabridiusculis. Spiculae plerumque numerosae 2–2.5 (3) mm longae, violaceae, rarius fuscidulo-violaceae. Glumae manifeste inaequilongae, superior inferiore multo longior, secus carinam breviter acuteque aculeolata, basi plerumque haud aculeolata. Lemma-gluma superiore ad 0.3–0.5 mm brevius, arista geniculata, e triente inferiore vel prope medium exeunte. Antherae 0.8–1.2 mm longae.

Typus. RSSA Tuva, declive occidentale jugi Academici Obruczevii, systema fl. Tapsa, in fluxu superiore affluxionis eius dextrae Karachem, 1300 m. s. m., pratum graminoso-varüherbosum fruticeto tectum, no. 732, 9 VII 1974, I. Krasnoborov, V. Chanaminzun.

Affinitas. Species *A. vinealis* Schreber et *A. trinii* Turcz. affinis, a quibus foliis radicalibus viridibus planis et sat longis differt, a priore praeterea gluma superiore prope apicem haud breviter aculeolata et paniculae ramulis subglabris vel vix scabridiusculis, a posteriore vero panicula densiore et plus minusve compressa, caespitulis majoribus necnon foliis planis viridibus sat copiosis distinguitur.

Rhizome short, with rather short subsurface shoots, latter sometimes lacking. Stems 15–55 cm tall, usually several, forming small mats together with numerous radical leaves. Leaves flat, narrow, 1–2 (3.5) mm broad, narrowly acuminate, light or yellowish green, rather scabrous (frequently subglabrous) along veins especially on lower surface; sharply scabrous along margin. Ligules of upper cauline leaves 1–2 mm long. Panicles 5–13 cm long, 1–3 cm broad, rather compressed before and after anthesis, with obliquely erect branches, glabrous or slightly scabrous in upper part. Spikelets usually many, 2–2.5 (3) mm long, violet, less often brownish-violet. Glumes distinctly unequal, lower slightly longer than upper, with short sharp spinules along keel; spinules usually absent at keel base. Lemma 0.3–0.5 mm shorter than lower glume; awn geniculate, arising in lower third or near middle. Anthers 0.8–1.2 mm long.

Middle and upper mountain belt, sand-pebble river banks, subalpine meadows and shrubs, mixed forests and exposed slopes. **West. Sib.:** AL—Go. **Cen. Sib.:** KR—Kha (Kamyzyak village, Ust'-Abakansk region; Kyzlas village, Askizsk region), Ve (Aginsk village), TU (Kara-Khem river on Akademika Obrucheva mountain range—class. hab.—and others). **East. Sib.**: IR—An (Kara-Buren' river on Udinsk mountain range), Pr (Baikalo-Patomsk upland), BU—Se (Stanovoe upland), Yuzh (Kashtak river estuary, Oki tributary; Shimki village on Irkut river upper courses; Uburt-Khongoldoi river on Tunkinsk mountain range), ChI—Shi, YAK—Vi, Al. —Nor. Mongolia. Map 135.

18. **A. × ussuriensis** Probat. 1984 in Bot. zhurn. 2: 254 (*A. stolonifera* L. × *A. trinii* Turcz.).

Differs from *A. stolonifera* in presence of straight awns in most florets emerging from spikelets, very lax panciles with relatively long branches; from *A. trinii* in long procumbent surface shoots rooting at joints and presence of palea half length of lemma. Anthers 0.7–1 mm long.

Wet meadows along river valleys. **West. Sib.**: TYU—Khm (Bel'sk mountain range around Surgut town). **East Sib.**: IR—An (Murino station on southeast. Baikal coast), ChI—Shi (Nerchinsk Zavod settlement). —Far East. Described from Ussuri river basin in Primorsk region.

19. **A. vinealis** Schreber 1771, Spicil. Fl. Lips.: 47. —*A. canina* auct. non L.

Rhizome short, with slender rather short subsurface shoots. Stems 20–70 cm tall, often several, rarely few, with few cauline leaves. Radical leaves convolute, 0.5–1.5 mm diam, grayish-green, narrowly acuminate, subglabrous on both surfaces, occasionally weakly scabrous on lower (outer) surface. Ligules of upper cauline leaves 0.5–1.5 mm long. Panicles 4–14 cm long, rather compressed or slightly spreading before and after anthesis, up to 5 cm broad, branches obliquely erect, slender, more or less sharply scabrous almost throughout length, rarely subglabrous in lower part. Spikelets 1.5–2.2 (2.5) mm long, several, violet. Glumes nearly equal or slightly unequal; lower usually slightly longer than upper, sharply scabrous almost all along keel and, additionally, covered with dense short spinules near tip. Lemma slightly shorter than glumes, with tufts of very short (about 0.2 mm long) hairs on callus; awn arising from lower third on back, geniculate. Anthers 0.8–1.3 mm long.

Steppe and forest steppe belt on wet solonetz and humid dry valley meadows, forest glades, steppified forests, sometimes fallow land. **West. Sib.**: TYU—Tb, KU, OM, TO (Zorkal'tsevo village), NO, KE, AL—Ba, Go. **Cen. Sib.**: KR—Kha, Ve. —Europe, Caucasus, North America (northeast). Described from GDR (environs of Leipzig). Map 137.

24. Hierochloë R. Br.

1. Lemma of staminate florets awnless or with short (not longer than 1 mm) subterminal or terminal awn. Plant with long rhizome .... 2.
+ Lemma of staminate florets with 5–8 mm long awn arising close to middle on back. Cespitose plant .... 1. *H. alpina*.
2. Panicles pyramidal or oblong, spreading or compressed, with many

spikelets (25–100 or more). Leaves flat or convolute, erect, rather coarse .... 3.
+ Panicles racemose, narrow, secund with few spikelets (3–15). Leaves narrow (up to 1–2 mm), convolute, slightly curved, relatively soft .... 9. *H. pauciflora.*
3. Callus of lower staminate florets with fairly stiff, sometimes sparse hairs at base .... 4.
+ Callus of lower staminate florets glabrous; sparse hairs extremely rare .... 6.
4. Leaves of vegetative shoots usually broad ((3) 5–8 (10) mm), relatively short (10–15 cm), rapidly acuminate, aggregated near base of reproductive stems. Rhizome relatively thick, 1–2 mm diam, yellow or creamy .... 5.
+ Leaves of vegetative shoots usually relatively narrow (2–4 (5) mm), long (up to 20 cm), acuminate, arising in tufts at some distance from flowering stems. Rhizome slender (up to 1 mm), light gray or dirty white .... 3. *H. arctica.*
5. Lemma of staminate florets with distinct spinules on back only in upper part, glabrous or subglabrous beneath (with sharp tubercles). Panicles with obliquely erect branches; spikelets many and almost wholly covering main rachis of panicle. Leaf blades grayish-green or glaucous, glabrous, with papillose protuberances (not spinules) along margin .... 10. *H. repens.*
+ Lemma of staminate florets glabrous and smooth on back only in lower fourth, with distinct spinules above. Panicles spreading at anthesis, later compressed, less dense; main rachis of panicle usually not covered with spikelets. Leaf blades green, with fine hairs on upper surface, rarely glabrous, with dense sharp spinules along margin .... 8. *H. odorata.*
6. Lemma usually without scarious fringe or very narrow one. Glumes as long or shorter than spikelet, acute or obtuse. Sheath of lower leaves rather densely pilose, rarely glabrous .... 7.
+ Lemma with distinctly visible but sometimes fairly narrow scarious fringe. Glumes almost invariably longer than spikelet or as long, usually narrowly acuminate. Sheath of lower leaves glabrous, smooth or scabrous due to short spinules, occasionally with diffuse short appressed hairs .... 9.
7. Ligules of upper cauline leaves 1.5–2.5 mm long, scarious, usually glabrous, rarely slightly pilose, undivided or lacerated at tip, sometimes with fine cilia .... 8.
+ Ligules of upper cauline leaves 0.5–1 (1.5) mm long, scarious, rather densely pilose on outer side (toward blade) ....
.... 6. *H. glabra* subsp. *chakassica.*
8. Panicles spreading at anthesis, later compressed; panicle branches

obliquely erect. Leaves usually green, glabrous or finely pilose on upper surface. Lemma of staminate florets acuminate, often with short terminal cusp, largely glabrous on back, with short spinules (very rarely with stray hairs) under tip .... 4. *H. glabra* s. str.

+ Panicles spreading even in fruit, with slender lower branches often nutant. Leaves grayish-green, glabrous. Lemma of staminate florets obtuse, awnless or with subapical short spinule not surpassing scarious fringe of lemma; glabrous, smooth on back ....
.... 8. 5. *H. glabra* subsp. *bungeana*.

9. Lemma of staminate florets with broad scarious fringe at tip and rather dark-colored midrib, awnless, very rarely with short up to 0.2 mm long cusp, with short spinules on back near tip .... 10.

+ Lemma of staminate florets with relatively narrow scarious fringe at tip and short (up to 0.6 mm) slightly recurved awn; coarsely scarious due to short spinules transforming into short hairs in upper half on back .... 11. *H. sibirica*.

10. Leaves and their sheaths glabrous. Sheath blade joints with band of dense hairs. Panicles compact, rather compressed. Spikelets 3–4 mm long. Lemma of staminate florets with short spinules in upper part, sometimes turning into cilia .... 2. *H. annulata*.

+ Leaf sheaths rather scabrous or with diffuse short hairs, sometimes subglabrous; leaf blades sometimes rather pilose on upper surface. Sheath blade joints glabrous or with short papillose protuberances. Panicles rather lax, somewhat compressed in fruit. Spikelets somewhat larger, 4–5 (6) mm long. Lemma of staminate florets subglabrous on back, slightly scabrous only under tip. ....
.... 7. *H. ochotensis*.

1. **H. alpina** (Sw.) Roemer et Schultes 1817, Syst. Veg. 2: 515.

Plant with short decumbent subsurface shoots, frequently forming fairly large mats. Stems 15–40 (60) cm tall, few. Radical leaves often convolute, rarely flat, glabrous and smooth outside (lower surface) and covered with short hairs within (upper surface) along veins. Cauline leaves with long slightly inflated sheath, scabrous due to short and sparse hairs, and short blade (3–15 mm long in upper leaf). Ligules up to 1–1.5 mm long, ciliate along margin. Panicles 1.5–5 (7) cm long, 1–2 cm broad, compressed, ovate or oblong. Glumes scarious, obtuse, glabrous, lustrous at tip. Lemma of staminate florets nearly as long as glumes, brown; diffusely hirsute or scabrous throughout surface; with long appressed hairs along margin; awn of lower floret short (1–3 mm long), straight, arising near tip; awn of upper floret long, surpassing lemma by 2–4 mm, geniculate, arising near middle on back or below it. Anthers up to 2.5 mm long. Putoran plateau (Talnakh settlement, Baselak Lake) and Stanovoi upland (source of Kotera-Maskit river), $2n = 56$.

Rocky, moss-lichen and dryad tundras, alpine grasslands, Japanese stone pine thickets, sparse forests below bald peaks, sand banks of rivers and lakes, rocky slopes, rock debris at high altitudes and Arctic region. **West. Sib.**: TYU—Yam, Khm, KE, Al—Go. **Cent. Sib.**: KR—Ta, Pu, Ve, TU. **East. Sib.**: IR—An, Pr, BU—Se, Yuzh, ChI—Shi, Ka, YAK—Ar, Vi, Al, Yan, Ko. —Arctic Europe, Nor. Urals, Far East, Mongolia, NE China, North America. Described from Sweden. Map 136.

2. **H. annulata** V. Petrov 1930, Fl. Yakutii 1: 131. —*H. odorata* auct. p. p. —*H. odorata* subsp. *kolymensis* Probat.

Rhizome 1–1.5 (2) mm diam, bright or slightly yellowish. Stems 20–45 cm tall, slender, glabrous, with 2 or 3 shortened (up to 2 cm) linear-lanceolate leaves, present in lower fourth or third of stem. Leaves of vegetative shoots 3–6 mm broad, flat, usually with convolute margin, long, acuminate, with diffuse fine hairs or glabrous on upper surface; sheaths glabrous, with ring of short dense hairs at sheath blade joints. Ligules of upper cauline leaves 1.5–2 mm long. Panicles 3–6 (8) cm long, 2.5–4 (6) cm broad, pyramidal or oblong, with short, rather declinate branches, compact, often compressed in fruit. Spikelets 3–4 mm long. Glumes nearly as long as spikelet or slightly exceeding it. Lemma of staminate florets gradually acuminate, with broad scarious fringe and fine and dark-colored prominent midrib, usually not produced into awn (very rarely rib in form of short cusp barely surpassing scarious fringe of lemma), scabrous due to short teeth or spinules in upper part on back. Callus glabrous but sometimes sparse stiff hairs seen on callus of some florets.

Sandy floodplains of rivers in larch forests, chosenia-poplar groves, willow reeds, along their borders, in meadows and along roadsides. **East. Sib.**: YAK—Ar (Kolyma river between Pokhodsk and Panteleevsk settlements—class. hab.—and others), Al (Nezhdaninsk settlement), Yan, Ko. —Far East. Map 125.

3. **H. arctica** C. Presl 1830, Reliq. Haenk. 1: 252. —*H. odorata* subsp. *arctica* (C. Presl) Tzvelev. —*H. odorata* auct. p. p.

Rhizome slender, 0.5–1 mm diam, colorless or light grayish. Stems 25–75 cm tall, with 2 or 3 flat linear-lanceolate 1.5–5 (7) cm long leaves. Leaves of vegetative shoots 2–4 (5) mm broad, narrow, acuminate, flat or convolute, glabrous or with very fine diffuse hairs on upper surface. Sheath glabrous, sheath blade joints with very short, subpapillose dense hairs not forming rings. Ligules of upper cauline leaves 1.5–4 (5) mm long, glabrous, lacerated. Panicles pyramidal at anthesis, up to 10 cm long and 7 cm broad, spreading, somewhat compressed in fruit, 2–4 cm broad, oblong or ovate, rather dense. Spikelets 3.5–5 mm long. Glumes distinctly longer than spikelet. Lemma of staminate florets gradually acuminate and with broad (0.5–1 mm) scarious fringe and dark-colored prominent midrib at tip; awn sometimes arising near tip, short and slender not surpassing or slightly exceeding (by 0.2–0.3 mm) scarious fringe of lemma; scabrous on back due to short teeth or spinules, frequently interspersed with stiff hairs in upper part.

Hill river valleys, sand-pebble riverbeds, wet meadows, scrubs and their borders; enters larch forests. **West. sib.**: TYU—Yam, Khm, Tb, TO. **Cen. Sib.**: KR—PU, Tn, Ve. **East. Sib.**: IR—An, PR, BU—Se, Yuzh, YAK—Ol, Vi. —Europe, Scandinavia, Far East, North America. Described from Canada (Vancouver Island). Map 141.

4. **H. glabra** Trin. s. str. 1821 in Sprengel, Neue Entdeck. 2: 6. —*H. odorata* subsp. *glabra* (Trin.) Tzvelev.

Rhizome decumbent, funicular, 1—2 mm diam, yellowish or light brown. Stems 10–40 (60) cm tall, with 2 or 3 lanceolate 1–2 cm long leaves. Leaves of vegetative shoots flat, rarely convolute, acuminate, glabrous on both surfaces, sometimes diffusely puberulent on upper surface, slightly scabrous beneath. Leaf sheaths pilose, most dense at sheath blade joint, sometimes glabrous. Ligules of upper cauline leaves 1–2.5 mm long, obtuse, undivided or slightly lacerated, frequently with fine cilia at tip. Panicles 2–10 cm long, 1.5–4 cm broad, ovate or oblong, often compressed in fruit. Spikelets 2.5–4.5 mm long. Glumes more or less as long or shorter than spikelet, usually unequal, lower 0.5–1 mm shorter than upper. Lemma of staminate florets with interrupted narrow scarious fringe at tip and subterminal or terminal cusp up to 0.2 mm long, nearly wholly glabrous, with barely visible spinules or sparse hairs only right at tip. Callus glabrous.

Lower hill-steppe belt in sandy and rocky steppes, sand-pebble bed and banks of rivers and water reservoirs, exposed steppe slopes, poplar groves, their borders, solonetz meadows and solonetzes; enters forests. **Cen. Sib.**: KR—Ve. **East. Sib.**: IR—An, Pr, BU—Se, Yuzh, ChI—Shi, YAK—Vi, Yan, Ko (Srednekolymsk town). —Mongolia, NE China and possibly Far East. Described from Transbaikal. Map 142.

5. **H. glabra** subsp. **bungeana** (Trin.) Peschkova comb. nova. —*H. bungeana* Trin. 1839 in Mém. Acad. Sci. Pétersb. sér. 6, 5 (2): 82. —*H. odorata* f. *pubescens* Krylov.

Leaves of vegetative shoots grayish-green, convolute or flat, glabrous, cauline leaves broadly lanceolate, 0.4–1 cm long, ligules of upper cauline leaves 1–2 mm long, glabrous, lacerated at tip. Panicles lax, with slightly nutant branches. Lemma of staminate florets obtuse, awnless but, less often, with subterminal cusp not surpassing lemma, glabrous and smooth on back. Altay (Kurai village and Chibit settlement), $2n = 42$.

Upper mountain belt (1,500–2,100 m above sea level), sand-pebble riverbeds, hill-slope steppes, floodplain meadows, southern desert-steppe regions—near lakes and in steppes on plains. **West. Sib.**: AL—Go. **Cen. Sib.**: TU (southern desert-steppe regions). —Central Asia. Described from China (Tien Shan province). Map 143.

6. **H. glabra** subsp. **chakassica** Peschkova subsp. nova.

Folia innovationum griseolo-viridia convoluta glabra, culmea lanceolata 1–2 cm longa, ligula folii superioris 0.5–1.5 mm longa, extus (a latere laminae folii vergente) plus minusve dense pilosa, apice fimbriata et ciliata. Panicula fructificatione plerumque compressa. Lemmata florum stamineorum obtusata vel acutata apice saepe breviter mucronata dorso glabra, prope apicem scabridiuscula.

Typus: Prov. Enissej, insula ad fl. Enissej, prope opp. Minus-sinsk 25 VI 1908, No. 2704 M. Martjanov (NS).

Affinitas. A subspecie typica foliorum culmeorum superiorum ligulis pilosis brevissimis, panicula plus minusve compressa necnon florum stamineorum lemmatibus subglabris prope apicem tantum scabridiusculis differt.

Leaves of vegetative shoots grayish-green, convolute, glabrous, cauline leaves lanceolate, 1–2 cm long. Ligules of upper cauline leaves 0.5–1.5 mm long, rather densely pilose outside (toward leaf blade), fimbriate-lacerate and ciliate at tip. Panicles usually rather compressed in fruit. Lemma of staminate florets obtuse or acuminate, often with terminal short cusp, glabrous on back, slightly scabrous near tip. Khakass (Borets village), $2n = 28$.

Solonetz, wet and marshy meadows along river valleys, sandy and rocky steppes, sometimes in pine groves. **Cen. Sib.**: KR—Kha, Ve, TU. **East. Sib.**: BU—Yuzh (Mondy village). —Nor. Mongolia. Map 144.

7. **H. ochotensis** Probat. 1984 in Bot. zhurn., 69 (2): 257. —*H. odorata* subsp. *sibirica* auct. non Tzvelev.

Rhizome 1–2 mm diam, slender, funicular, bright or slightly yellowish. Stems (20) 30–65 cm tall, glabrous, with 2 or 3 lanceolate short (1–2 cm) leaves. Leaves of vegetative shoots 2–3 mm broad, long, narrowly acuminate, with convoluted edges, often with fine hairs on upper surface. Leaf sheaths usually scabrous or covered with rather sparse, short decurved hairs, sometimes subglabrous or glabrous. Sheath blade joints glabrous or with narrow band of papillose protuberances. Ligules of upper cauline leaves 1–2 mm long. Panciles 3–6 (9) cm long, (1.5) 2–4 (5) cm broad, oblong or ovate, initially rather spreading, more often compressed in fruit. Spikelets (3) 4–6 mm long. Glumes generally longer than spikelets, rarely nearly as long or slightly shorter. Lemma of staminate florets glabrous, smooth, with barely visible spinules only in uppermost part, gradually narrowed at tip, with rather narrow scarious fringe and thick brown midrib not transformed into awn (very rarely, rib surpasses scarious fringe in form of short cusp). Callus glabrous, sometimes sparse hairs present. Baikal region (Chivyrkuisk strait), $2n = 56$.

Sand-pebble riverbeds, meadows and banks of water reservoirs, valley scrubs; subalpine meadows in upper mountain belt of southern

118

regions, rock debris, forests around rivers, frequently along roadsides.
Cen. Sib.: TU (east). **East. Sib.**: IR—An, Pr, BU—Se, Yuzh, ChI—Ka, Shi,
YAK—Al, Vi, Yan (Tompo river near Khunkhada river estuary). —Far
East. Described from Magadan region (between Ola settlement and
Magadan town). Map 147.

8. **H. odorata** (L.) Beauv. 1812. Ess. Agrost.: 164.

Rhizome long, decumbent, funicular (1–2 mm diam), yellowish. Stems
25–70 cm tall, with 1–3 short lanceolate leaves. Leaves of vegetative
shoots 3–6 (8) mm broad, flat, acuminate, with diffuse very fine hairs on
upper surface, occasionally glabrous, with acute teeth along margin.
Ligules 2–4 mm long, scarious, glabrous, lacerated at tip. Panicles 5–10
cm long, 3–7 cm broad, pyramidal or oblong-ovate, spreading at anthesis,
with subhorizontal spreading branches; rather compressed before and
after anthesis. Spikelets about 5 mm long, glumes scarious, gradually
acuminate, often lacerated along fringes, 1–1.5 mm longer than spikelet,
totally covering it. Lemma of staminate florets with short hairs on callus, scabrous or with short spinules-hairs on back in upper half, obtuse,
without cusp and awn, scarious margin narrow, midrib poorly visible
and rarely surpassing lemma in form of fine cusp. West. Siberia (introduced in sections of Novosibirsk university township), $2n = 12$.

Loamy and sandy soils along valley meadows, meadow slopes, sparse
forests, sand-pebble beds, marsh fringes; ascends to altitude of 2,000 m.
West. Sib.: TYU—Tb (Bogondinsk village), KU, OM, NO, KE, AL—Ba
(Savvushka village in Zmeinogorsk region), Go (Kurai village; Ak-kol
river on Ukok upland). **Cen. Sib.**: KR—Kha, TU. **East. Sib.**: IR—An,
BU—Se, Yuzh, ChI—Shi (Arei Lake in Uletovsk region). —Europe, Mid.
Asia, West. China, North America. Described from Europe. Map 145.

In Khakassia (Beregovoe village of Bogradsk region), hybrids with *H. sibirica*
(Tzvelev) Czer. have been reported.

9. **H. pauciflora** R. Br. 1824 in Parry, Voy. N.W. Pass., Suppl. App.:
293.

Rhizome slender, decumbent, projecting foliated stray shoots and
ascending stems 10–20 (35) cm tall. Radical leaves up to 10 cm long, 1–
2 mm broad, narrow, convolute, together with sheath glabrous; cauline
leaf blades 2–10 mm long, flat, lanceolate, ligules 0.2–0.7 (1.5) mm long,
rounded, glabrous. Panicles 1–2 (3.5) cm long, narrow, secund, racemose, with 3–10 (15) spikelets. Spikelets about 4 mm long, oval. Glumes
scarious, slightly lustrous. Lemma of staminate flowers as long or longer
than glumes, brownish, scabrous on back due to short spinules, with
cilia along margin, slightly laciniate along midrib at tip, awnless, or
with short (up to 1 mm) awn slightly surpassing lemma. Callus with
few hairs. Central (Taimyr, Bogatyr Lake on Putoran plateau) and Eastern Siberia (Shandrin river in Yakutia), $2n = 28$.

Arctic moss-sedge tundras, fringes of water reservoirs, sometimes in sparse larch forests. **West. Sib.**: TYU—Yam. **Cen. Sib.**: KR—Ta, Pu (Bokovoe and Bogatyr' Lakes). **East. Sib.**: YAK—Ar. —European Arctic, Far East, North America (Alaska and Northern Canada). Described from Canada. Map 146.

10. **H. repens** (Host) Beauv. 1812, Ess. Agrost.: 164. —*H. stepporum* P. Smirnov.

Rhizome funicular, decumbent, coarsely flexuous, dark brown. Stems 25–65 cm tall, single, with dense tuft of stiff glaucescent or grayish-green leaves at base. Leaf blades 4–8 (10) mm broad, glabrous on both surfaces, with very short obtuse papillose teeth along margin. Ligules of upper cauline leaves 2–3.5 mm long, scarious, glabrous, lacerate at tip. Panicles 4–10 cm long, 1–4 cm broad, oblong or elongated-ovate, dense; branches short, obliquely erect, spikelets several (up to 100 or more), often overlapping and nearly concealing whole main rachis of panicle. Glumes distinctly unequal, gradually narrowed toward tip, undivided, rarely slightly laciniate, longer than adjoining florets. Lemma of staminate florets brownish-cinnamon, lustrous, obtuse or acute, with narrow scarious fringe, awnless or with short subterminal cusp, with distinct spinules, sometimes with sparse hairs in upper third on back; glabrous below. West. Siberia (Popovka town in Kulundinsk steppe), $2n = 28$.

Steppes, pine forests, steppified meadows. **West. Sib.**: OM (Novotsaritsino village and Ebeita Lake in Moskalensk region), AL—Ba. —Europe (southeast), Fore Caucasus, Mid. Asia (north). Described from Hungary. Map 138.

11. **H. sibirica** (Tzvelev) Czer. 1981, Sosud. rast. SSSR: 362. —*H. odorata* subsp. *sibirica* Tzvelev. —*H. glabra* subsp. *sibirica* (Tzvelev) Tzvelev.

Rhizome slender, grayish or light brownish. Stems 35–65 cm tall, with 1 or 2 short (0.5–1.5 cm long, rarely more) lanceolate leaves in lower third or half. Leaves of vegetative shoots 2–5 mm broad, flat, usually with convolute margin, long, acuminate, with grayish hue, glabrous on upper surface, rarely with fine hairs, sheaths glabrous or with diffuse short decurved spinules, sheath blade joint with poorly developed ring of very short, subpapillose hairs. Ligules of upper cauline leaves 1.5–3 mm long, glabrous, lacerated at tip, sometimes with short cilia. Panicles 3–7 cm long, as broad, pyramidal or oblong, rather spreading before anthesis, somewhat compressed in fruit. Glumes nearly as long as spikelets, sometimes slightly longer or shorter. Lemma of staminate florets glabrous, lustrous in lower half, scabrous in upper third due to short spinules, sometimes passing into hairs, with narrow scarious fringe and up to 0.6 mm long terminal awn; usually slightly recurved outward.

River valleys in shrubby forest meadows, forests, sand-pebble riverbeds, marshes; ascends into upper mountain belt where it is found in
alpine meadows, marshy tundras, exposed southern slopes, rock debris.
West. Sib.: NO, KE (Novochuvashka village, Topki town), AL—Ba, Go.
Cen. Sib.: KR—Tn (Nidym settlement, Turukhansk town), Kha, Ve
(Rybinsk village in Kansk region—class. hab.—and others). **East. Sib.**:
IR—An, Pr (western part), BU—Yuzh (Munku-Sardyk hill, Zabit river
estuary). —Mid. Asia, West. China (Junggar, Kashgar), Mongolia. Map
148.

25. Anthoxanthum L.

1. Spikelet stalks glabrous and smooth. Leaves invariably glabrous
.... 1. *A. alpinum.*
+ Spikelet stalks usually covered with distant hairs or elongated
spinules. Leaves rather pilose .... 2. *A. odoratum.*

1. **A. alpinum** A. et D. Löve 1948 in Rep. Depart. Agric. Univ.
Reykjavik, ser. B, 3: 105. —*A. odoratum* subsp. *alpinum* (A. et D. Löve) B.
Jones et Meld. —*A. odoratum* var. *glabrescens* Čelak.

Rather low (20–40 cm tall) odorous plant with glabrous lustrous
straight stems aggregated into loose mat. Leaves bright green, glabrous.
Panicles spicate, lustrous, with short glabrous branches. Spikelet stalk
glabrous, sometimes covered with sparse short hairs. Spikelets golden-
green, with one bisexual and 2 reduced florets. Glumes lanceolate, acuminate, strongly unequal. Lemma of reduced florets brown, bilobed at
tip, covered along margin with long stiff appressed hairs; firm, convolute, geniculate awn arising from lower third of lemma. Stamens 2; anthers 3–4.5 mm long. In Western Sayan (Kashareta river), in Buryat
(northeastern Baikal coast and Tunkinsk mountain range), $2n = 10$; on
Stanovoi upland (South Muisk mountain range), $2n = 20$.

Subalpine and alpine meadows, around snowfields at altitude up to
2,400 m. **West. Sib.**: TO, KE, AL—Ba, Go. **Cen. Sib.**: KR—Pu, Tn
(Turukhansk town, Uchami river basin), Kha, Ve, TU. **East. Sib.**: IR—
An, Pr, BU—Se, Yuzh, ChI—Shi. —Europe, Caucasus, Mid. Asia, Mediterranean, West. Asia, West. China, Mongolia, North America. Described
from Nor. Sweden. Map 149.

2. **A. odoratum** L. 1753, Sp. Pl.: 28.

Rather low, 20–60 cm tall, odorous plant, with glabrous and straight
stems aggregated into mat. Leaves green, more or less uniformly pubescent. Panicles spicate, lustrous, with short pilose branches. Spikelet stalks
densely covered with hairs or elongated spinules. Spikelets yellowish-
green with one bisexual and 2 reduced florets. Lemma of reduced florets similar, brown, half as long as spikelet, bilobed at tip, covered with
long, stiff, appressed hairs along margin. Awn convolute, geniculate, 2–
3 times longer than spikelet. Glumes lanceolate, acute, strongly unequal.

Lemma of fertile florets glabrous, scarious, shorter than glumes. Stamens 2; anthers 3–4 mm long.

Sand-pebble riverbeds, forests in mountain belt. West. Sib.: TO (environs of Tomsk). **East. Sib.**: BU—Yuzh (Snezhnaya river). —Europe, Caucasus, Far East, Mediterranean, North America. Described from Europe.

26. Phalaroides Wolf

1. Glumes with wingless keel, rarely with poorly developed wing less than 0.2 mm broad. Panicle branches scabrous ....
.... 1. *P. arundinacea.*
+ Glumes in upper half with winged keel 0.2–0.3 mm broad. Panicle branches glabrous .... 2. *P. japonica.*

1. **P. arundinacea** (L.) Rausch. 1969 in Feddes Repert. 79 (6): 409. — *Phalaris arundinacea* L. —*Digraphis arundinacea* (L.) Trin. —*Typhoides arundinacea* (L.) Moench.

Perennial, with long, decumbent rhizome. Stems up to 1.5 m tall, erect, glabrous, smooth. Leaves scabrous on upper surface, glabrous beneath. Panicles rather lax, 10–20 cm long, 1.5–2 cm broad, with obliquely erect branches. Spikelets 3-flowered, of which one developed, bisexual; other 2 florets with narrowly lanceolate lemma comprising 2 parts: lower cartilaginous, glabrous, lustrous; upper coriaceous-membranous. Glumes similar, 5–6 mm long, lanceolate, with wingless keel. Lemma cartilaginous, lustrous, glabrous, with sparse hairs at base. Novosibirsk region (Kotorovo village), $2n = 28$.

Wet meadows, banks of water reservoirs. **West. Sib.**: TYU—Khm (Sev. Sos'va river), Tb, KU, OM, TO, NO, KE, AL—Ba, Go. **Cen. Sib.**: KR—Kha, Ve, TU. **East. Sib.**: IR—An, Pr, BU—Se, Yuzh, ChI—Shi, YAK—Vi, Al, Yan, Ko. —Europe, Caucasus, Mediterranean, Asia, North America. Described from Europe.

2. **P. japonica** (Steudel) Czer. 1981, Sosud. rast. SSSR: 371. —*P. arundinacea* subsp. *japonica* (Steudel) Tzvelev.

Perennial, with long, decumbent rhizome. Stems up to 1 m tall, erect. Panicles spicate, narrow, 10–15 cm long. Spikelets 3-flowered, of which one developed, bisexual, other 2 florets with narrowly lanceolate lemma, comprising 2 parts: lower cartilaginous, glabrous, lustrous; upper coriaceous-membranous, covered with long hairs. Glumes similar, lanceolate, with winged keel 0.2–0.3 mm long.

Wet meadows, banks of water reservoirs. **East. Sib.**: ChI—Shi. —Far East, East. Asia. Described from Japan. Map 253.

27. Phalaris L.

1. **P. canariensis** L. 1753, Sp. Pl.: 54.

Stems 20–60 (70) cm tall, usually branched from base. Leaves 5–7 mm broad, flat, scabrous. Sheath of upper leaves highly inflated, weakly scabrous. Panicles 2–4 cm long, dense, oval or broadly elliptical. Spikelets 5–9 mm long, pale green, laterally compressed. Lower glume 7–8 mm long, sharply keeled, with broad wing along keel entire or slightly crispate along margin and green longitudinal band on back. Upper glume about 3 mm long, narrowly lanceolate, without keel. Lemma of fertile floret 4.5–6 mm long, coriaceous, lustrous, appressed-pubescent; of reduced floret narrowly lanceolate, membranous, half length of fertile floret. Anthers 2.3–3.5 mm long.

Wastelands and roadsides. **West. Sib.**: AL—Ba. **East. Sib.**: IR—An (environs of Irkutsk). —Europe, Caucasus, Far East, Mediterranean. Described from South. Europe and Canary Islands.

28. Beckmannia Host

1. Spikelets 1-flowered, glumes slightly inflated, anthers 0.3–0.8 (1) mm long .... 2.
+ Spikelets 2-flowered, glumes strongly inflated, anthers 1.2–1.8 mm long .... 3.
2. Glumes profusely pilose .... 3. *B. hirsutiflora.*
+ Glumes glabrous .... 4. *B. syzigachne.*
3. Plant with short decumbent shoots forming relatively loose mat. Lower stem internodes with tuberous thickening ....
.... 1. *B. eruciformis.*
+ Plant without decumbent subsurface shoots forming dense mats; lowermost stem internodes not thickened ....
.... 2. *B. eruciformis* subsp. *borealis.*

1. **B. eruciformis** (L.) Host 1805 in Gram. Austr. 3: 5.

Perennial, loosely cespitose plant with short decumbent shoots. Stems 30–100 cm tall, with internodes with tuberous thickening at base. Leaves grayish-green, scabrous along margin. Panicles 8–30 cm long, with appressed, simple recurved branches bearing oblong and secund spikes. Spikelets 2-flowered, 3–3.5 mm long, broadly pyriform, approximate in 2 rows. Glumes cymbiform, glabrous, highly inflated. Lemma with rather long (about 0.5 mm) terminal beak.

Wet and solonchak meadows, banks of rivers and water reservoirs. **West. Sib.**: TYU—Yam (Shuryshkar region), Tb, KU, OM, TO, NO, KE, AL—Ba, Go. **Cen. Sib.**: KR—Ve, Tn (Uchami settlement), Kha, TU. **East. Sib.**: IR—An, Pr, BU—Yuzh (Onokhoi settlement), YAK—Vi, Al. —Europe, Caucasus, Mid. Asia, Mediterranean. Described from South. Europe. Map 251.

2. **B. eruciformis** subsp. **borealis** Tzvelev 1973 in Novosti sist. vyssh. rast. 10: 81.

Loosely cespitose perennial, without subsurface shoots. Stems up to
124 100 cm tall, without tuberous thickened internodes at base. Leaves green,
scabrous along margin. Inflorescence a compound spike 10–25 cm long,
with highly appressed secund recurved branches; branches in lower part
of inflorescence poorly branched. Spikelets 2 (1)-flowered, 2.5–3 mm
long, broadly pyriform, with flattened sides. Glumes cymbiform, glabrous, slightly inflated. Lemma with small beak (about 0.3 mm long)
scarcely exserted from spikelet. Anthers 1.2–1.8 mm long.

Wet meadows, banks of lakes and rivers. **West. Sib.**: TYU—Tb (introduced), KU, AL—Ba. —Europe, Urals. Described from Komi ASSR.

N.S. Probatova (Novosti sist. vyssh. rast., 1976: 33) reports this subspecies from the
environs of Yakutsk. We did not find herbarium specimens from this site. This subspecies was perhaps either introduced there or its identification erroneous.

3. **B. hirsutiflora** (Roshev.) Probat. 1981 in Bot. zhurn. 66 (11): 1588.
—*B. syzigachne* var. *hirsutiflora* Roshev. —*B. syzigachne* subsp. *hirsutiflora*
(Roshev.) Tzvelev.

Compactly caespitose perennial up to 100 cm tall. Stems without
tuberous thickened internodes at base. Leaves light green, weakly scabrous. Inflorescence a compound spike 7–20 cm long, with appressed
branches further branched and bearing sessile, secund spikes. Spikelets
2.2–3 mm long, single-flowered. Glumes cymbiform, slightly inflated,
densely setose throughout surface. Tip of lemma drawn into beak distinctly exserted from glumes. Anthers 0.4–0.8 mm long.

Wet floodplain meadows. **East. Sib.**: ChI—Shi. —Far East, East. Asia.
Described from Amur basin.

4. **B. syzigachne** (Steudel) Fern. 1928 in Rhodora 30: 27. —*B.
eruciformis* var. *baicalensis* V. Kuznezov.

Light green, caespitose perennial. Stems 30–120 cm tall, without
thickened internodes at base. Leaves grayish-green, flat, scabrous. Inflorescence 10–30 (35) cm long, with appressed branches further branched,
especially in lower part. Spikelets single-flowered, 2.5–3 mm long, round-
cuneate, poorly inflated, flattened; glumes glabrous, with prominent
green nerves. Tip of lemma drawn into sharp beak not exserted from
glumes. In Eastern Sayan (Tunkinsk mountain range), Northern Baikal
region (Kotera river), Yakutia (Pokhodsk settlement), $2n = 14$.

Banks of rivers, marshes, lakes overgrown with vegetation, wet
meadows. **West. Sib.**: TYU—Yam (Salekhard), Khm, Tb, KU, OM, NO,
KE, AL—Ba, Go. **Cen. Sib.**: KR—Tn, Kha, Ve, TU. **East. Sib.**: IR—An, Pr,
BU—Se, Yuzh, ChI—Ka, Shi, YAK—Vi, Al, Yan, Ko. —Europe, Mid. Asia,
Far East, East. Asia, North America. Described from Japan.

29. Phleum L.

1. Branches of spicate panicles adnate to panicle rachis. Glumes with

long cilia along keels .... 2.
+ Branches of spicate panicles not adnate to panicle rachis. Glumes
without cilia along keels .... 2. *P. phleoides.*
2. Sheaths of upper cauline leaves highly inflated. Panicles short-
cylindrical, 1.5–2.5 (3) cm long .... 1. *P. alpinum.*
+ Sheath of upper cauline leaves not inflated. Panicles long-cylindri-
cal, 8–10 cm long .... 3. *P. pratense.*
1. **P. alpinum** L. 1753 Sp. Pl.: 59. —*P. commutatum* Gaudin.

Plant low, 10–30 (40) cm tall, glaucous-violet. Stems with short decumbent rhizome, with many dead leaves at base. Leaves broadly linear, glabrous, 4–7 mm broad; sheath of upper leaves greenish-violet,
highly inflated. Panicles 1.5–2.5 cm long, 8–10 mm broad, dense, short-
cylindrical or oval-oblong, violet. Glumes 3–4 mm long, with horny awns
at tip, linear-oblong, with obtuse incision along margin and long stiff
cilia along keel. Awns long, subulate. Lemma nearly half length of
glumes, scarious, with oblong nerves. Anthers about 1 mm long. Western Sayan (Olen' river upper course), $2n = 28$.

Grasslands in Arctic and bald peaks, sand-pebble riverbeds. **West.
Sib.**: KE (Verkh. Ters' river upper course), AL—Ba (Korgonsk mountain
range), Go. **Cen. Sib.**: KR—Pu, Kha, Be, TU. —Nor. Eurasia, North
America, South America. Described from Europe. Map 151.

2. **P. phleoides** (L.) Karsten 1880 in Deutsche Fl.: 374. —*P. boehmeri*
Wibel.

Stems 30–70 cm tall, several, erect, aggregated into loose mats. Leaves
mostly radical, linear, with thickened scabrous marginal veins. Panicles
narrowly cylindrical, 3–16 (19) cm long, with glabrous or scabrous
branches, densely appressed to rachis (but not adnate), 1–2 cm long.
Glumes 2–2.5 mm long, glabrous, or scabrous along nerves, without
cilia along keels, with sharply angled incision at tip terminating in short
awnlike 0.2–0.3 mm long cusp. Lemma awnless, 1.5–1.7 mm long,
scarious, with faint nerves. Tuva (Sangilen upland), $2n = 14 + 1–2B$.

Steppes, sparse forests, dry valley meadows. **West. Sib.**: TYU—Tb,
KU, OM, TO, NO, KE, AL—Ba, Go. **Cen. Sib.**: KR—Kha, Ve, TU. **East.
Sib.**: IR—An, Pr, BU—Se (Ushkanii island on Baikal; Barguzin settlement), Yuzh, YAK—Vi (environs of Olekminsk). —Eurasia. Described
from Europe. Map 152.

3. **P. pratense** L. 1753, Sp. Pl.: 59.

Stems 50–100 cm tall, erect, with tuberous thickening at base. Leaves
broadly linear, 3–10 mm broad, scabrous, without inflated sheaths. Panicles compact, 6–8 mm broad, 8–14 cm long, branches adnate to rachis.
Glumes 2.5–3 mm long, with long cilia separated horizontally, with
obtuse-angled incision at tip terminating in lateral long awnlike cusp.
Lemma weakly keeled, truncate at tip, nearly half length of glume,
scarious, with oblong nerves.

Forest and dry valley meadows, along roadsides. **West. Sib.**: TYU—Tb, KU, OM, TO, NO, KE, AL—Ba, Go. **Cen. Sib.**: KR—Kha, Ve, TU. **East. Sib.**: IR—An, Pr, BU—Yuzh, ChI—Shi (Maleta village), YAK—AL (environs of Aldan). —Europe, Caucasus, Mid. Asia, Mediterranean, West. Asia. Described from Europe. Map 153.

30. Limnas Trin.

1. Mats compact, rhizome absent or very short, without subsurface shoots. Radical leaves numerous, setaceous, 2 or 3 times shorter than stem .... 1. *L. malyschevii.*
+ Mats loose, rhizome long, branched, with subsurface shoots. Radical leaves few, flat or longitudinally folded, reaching panicle or surpassing it .... 2. *L. stelleri.*

1. **L. malyschevii** Nikiforova 1987 in Bot. zhurn. 72 (3): 391.

Compactly cespitose perennial 20–30 cm tall. Rhizome short, without subsurface shoots. Radical leaves 0.5–0.7 mm broad, several, longitudinally convolute, setaceous, 2 or 3 times shorter than stem, usually glabrous on lower surface, rarely covered with spinules. Panicles 1.5–3 cm long, 0.5–0.7 cm broad, lax, with scabrous recurved branches. Spikelets light violet; ultimately yellowish. Glumes 2.7–3.2 mm long, lanceolate, with 2 distinct lateral nerves, scabrous on back, with short spinules along keel, particularly abundant near tip. Lemma broadly lanceolate, with indistinct lateral nerves, glabrous, lustrous, as long as spikelet, with firm geniculate awn surpassing spikelet by 6–7 mm. Palea narrowly lanceolate, scarious, shorter than lemma. Anthers 1.9–2 mm long.

Belt below bald peaks on rock debris, rubble semiturfed arid southern slopes, mostly on limestone outcrops. **Cen. Sib.**: KR—Ta (Syndasko settlement), Pu (Khaya-Kyuël' Lake—class. hab.—and others). **East. Sib.**: YAK—Ol (Arga-Sala river; Mogdy river), Al (Teplyi Klyuch settlement), Yan (Dogdo and Ulokan-Sala rivers). —Far East. Map 140.

2. **L. stelleri** Trin. 1820, Fund. Agrost.: 116.

Loosely cespitose perennial. Stems 30–60 cm tall, covered with light brown remnants of leaf sheaths at base. Radical leaves rather few, 0.7–1.5 mm broad, flat, sometimes longitudinally folded, profusely covered with spinules on lower surface. Panicles (3) 3.5–4.5 cm long, 0.5–1 cm broad, rather lax. Spikelets single-flowered, yellowish-green. Glumes (3.5) 4–4.5 mm long, lanceolate, glabrous along keel, rarely with few short bristles. Lemma broadly lanceolate, with indistinct lateral nerves, glabrous, lustrous, as long as spikelet, with firm geniculate awn surpassing spikelet by 7–8 mm. Palea narrowly lanceolate, shorter than lemma. Anthers about 2 mm long.

Sparse pine and larch forests. **Cen. Sib.**: KR—Tn (Baikit and Taimba settlements). **East. Sib.**: IR—An, Pr, BU—Se, YAK—Vi. Described from region between Yakutsk and Okhotsk. Map 150.

126

31. Alopecurus L.

1. Perennial plant with erect stems not rooting at lower nodes. Anthers 2–3 mm long .... 2.
+ Annual with several geniculate stems rooting at lower nodes. Anthers 0.4–1.0 mm long .... 1. *A. aequalis.*
2. Inflorescence long, cylindrical, 5–8 (10) cm long .... 6.
+ Inflorescence short, oval, oval-elliptical or oval-cylindrical, 1–3 (4) cm long .... 3.
3. Sheaths of all leaves or of upper one inflated. Stems short, erect, poorly foliated .... 4.
+ Leaf sheaths not inflated. Stems weak, slender, densely foliated .....
.... 6. *A. roshevitzianus.*
4. Plant low, 15–35 cm tall. Awns of lemma slender at base, not convolute, not lustrous, shorter than or as long as spikelet; if longer, black-violet and geniculate .... 5.
+ Plant up to 50 (90) cm tall. Awns of lemma thick, convolute, geniculate, dark brown, lustrous, 2–2.5 times longer than spikelet ....
.... 4. *A. brachystachyus.*
5. Inflorescence oval-elliptical, densely pilose, 1–2 cm long, 5–10 mm broad. Spikelets dense, imbricate. Awns of lemma slender, weak, not longer than spikelet, sometimes lacking .... 2. *A. alpinus.*
+ Inflorescence cylindrical, 2–3 cm long, 1–1.5 cm broad. Spikelets sparsely hairy, loosely imbricate. Awns of lemma firm, black-violet, 1.5–2 times longer than spikelet .... 7. *A. turczaninovii.*
6. Plant grayish white-green. Glumes covered with soft sinuate hairs along and between nerves. Glumes obtuse, diverging laterally (spikelets urceolate). Awns of lemma slender, straight, barely longer than spikelets (sometimes by 1–2 mm) .... 3. *A. arundinaceus.*
+ Plant green. Glumes covered with stiff straight hairs only along main nerves. Glumes acuminate, convergent. Awns of lemma thick, geniculate, considerably surpassing spikelets ....
.... 5. *A. pratensis.*

1. **A. aequalis** Sobol. 1799 in Fl. Petropol.: 16. —*A. amurensis* Kom. —*A. fulvus* Smith.

Low (15–30 cm) annual. Stems several, greenish-glaucous, spreading, geniculate in lower part, rooting at nodes. Leaves narrowly linear, scabrous, sheath somewhat inflated. Inflorescence 2–5 cm long, 3–5 mm broad, narrowly cylindrical. Spikelets small, 1.8–2.5 mm long, elliptical. Glumes obtuse or rounded at tip, with several soft cilia along keel. Lemma glabrous, with slender straight awn, scarcely exserted from spikelet. Anthers orange, 0.4–1 mm long. Northeastern coast of Lake Baikal, $2n = 14$.

Banks of rivers and water reservoirs, sand-pebble riverbeds. **West. Sib., Cen. Sib., East. Sib.**: all regions. —Europe, Asia, North America. Described from Leningrad region.

2. **A. alpinus** Smith 1809, Fl. Brit. 3: 1386.

Low (15–25 (30) cm) plant with long rhizome. Stems with 2 internodes, covered with brown leaves at base. Cauline leaves usually 2, upper one at middle of stem. Leaf blades shorter than highly inflated sheaths. Inflorescence short, 1–1.5 (2) cm long, 5–10 mm broad, oval-elliptical, densely hairy. Spikelets 2.7–3 (3.5) mm long, dense and imbricate. Glumes covered with slender silky sinuate hairs along keel, nerves and between them. Lemma shorter than glumes, with slender, weakly developed awn, usually not exserted from spikelet. Anthers 1.6–2.2 mm long. Putoran plateau (Baselak Lake), $2n = 100$.

Grasslands and pebble beds, slopes and rock debris. **West. Sib.**: TYU—Yam. **Cen. Sib.**: KR—Ta, Pu. **East. Sib.**: YAK—Ar, Ol, Vi (Segyan-Kyuël village), Ko. —Arctic Europe, Far East (north), North America. Described from Scotland. Map 154.

3. **A. arundinaceus** Pioret 1808 in Lam., Encycl. Meth. Bot. 8: 776. — *A. ventricosus* Pers.

Perennial, 50–100 cm tall, with decumbent subsurface shoots. Leaves glabrous, grayish-green, 3–5 mm broad, with inflated sheaths. Inflorescence 5–10 cm long, cylindrical, pale green, nigrescent on maturing. Spikelets urceolate, 3.5–4 mm long, densely hairy along and between nerves. Tips of glumes diverging laterally, forming obtuse incision among themselves. Lemma shorter than glumes, obtusely abscised, with straight slender awn not exserted from spikelet, rarely longer than it by 1–2 mm. Anthers 2–3 mm long. Eastern Sayan (Tunkinsk mountain range), $2n = 28$.

Banks of rivers and water reservoirs, meadows and solonchaks. **West. Sib.**: TYU—Khm, KU, OM, TO, NO, KE, AL—Ba, Go. **Cen. Sib.**: KR—Tn, Kha, Ve, TU. **East. Sib.**: IR—An, Pr, BU—Se, Yuzh, ChI—Shi, YAK—Vi, Al. —Europe, Caucasus, Mid. Asia, Mediterranean, West. Asia, Far East, West. China, Mongolia. Described from cultivated specimens probably from Fore Asia.

4. **A. brachystachyus** Bieb. 1819 in Fl. Taur.-Cauc. 3: 56.

Perennial, 50–60 (90) cm tall, with slender branched rhizome. Stems solitary, with 2 or 3 internodes, few long radical leaves with uninflated sheaths. Blades of cauline leaves short, glabrous. Sheaths of upper leaves inflated. Inflorescence 2.5–3.5 cm long, 1–1.5 cm broad, oval or broadly cylindrical. Spikelets yellow-green, with violet tip, finally turning brown. Glumes 3.5–4 (5) mm long, covered with long, straight, silky hairs along lateral nerves and partly between them, upper ends interrupted laterally, forming obtuse incision equal to 1/3 length of spikelet. Lemma sheared obliquely at tip, shorter than spikelet, with thick, geniculate, convolute dark brown awn, surpassing spikelets by 5–8 mm. Anthers 2.3–2.5 mm long.

Solonetz meadows, water reservoirs in steppe sections. **Cen. Sib.**: KR—Ve (Solyanoozernoe and Kamchatki villages). **East. Sib.**: IR—

An, Pr, BU—Se, Yuzh, ChI—Shi, YAK—Al (Nagornyi settlement; Evota pass). —Far East, Mongolia, East. Asia. Described from Transbaikal. Map 155.

5. **A. pratensis** L. 1753, Sp. Pl.: 60.

Loosely cespitose plant up to 100 cm tall, with short decumbent rhizome. Leaves green, glabrous, 4–10 mm broad; leaf sheaths long, poorly inflated. Inflorescence cylindrical, up to 10 cm long, with short branches densely appressed to rachis. Spikelets elliptical, large, 5–6 mm long. Glumes with straight, convergent, acuminate tips with 3 greenish nerves, pubescent only along keel, rarely along nerves. Lemma acuminate, nearly as long as glumes, whitish, with firm geniculate awn considerably surpassing spikelet. Anthers 3–4 mm long. Putoran plateau (Talnakh settlement), $2n = 28$.

Meadows, banks of water reservoirs, among shrubs. **West. Sib.**: all regions. **Cen. Sib.**: KR—Pu (Talnakh settlement, Potakovo village), Tn, Kha, Ve, TU. **East. Sib.**: IR—An, Pr, BU—Se, Yuzh, ChI—Ka, Shi, YAK—Vi, Al. —Eurasia. Described from Europe.

6. **A. roshevitzianus** Ovcz. 1934 in Fl. SSSR 2: 154, 745. —*A. glaucus* auct. non Less. —*A. alpinus* subsp. *glaucus* auct. non Hultén.

Loosely cespitose green plant with short decumbent rhizome. Stems weak, 40–90 cm tall, with 3 or 4 internodes. Cauline leaves 10–15 cm long, many, reaching or exceeding inflorescence, green, glabrous, flat; leaf sheaths not inflated. Inflorescence 2–3 cm long, 0.7–0.8 cm broad; spikelets loosely arranged, 3–3.5 mm long, flattened, greenish-gray. Glumes densely adhering to each other; acute-acuminate, with diffuse straight hairs along and between nerves. Lemma glabrous, as long as glumes, with unconvoluted slender awn exserted from lower third and surpassing spikelets by 1–1.5 mm. Anthers up to 3 mm long.

Marshy meadows, turfed grassy slopes. **East. Sib.**: YAK—Ko (Kolyma district, on Pokhodsk to Panteleevsk settlement road—class. hab.—and others). —Endemic. Map 157.

7. **A. turczaninovii** Nikiforova 1988 in Bot. zhurn. 73 (11): 1601. —*A. glaucus* auct. non Less.

Low, loosely cespitose, glaucous-green plant with short decumbent rhizome. Stems up to 30 cm tall, solitary, with 2 or 3 internodes, with few dead leaf sheaths at base. Cauline leaves 2 or 3, glabrous, leaf blades 2 or 3 times shorter than highly inflated leaf sheaths. Upper cauline leaves above middle of stem. Inflorescence 2–3 cm long, 1–1.5 cm broad, glaucous-violet, cylindrical. Spikelets on short scabrous branches, loosely imbricate. Glumes 4–4.5 mm long, acutely abscised, hardly longer than lemma, dark violet along nerves, keel and at tip, densely covered with long straight silky hairs. Lemma glabrous, obtuse, with 2 poorly visible lateral nerves. Awns black-violet; firm, more slender at base, not convoluted, longer than spikelet by 5–6 mm, arising from lower third of lemma.

In alpine meadows. **West. Sib.:** AL—Ba, Go. **Cen. Sib.:** KR—Kha, TU. **East. Sib.:** IR—An, Pr, BU—Se, Yuzh (Posol'sk settlement—class. hab.—and others). —North. Mongolia. Map 156.

32. Scolochloa Link

1. S. festucacea (Willd.) Link 1827, Hort. Bot. Berol. 1: 137.

Rhizome decumbent, long and thick, bearing shoots. Leaves stiff, smooth, scabrous along margin, up to 10 mm broad. Stems straight, thick, up to 2 m tall, rooting at lower node. Panicles broad and lax, 20–35 cm long and 5–15 cm broad, with scabrous triquetrous recurved branches. Spikelets with 2–5 florets, 7–10 mm long, slightly laterally compressed, disintegrating. Glumes lanceolate, acute, slightly unequal. Lemma with 7 sharp nerves, apically obtuse and crenate, glabrous, with 2 tufts of stiff, straight hairs on callus. Yakutia (environs of Yakutsk), $2n = 28$.

Banks of rivers and lakes overgrown with vegetation, swamps, marshy meadows. **West. Sib.:** TYU—Tb, KU, OM, TO, NO, KE, AL—Ba. **Cen. Sib.:** KR—Kha, Ve, TU. **East. Sib.:** IR—An, Pr (Tutura village), ChI—Shi, YAK—Vi). —Europe, Caucasus, Mid. Asia, Mongolia, North America. Described from environs of Berlin. Map 159.

33. Festuca L.

1. Spikelets in panicle mostly viviparous ....2.
+ Spikelets not viviparous .... 3.
2. Lemma of all florets awnless. Leaf blades with (3) 5–7 (9) sclerenchyma strands .... 45. *F. viviparoidea.*
+ Lemma of lower florets with 0.6–1.8 mm long awn. Leaf blades with 3 poorly developed sclerenchyma strands ....
.... 19. *F. chionobia.*
3. Sheath blade joints along edges with lanceolate, usually falcate auricles .... 4.
+ Sheath blade joints without auricles .... 6.
4. Lemma awnless or with up to 3 (5) mm long awn. Leaf blades 3–8 mm broad, with distinctly developed veins on upper surface .... 5.
+ Lemma with 10–18 mm long awn. Leaf blades 8–15 mm broad, without distinctly developed veins on upper surface ....
.... 3. *F. gigantea.*
5. Sheath blade joints and auricles with cilia (sometimes only stray) along edges. Lemma more often awned ....
....4. *F. arundinacea.*
+ Sheath blade joints and auricles without cilia. Lemma awnless .... 5. *F. pratensis.*

6 (3). Glumes almost wholly membranous, distinctly differing in texture and frequently even in color from rather coriaceous lemma; latter keeled .... 7.

\+ Glumes membranous only along margin, elsewhere rather coriaceous, similar to lemma in texture and color; latter without keels, rounded on back .... 13.

7. Blades of all leaves flat, without distinct veins on upper surface .... 8.

\+ Blades of all or most leaves longitudinally folded or convolute, with distinct veins on upper surface .... 10.

8. Plant compactly cespitose, without decumbent subsurface shoots. Lemma with 3 strong and 2 very weak nerves .... 9.

\+ Plant loosely cespitose, with short decumbent subsurface shoots. Lemma with 5 similar nerves .... 6. *F. hubsugulica.*

9. Plant dioecious. Ligules of cauline leaves shorter than 0.5 mm. Panicles poorly spreading. Lemma uniformly covered with spinules on back, acute (cusp less than 0.4 mm long). Ovary densely pilose at tip .... 7. *F. komarovii.*

\+ Plant monoecious. Ligules of cauline leaves 0.5–2 mm long. Panicles broadly spreading. Lemma with scattered spinules only in upper third on back; short-awned (cusp 0.4–1.2 mm long). Ovary with few hairs at tip .... 11. *F. popovii.*

10 (7). Plant monoecious. Leaf blade highly scabrous on outer surface. Panicles rather spreading, with extremely long branches. Lemma pinkish-violet, very rarely yellowish or green. Ovary with few hairs at tip .... 11.

\+ Plant dioecious. Leaf blades glabrous on outer surface. Panicles compressed, with short branches. Lemma whitish, rarely pinkish. Ovary densely pilose at tip .... 8. *F. sibirica.*

11. Lemma with 1.2–1.8 mm long awn. Leaf blades 0.3–0.6 mm diam, with 5 vascular bundles, with 3 veins within, sparsely covered with trichomes 0.02–0.05 mm long; sclerenchyma strands arranged on margin of leaf blades and opposite vascular bundles only under lower epidermis .... 12.

\+ Lemma awnless, sometimes with cusp not longer than 1 mm. Leaf blades 0.8–1.5 mm diam, with 7–11 vascular bundles, with 5–7 veins on inner surface, densely covered with trichomes 0.07–0.12 mm long; sclerenchyma strands arranged on margin of leaf blades and opposite vascular bundles under upper and lower epidermis .... 9. *F. altaica.*

12. Lemma with few spinules in lower third on back. Anthers 2.5–4 mm long .... 12. *F. tristis.*

\+ Lemma densely covered with spinules on back up to base. Anthers 1.8–2.3 mm long .... 10. *F. bargusinensis.*

13 (6).   Blades of all leaves flat, over 3 mm broad, without distinct veins on upper surface. Ligules longer than 1 mm. Ovary densely pilose at tip .... 14.

+   Blades of all leaves or at least those of vegetative shoots double folded longitudinally, less than 2.5 mm broad (1.25 mm diam), invariably with distinct midrib on upper surface; usually also with lateral veins. Ligules shorter than 0.5 mm. Ovary glabrous or rather pilose .... 15.

14.   Lemma awnless, with 3 nerves, with spinules on back throughout surface. Shoots with coriaceous scale leaves at base. Sheath laciniate almost to base. Anthers longer than 2 mm ....
.... 1. *F. altissima.*

+   Lemma with 4–8 mm long awn, 5 nerves; diffuse spinules only in upper half on back. Shoots without coriaceous scale leaves at base. Sheaths closed for not less than 1/2 length from base. Anthers 0.7–1.2 mm long .... 2. *F. extremiorientalis.*

15.   Anthers 0.5–1 (1.25) mm long .... 16.

+   Anthers longer than 1.5 mm .... 19.

16.   Compactly cespitose plant with intravaginal shoot regeneration, without decumbent subsurface shoots. Panicles compressed. Spikelets with (3) 4–6 florets. Ovary glabrous or with few hairs at tip. Blades of all leaves double folded longitudinally, oval or obovate in cross section; sclerenchyma strands arranged along margin of leaf blades opposite central and, quite often, opposite other vascular bundles only under lower epidermis .... 17.

+   Plant loosely cespitose or not forming mats, with extravaginal shoot regeneration and short decumbent subsurface shoots. Panicles spreading. Spikelets usually with 2 or 3 florets. Ovary pilose at tip. Leaf blades of vegetative shoots double folded longitudinally, 4–6-angled in cross section, of flowering shoots flat or longitudinally convolute; sclerenchyma strands arranged along margin of leaf blades and opposite vascular bundles under upper and lower epidermis ....
.... 44. *F. venusta.*

17.   Stems glabrous or covered with scattered spinules only for 1–2 cm under panicle. Ovary glabrous .... 18.

+   Stems densely puberulent almost throughout length. Ovary with few hairs at tip .... 14. *F. baffinensis.*

18.   Panicle branches rather branched, scabrous, covered with more or less numerous spinules, sometimes transformed into hairs. Leaf blades without glaucous bloom. shoots usually (but not always) with few dead leaf sheaths at base ....
.... 18. *F. brachyphylla.*

+ Panicle branches usually unbranched, glabrous, smooth or covered with stray spinules. Leaf blades with rather distinct glaucous bloom. Shoots with many dead leaf sheaths at base .... 21. *F. hyperborea.*

19 (15). Lemma awnless, sometimes with up to 0.8 mm long cusp .... 20.

+ Lemma with awn longer than 1 mm .... 26.

20. Leaf blades 4–6-angled in cross section, with 3–5 (7) veins within, with 7–9 (11) sometimes connective sclerenchyma strands under lower epidermis. Leaf sheaths of vegetative shoots closed for not less than 1/2 length from base. Spikelets green or pinkish-violet .... 21.

+ Leaf blades oval or rounded in cross section, rarely obovate, with solitary midrib within (sometimes lateral veins present but not prominent), with different number of generally connective sclerenchyma strands under lower epidermis. Leaf sheaths of vegetative shoots closed for 1/6 to 1/4 length from base. Spikelets green or brownish .... 20. *F. dahurica.*

21. Leaf blades totally smooth on outer surface, covered with sparse, 0.02–0.05 mm long trichomes within. Plant with extravaginal or mixed shoot regeneration, with rather long decumbent subsurface shoots. Leaf sheaths of vegetative shoots closed for 3/4 to 4/5 length from base .... 22.

+ Leaf blades scabrous on outer surface (at least in upper half), densely covered with 0.06–0.08 mm long trichomes within. Plants with exclusively intravaginal shoot regeneration, without decumbent subsurface shoots. Leaf sheaths of vegetative shoots closed for nearly 1/2 length from base .... 25.

22. Panicle branches pilose or scabrous. Leaf blades over 0.4 mm diam, with 7–9 (11) vascular bundles, with 3–5 (7) veins within. Palea covered with spinules all along keels. Lemma 4–5.3 mm long, villous, rarely glabrous .... 23.

+ Branches glabrous, smooth. Leaf blades 0.35–0.4 mm diam, with 5 vascular bundles, with 3 veins within. Palea glabrous or subglabrous along keels. Lemma 5.5–5.8 mm long, glabrous .... 23. *F. karavaevii.*

23. Panicle branches scabrous. Lemma lanceolate to broadly lanceolate, 4–7 mm long, rather pilose or glabrous, with 0.7–3.5 mm long awns .... 24.

+ Panicle branches pilose or scabrous. Lemma lanceolate-ovate, 4–5.3 mm long, villous, rarely subglabrous, with 0–1.5 mm long awns .... 37. *F. rubra* subsp. *arctica.*

24. Plant loosely cespitose, widely distributed. Sheaths of outer leaves of vegetative shoots puberulent, rarely glabrous. Lemma

rather pilose or glabrous. Anthers longer than 2.5 mm ....
.... 36. *F. rubra* s. str.

+ Plants on Baikal Lake coast not forming mats, with extremely long, decumbent subsurface shoots. Sheaths of all leaves usually glabrous. Lemma usually glabrous, rarely puberulent. Anthers 2–3 mm long .... 38. *F. rubra* subsp. ***baicalensis.***

25. Palea covered with extremely short spinules along keels in upper half. Spikelets usually pale green. Panicles not nutant .... 22. *F. jacutica.*

+ Palea glabrous or subglabrous along keels. Spikelets violet. Panicles more or less nutant .... 29. *F. malyschevii.*

26 (19). Leaf blades 4–6-angled in cross section, with nearly equal sclerenchyma strands, arranged along margin and opposite vascular bundles along lower and sometimes even under upper epidermis. Plants with mixed or exclusively extravaginal shoot regeneration, frequently with decumbent subsurface shoots, loosely cespitose or not forming mats .... 27.

+ Leaf blades oval, round or obovate in cross section, with compact or interrupted sheath of sclerenchyma or 3 (5) sclerenchyma strands; further, lateral strands, if present, less wide than carinal. Plants with intravaginal shoot regeneration, without decumbent subsurface shoots, compactly caespitose .... 28.

27. Leaf blades covered with very sparse, 0.03–0.05 mm long trichomes within, with sclerenchyma strands usually only under lower epidermis. Lemma pilose or glabrous. Palea covered with spinules all along keels. Spikelets green or pinkish-violet .... 36. *F. rubra* s. str.

+ Leaf blades covered with dense, 0.07–0.1 mm long trichomes within, invariably with sclerenchyma strands under lower as well as upper epidermis. Lemma glabrous; palea covered with spinules for 1/3 to 1/2 along keels. Spikelets glaucous green .... 40. *F. skrjabinii.*

28. Leaf blade cross section with continuous or rarely interrupted sheath of sclerenchyma, wholly uniform in width, under lower epidermis .... 29.

+ Leaf blade cross section with isolated sclerenchyma strands under lower epidermis, frequently coalescing into continuous or interrupted sheath; width of latter opposite midrib considerably more than along sides of leaf blades .... 33.

29. Spikelets green or pinkish-violet. Plants of plains and all hill belts .... 30.

+ Spikelets brownish. High-altitude plants ....
.... 32. *F. ovina* subsp. ***sphagnicola.***

30.   Leaf blades with 1–7 veins within, densely covered with 0.06 –0.1 (0.15) mm long trichomes .... 31.

+   Leaf blades with solitary vein within, sparsely covered with 0.02–0.05 mm long trichomes .... 31. *F. ovina* s. str.

31.   Roots with sheaths of conglutinate sand grains. Leaf blades with (1) 3–7 veins within. Widely distributed plant .... 32.

+   Roots without sheaths of conglutinate sand grains. Leaf blades with 1 (3) veins within. Plant on coasts of Baikal Lake .... 33. *F. ovina* subsp. *vylzaniae.*

32.   Leaf blades abruptly acuminate toward tip, (0.4) 0.65–0.85 (1.1) mm diam, with (7) 9–11 (15) vascular bundles, scabrous or smooth outside, with 5–7 (9) veins within. Lemma (2.8) 4–4.8 (5.2) mm long; palea covered with spinules along keels usually in upper third. Stems often pubescent throughout length.... .... 16. *F. beckeri* subsp. *polesica.*

+   Leaf blades gradually acuminate toward tip, (0.3) 0.4–0.6 (0.7) mm diam, with 7–9 vascular bundles, smooth on outer surface, with (1) 3–5 veins within. Lemma glabrous or subglabrous along keels. Stems pubescent or glabrous .... 15. *F. beckeri* s. str.

33.   Dry leaf blades grooved, with 3 (5) sclerenchyma strands; lateral strands, if present, much narrower than carinal and marginal .... 34.

+   Dry leaf blades usually not grooved, with continuous or interrupted sheath broadened only opposite midrib or 5–7 sclerenchyma strands, of which lateral ones much narrower than carinal and nearly as broad as marginal .... 47.

34.   Mat divided into rather isolated tufts of shoots enveloped by dead leaf sheaths at base. Leaf blades with middle and frequently 1 or 2 nearly invisible lateral veins within; upper vascular bundles of 1/3 size almost invariably present and nearly equal to lower .... 35.

+   Mat not divided into shoot tufts, dead leaf sheaths at base of shoots usually few. Leaf blades with 3–5 highly prominent veins or with solitary vein; in latter case, upper vascular bundles 1/3 size lacking or, if present, usually smaller than lower .... 37.

35.   Carinal sclerenchyma strands highly developed in leaf blades. Latter covered outside with tiny spinules, scabrous, rarely glabrous, smooth or pilose. Spikelets brownish, green or pinkish-violet .... 36.

+   All 3 sclerenchyma strands in leaf blades poorly developed, barely visible. Leaf blades glabrous, smooth, occasionally pilose on outer surface. Spikelets green or pinkish-violet ....

.... 13. *F. auriculata.*

36. Carinal and marginal sclerenchyma strands in leaf blades nearly equally developed. Leaf blades usually not sinuate at tip .... 27. *F. lenensis.*

+ Carinal sclerenchyma strands in leaf blades developed far better than marginal. Leaf blades rather sinuate at tip ....
.... 41. *F. tschujensis.*

37 (34). Leaf blades scabrous on outer surface throughout length .... 38.

+ Leaf blades smooth on outer surface throughout length, rarely weakly scabrous only in upper half .... 46.

38. Spikelets brownish .... 39.

+ Spikelets green or pinkish-violet .... 43.

39. Leaf blades with 5 (7) vascular bundles, with 1–3 veins within. Upper vascular bundles of 1/3 size, if present, smaller than lower .... 40.

+ Leaf blades with 7 (rarely only in case of extremely narrow ones, 5) vascular bundles, with 3–5 veins within. Upper vascular bundles of 1/3 size nearly equal to lower ....
.... 25. *F. kryloviana.*

40. Leaf sheaths of vegetative shoots closed for not more than 1/2 length from base .... 41.

+ Leaf sheaths of vegetative shoots closed for 2/3 to 4/5 length from base .... 26. *F. kurtschumica.*

41. Leaf blades with 3 highly prominent veins. Leaf sheaths of vegetative shoots closed for 1/4 to 1/2 length from base ...
.... 42.

+ Leaf blades with solitary midrib within and sometimes with 1 or 2 poorly visible lateral veins. Leaf sheaths of vegetative shoots laciniate almost to base .... 24. *F. kolymensis.*

42. Leaf blades (0.4) 0.55–0.85 (1) mm diam. Lemma (4.5) 4.8–5.5 (6) mm long; palea covered with spinules for 1/2 to 3/4 length along keels .... 25. *F. kryloviana.*

+ Leaf blades 0.3–0.6 (0.7) mm diam. Lemma 3.2–4.2 (4.6) mm long, palea covered with spinules for 1/3 length along keels
.... 43. *F. valesiaca* subsp. *hypsophila.*

43 (38). Leaf blades with 3 highly prominent veins within. Leaf sheaths of vegetative shoots laciniate almost to base .... 44.

+ Leaf blades with 1 midrib within and sometimes with 1 or 2 lateral veins; latter almost not visible. Leaf sheaths of vegetative shoots closed for 1/4 to 1/3 length from base ....
.... 34. *F. pseudosulcata.*

44. Leaf blades green .... 45.

+ Leaf blades with glaucous bloom .... 42. *F. valesiaca.*

136

45. Spikelets (4) 4.5–5.5 (6) mm long. Lemma (2.3) 2.8–3.8 (4.2) mm long. Leaf blades (0.3) 0.4–0.6 (0.75) mm diam .... 35. *F. pseudovina.*
+ Spikelets (5.5) 6.5–8.5 (10) mm long. Lemma (4.5) 4.8–5.5 (6) mm long. Leaf blades (0.4) 0.55–0.85 (1.1) mm diam .... 39. *F. rupicola.*
46 (37). Panicle branches smooth. Leaf blades with 3–5 veins within. Leaf sheaths of vegetative shoots closed for 2/3 to 4/5 length from base. Spikelets brownish .... 17. *F. borissii.*
+ Panicle branches scabrous. Leaf blades with solitary midrib within and sometimes with 1 or 2 poorly visible lateral veins. Leaf sheaths of vegetative shoots closed for 1/4 to 1/3 length from base. Spikelets green, with glaucous bloom or brownish .... 28. *F. litvinovii.*
47 (33). Leaf blades with glaucous bloom, with 3–5 veins within, covered with 0.06–0.1 mm long trichomes, usually with continuous sclerenchyma sheath under lower epidermis. Leaf sheaths of vegetative shoots laciniate almost to base .... 46. *F. wolgensis.*
+ Leaf blades green, with 1 (3) veins within, covered with 0.03–0.06 mm long trichomes, with continuous or interrupted sheath or individual sclerenchyma strands under lower epidermis. Leaf sheaths of vegetative shoots closed for roughly 1/3 length from base .... 30. *F. olchonensis.*

Section P h a e o c h l o a Griseb. s. str.

Plant monoecious. Shoots extravaginal, with coriaceous scale leaves at base. Sheath laciniate almost to base. Sheath blade joint without auricles. Leaf blades flat, veinless on upper surface. Glumes rather coriaceous, similar in texture to lemma; latter without keels, awnless. Ovary rather pilose at tip. Caryopsis conglutinate with lemma and palea at base, grooved ventrally, with long linear hilum.
1. **F. altissima** All. 1789 in Auct. Fl. Pedem.: 43. —*F. sylvatica* (Poll.) Villar non Hudson.
Sheath scabrous. Panicles broadly spreading, with multispicate highly scabrous branches. Lemma 5–6 mm long, scabrous throughout surface on back, with 3 nerves. Anthers 2.5–3.5 mm long. Ovary densely pilose at tip. Plate I (1).
Deciduous and mixed forests, up to midmountain belt. **West. Sib.:** KE (Kuzedeevo village), AL—Go. **Cen. Sib.:** KR—Ve (Kazyr river valley). **East. Sib.:** BU—Yuzh (Khamar-Daban mountain range, Snezhnaya river). —Europe, Caucasus, Kazakhstan Altay, West. Asia. Described from Nor. Italy. Map 160.

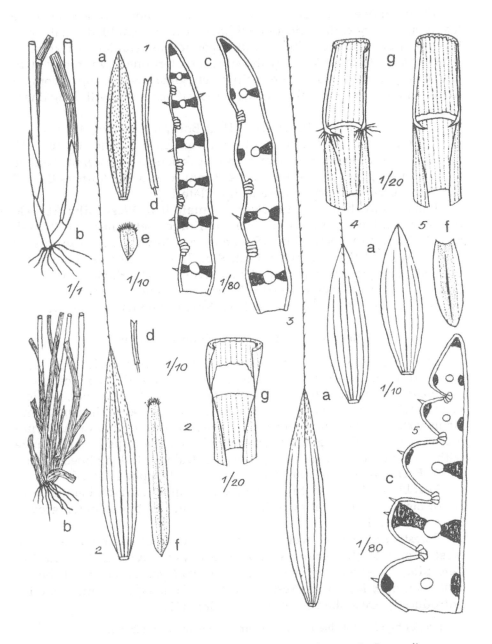

Plate I. 1—*Festuca altissima*; 2—*F. extremiorientalis*; 3—*F. gigantea*; 4—*F. arundinacea*;
5—*F. pratensis*:
a—lemma, b—shoot base, c—leaf blade cross sections, d—stamens, e—ovary,
f—caryopsis, g—sheath blade joint.

Section Subulatae Tzvelev

Plant monoecious. Shoots extravaginal, without coriaceous scale leaves at base. Sheath blade joint without auricles. Leaf blades flat, veinless on upper surface. Glumes rather coriaceous, similar in texture to lemma; latter without keels, with rather long terminal awn, rarely awnless. Ovary more or less pilose at tip, rarely glabrous. Caryopsis conglutinate with lemma and palea at base, grooved ventrally, with long linear hilum.

2. **F. extremiorientalis** Ohwi 1931 in Bot. Mag. Tokyo, 45: 194.

Sheath closed for 1/2 to 2/3 length from base. Ligules up to 2–2.5 mm long. Leaf blades 6–15 mm broad. Spikelets 4–6-flowered. Lemma 6–6.5 mm long, with 5 nerves and 4–8 mm long awn. Anthers 0.7–1.2 mm long. Ovary densely pilose at tip. Plate I (2).

Forests and shrubs up to lower mountain belt. **West. Sib.**: AL—Go (Uznezya settlement on Katun' river and Korgon settlement on Charysh river). **Cen. Sib.**: KR—Kha (Sayanogorsk town), Ve (environs of Minusinsk), TU (Derzik river). **East. Sib.**: IR—An (Kultuk settlement), BU—Se (Chivyrkuisk gulf and Almat river), Yuzh (Kamennaya river, Dzhida river tributary), ChI—Shi (Nerchinsk, Sretensk, Undurga river). —Far East, East. Asia. Described from Korean peninsula. Map 164.

Section Plantynia (Dum.) Tzvelev

Plant monoecious. Shoots extravaginal, without coriaceous scale leaves at base. Sheath laciniate almost to base. Sheath blade joint with lanceolate, usually falcate auricles along edges. Leaf blades flat, veinless or somewhat veined on upper surface. Glumes rather coriaceous, similar in texture to lemma; latter without keel, with terminal or subterminal awn longer than glumes. Ovary glabrous. Caryopsis conglutinate with lemma and palea at base, grooved ventrally, with long linear hilum.

3. **F. gigantea** (L.) Villar 1787 in Hist. Pl. Dauph. 2: 110.

Leaf blades 8–15 mm broad, veinless on upper surface. Lemma with 10–18 mm long terminal awn. Kuznets Alatau (Tel'bessk linden grove), $2n = 42$. Plate I (3).

Forests and shrubs, wet meadows; up to subalpine mountain belt. **West. Sib.**: OM (Ekaterininsk village on Irtysh), NO, KE, AL—Ba, Go. **Cen. Sib.**: KR—Ve (Amyl and Tabrat river valleys, Kazyr river tributaries). —Europe, Mid. and West. Asia, West. Himalayas, North America (introduced). Described from Europe. Map 161.

Species highly probable in all southern regions of West. Siberia.

Section S c h e d o n o r u s (Beauv.) Koch

Plant monoecious. Shoots extravaginal, without coriaceous scale leaves at base. Sheaths laciniate almost to base. Sheath blade joint with lanceolate, usually falcate auricles along edges. Leaf blades flat or partly longitudinally convolute, veined on upper surface. Glumes rather coriaceous, similar in texture to lemma. Latter without keels, awnless or with awn shorter than lemma, emerging from undivided or shortly bidentate tip. Ovary glabrous. Caryopsis conglutinate with lemma and palea at base, grooved ventrally, with long linear hilum.

4. **F. arundinacea** Schreber 1771 in Spicil. Fl. Lips.: 57. —*F. orientalis* (Hackel) V. Krecz. et Bobrov non B. Fedtsch. —*F. arundinacea* subsp. *orientalis* (Hackel) Tzvelev.

Sheath blade joint and auricles with cilia along edges. Lemma awned, rarely awnless. Plate I (4).

More or less solonetz meadows, banks of brooks and rivers, limestone and chalk exposures; sporadic. **West. Sib.**: KU, OM (Tara settlement). —Europe, Mid. Asia, Nor. Africa, East. Asia (introduced), North and South America. Described from GDR (environs of Leipzig). Map 162.

This species is represented in southern European and Asian USSR by the eastern race of subsp. *orientalis* (Hackel) Tzvelev, differing from subsp. *arundinacea* usually in aristate, undivided (not apically shortly bidentate) lemma. The extent of isolation of these subspecies is not entirely clear.

5. **F. pratensis** Hudson s. str. 1762 in Fl. Angl.: 37.

Sheath blade joints and auricles glabrous. Lemma awnless. Krasnoyarsk region (Goryachegorsk settlement), $2n = 14$. Novosibirsk region (environs of Akademgorodka), $2n = 14 + 2B$. Plate I (5).

Meadows, forest glades, sparse forests, near roads, inhabited sites, gardens and parks, various secondary and disturbed habitats, frequently abundant. **West. Sib.**: TYU—Khm (Salym river valley), Tb, KU, OM, TO, NO, KE, Al—Ba, Go. **Cen. Sib.**: KR—Pu (Noril'sk, Dudinka, Igarka), Tn, Kha, Ve, TU (foothills of Uyuksk mountain range). **East. Sib.**: IR—An, Pr, BU—Se (Ust'-Barguzin settlement), Yuzh, YAK—Vi (Peledui, Lensk, Olekminsk), Al (Aldan settlement). —Distributed in almost all extratropical countries of Nor. Hemisphere. Described from Great Britain. Map 165.

In Siberia, this species continues to spread actively northerly and easterly.

Section L e u c o p o a (Griseb.) Krivot.

Plants dioecious or monoecious. Shoots without coriaceous scale leaves at base. Sheath laciniate almost to base. Sheath blade joint without auricles. Leaf blades flat or longitudinally convolute. Glumes almost wholly membranous, distinctly different in texture and usually even in color from more or less coriaceous lemma; latter somewhat keeled. Ovary mainly pilose, rarely glabrous at tip. Caryopsis free, grooved ventrally, with long linear hilum.

6. **F. hubsugulica** Krivot. 1955 in Bot. mat. (Leningrad), 17: 77.

Plant dioecious, loosely cespitose, with extravaginal shoot regeneration and short decumbent subsurface shoots. Ligules of cauline leaves 0.5–1 mm long. Leaf blades flat, veinless on upper surface. Panicles weakly spreading, with scabrous branches. Lemma pinkish-violet, uniformly covered with spinules on back, with 5 nearly uniformly developed nerves, awnless. Ovary densely pilose at tip. Plate II (1).

Rocky slopes, rocks, debris and pebble beds; predominantly in upper mountain belt. **Cen. Sib.:** TU (Sangilen mountain range). **East Sib.:** BU—Se (Bogdarino settlement), Yuzh, ChI—Shi (Nerchinsk Zavod settlement). —Nor. Mongolia. Described from Mongolia. Map 167.

In Transbaikal, this species is quite likely to be found at other sites as well.

7. **F. komarovii** Krivot. 1955 in Bot. mat. (Leningrad), 17: 80.

Plant dioecious, compactly cespitose, with intravaginal shoot regeneration, without decumbent subsurface shoots. Shoots with many brownish dead leaf sheaths at base. Ligules of cauline leaves 0.1–0.5 mm long. Leaf blades flat, without distinct veins on upper surface. Panicles weakly spreading, with more or less scabrous branches. Lemma uniformly covered with spinules on back, with 3 stout and 2 very weak nerves and terminal cusp not longer than 0.4 mm. Ovary densely pilose at tip. East. Sayan (Tunkinsk mountain range), $2n = 28$. Plate II (2).

Rocky slopes, rocks, debris and pebble beds; middle and upper hill belts. **East. Sib.:** IR—An, BU—Yuzh (Tunkinsk bald peaks—class. hab.—and others). —Nor. Mongolia. Map 163.

8. **F. sibirica** Hackel ex Boiss. 1884 in Fl. Or. 5: 626, quoad nom. — *Leucopoa albida* (Turcz. ex Trin.) V. Krecz. et Bobrov.

Plants dioecious, compactly cespitose, with intravaginal shoot regeneration. Shoots with light gray or brownish dead leaf sheaths at base, partly disintegrating into oblong fibers. Ligules of cauline leaves 0.2–0.8 mm long. Panicles weakly spreading, with rather scabrous branches. Lemma uniformly covered with spinules on back, 3 stout and 2 very weak nerves, acuminate. Ovary densely pilose at tip. Plate III (1).

Rocky slopes, rocks, debris and pebble beds, sometimes sand; up to upper mountain belt. **Cen. Sib.:** KR—Kha, Ve, TU (Sangilen and East. Tannu-Ol mountain ranges). **East. Sib.:** IR—An, Pr, BU—Se, Yuzh, ChI—Ka (Udokan mountain range), Shi, YAK—Vi (Lena bank near Olekminsk; opposite Russkaya river estuary; Dapparai settlement). —Mongolia, Manchuria. Described from Baikal. Map 168.

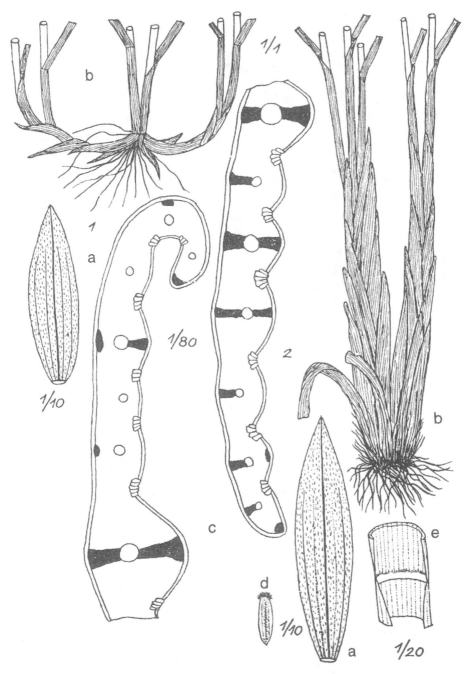

Plate II. —*Festuca hubsugulica*; 2—*F. komarovii*:
a—lemma, b—shoot bases, c—leaf blade cross sections, d—ovary, e—sheath
blade joint.

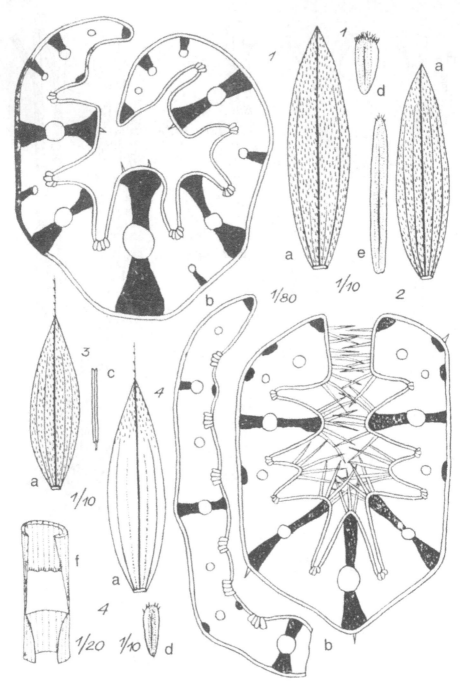

Plate III. 1—*Festuca sibirica;* 2—*F. altaica;* 3—*F. bargusinensis;* 4—*F. popovii:*
a—lemma, b—cross sections of leaf blades, c—stamens, d—ovary,
e—caryopsis, f—sheath blade joint.

Section B r e v i a r i s t a t a e Krivot.

Plant monoecious. Shoots without coriaceous scale leaves at base. Sheath laciniate almost to base. Sheath blade joints without auricles. Leaf blades double folded longitudinally, rarely flat. Glumes almost wholly membranous, distinctly different in texture from somewhat coriaceous lemma, latter somewhat keeled. Ovary usually with few hairs at tip, rarely glabrous. Caryopsis conglutinate at base with lemma and palea, ventrally grooved, with long linear hilum.

9. **F. altaica** Trin. 1829 in Ledeb. Fl. Alt. 1: 109.

Compactly cespitose plant with intravaginal shoot regeneration. Shoots with many brownish scabrous dead leaf sheaths at base. Ligules of cauline leaves 0.3–0.8 mm long. Leaf blades double folded longitudinally, 0.8–1.5 mm diam, with 7–11 vascular bundles, with 5–7 veins within; densely covered with 0.07–0.12 mm long trichomes; sclerenchyma strands arranged along margin of leaf blades and opposite vascular bundles under lower and upper epidermis. Lemma uniformly covered with spinules on back, with 5 firm nerves, with up to 1 mm long terminal cusp. Ovary with few hairs at tip. Putoran plateau (Talnakh settlement; Baselak Lake), $2n = 28$. Plate III (2).

Rocky meadows, slopes, rocks, debris and pebble beds; up to upper hill belt. **West. Sib.**: AL—Go. **Cen. Sib.**: KR—Pu, Kha, Ve, TU. **East. Sib.**: IR—An, Pr, BU—Se, Yuzh, ChI—Ka, YAK—Ar, Ol, Vi, Al, Yan, Ko. —Kazakhstan (Altay and Saur mountain range), Far East, Mongolia, North America. Described from Altay. Map 166.

10. **F. bargusinensis** Malyschev 1971 in Novosti sist. vyssh. rast. 7: 296.

Compactly cespitose plant with intravaginal shoot regeneration. Ligules of cauline leaves 0.2–0.5 mm long. Leaf blades double folded longitudinally, 0.3–0.6 mm diam, with 5 vascular bundles, scabrous on outer surface, with 3 veins within; sparsely covered with 0.02–0.05 mm long trichomes; sclerenchyma strands arranged along margin of leaf blades and opposite vascular bundles under lower epidermis. Lemma uniformly covered with spinules on back, with 5 firm nerves; terminal awn 1.2–1.8 mm long. Anthers 1.8–2.3 mm long. Ovary with few hairs at tip. Plate III (3).

Grasslands, rocky slopes and rocks, bald-peak hill belt. **East. Sib.**: BU—Se (Barguzinsk mountain range, Shegnanda river upper course— class. hab.—and others). —Endemic.

11. **F. popovii** E. Alexeev 1978 in Byul. Mosk. o-va isp. prir. otd. biol. 83 (5): 94. —*F. insularis* M. Popov non Steudel.

Plant compactly cespitose, with intravaginal shoot regeneration. Shoots with many brownish dead leaf sheaths at base. Ligules of cauline leaves 0.5–2 mm long. Leaf blades flat, veinless on upper surface. Panicles broadly spreading. Lemma pinkish-violet, with diffuse spinules only in uppr third on back, with 3 stout and 2 very weak nerves; 0.4–1.2 mm long terminal cusp. Ovary with few hairs at tip. Plate III (4).

Rocks and rocky slopes, larch forests; up to midmountain belt. **East. Sib.**: BU—Se (NE Baikal coast, Mokhnatyi Kaltygei Island—class. hab.—and others). —Endemic.

Undoubtedly, this species originated through hybridization F. *altaica* × F. *hubsugulica* but it is totally isolated from both parent species.

12. **F. tristis** Krylov et Ivanitzk. 1928 in Sist. zam. Gerb. Tomsk. un-ta. 1: 1—F. *sajanensis* Roshev. —F. *tristis* subsp. *sajanensis* (Roshev.) Tzvelev.

Compactly cespitose plant with intravaginal shoot regeneration. Ligules of cauline leaves 0.2–0.5 mm long. Leaf blades double folded longitudinally, 0.3–0.6 mm diam, with 5 vascular bundles, scabrous on outer surface, with 3 veins within; sparsely covered with 0.02–0.05 mm long trichomes; sclerenchyma strands arranged along margin of leaf blades and opposite vascular bundles under lower epidermis. Lemma glabrous or subglabrous in lower third on back, elsewhere uniformly covered with spinules, with 5 firm nerves; terminal awn 1.2–1.8 mm long. Anthers 2.5–4 mm long. Ovary with few hairs at tip. Altay (Kuraisk mountain range, Aktash settlement), $2n = 14$. Plate IV (1).

Grasslands, rocky slopes and rocks; bald-peak hill belt. **West. Sib.**: AL—Go. **Cen. Sib.**: KR—Kha (Ona river valley, tributary of Abakan river), TU (Uyuksk and Sangilen mountain ranges). **East. Sib.**: IR—An (East. Sayan), BU—Yuzh (East. Sayan). —Kazakhstan and Mongolian Altay. Described from Kazakhstan. Map 170.

Section Festuca

Plant monoecious. Shoots without coriaceous scale leaves at base. Sheath blade joint without auricles. Blades of all leaves or at least of vegetative shoots double folded longitudinally. Glumes usually membranous only along margin, elsewhere rather coriaceous, invariably similar in texture to lemma. Latter usually without keels. Caryopsis ventrally grooved, with long linear hilum.

13. **F. auriculata** Drobov 1915 in Tr. Bot. muz. Akad. nauk, 14: 163.

Plant densely cespitose, with intravaginal shoot regeneration. Mats divided into rather isolated shoot tufts, enveloped by dead leaf sheaths at base. Sheath closed for 2/5 to 3/4 length. Leaf blades usually with 7 vascular bundles, smooth outside, glabrous, sparsely hairy, with solitary midrib and sometimes 1 or 2 almost invisible lateral veins within, and 3 very weak sclerenchyma strands. Spikelets pinkish-violet or green. Lemma with 1–2.5 mm long awn. Yakutia (Tiksi bay) and Putoran plateau (Talnakh settlement, Baselak Lake), $2n = 14$. Plate IV (2).

Rocky tundras, grasslands, rocks, debris, pebble beds; Arctic and bald-peak hill belt. **Cen. Sib.**: KR—Ta, Pu. **East. Sib.**: BU—Yuzh

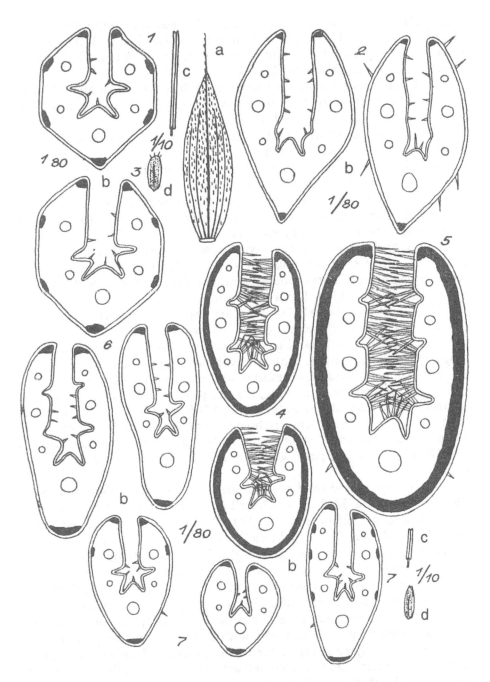

Plate IV: 1—*Festuca tristis*; 2—*F. auriculata*; 3—*F. baffinensis*; 4—*F. beckeri* s. str.; 5—*F. beckeri* subsp. *polesica*; 6—*F. borissii*; 7—*F. brachyphylla*:
a—lemma, b—leaf blade cross sections, c—stamens, d—ovaries.

146

(Dzun-Murinsk mountain range), ChI—Ka (Kalarsk mountain range), Shi (Sokhondo bald peak), YAK—Ar (Panteleikha hill in Kolyma river lower courses—class. hab.—and others), Yan, Ko. —Nor. Urals, Far East, East. China, North America. Map 169.

14. **F. baffinensis** Polunin 1940 in Bull. Nat. Mus. Canada, 92, Biol. 24: 91. —*F. brevifolia* auct. p. p. non Muhl. nec R. Br. —*F. brachyphylla* auct. p. p. non Schultes et Schultes fil.

Compactly cespitose plant with intravaginal shoot regeneration. Shoots with few dead leaf sheaths at base. Sheath closed for 1/2 to 3/4 length. Leaf blades green, with isolated sclerenchyma strands under lower epidermis. Stems densely pilose almost throughout length. Panicles with densely pilose branches rather declinate from rachis, short and somewhat branched. Spikelets usually violet. Anthers 0.7–0.9 mm long. Ovary with few hairs at tip. Plate IV (3).

Rocky and sandy tundras, usually near limestone outcrops. **East. Sib.:** YAK—Ar (Chetyrekhstolbovoi island in Medvezh'ikh island group), Ko (Darpir Lake in Chersk mountain range). Distributed almost throughout Circumpolar region. Described from Canada.

15. **F. beckeri** (Hackel) Trautv. s. str. 1884 in Tr. Peterb. bot. sada 9 (1): 325. —*F. duriuscula* auct. non L. —*F. dahurica* auct. fl. prov.

Roots covered with sand grains glued with secretions of root hairs. Plant compact caespitose, with intravaginal short regeneration. Sheath laciniate almost to base. Leaf blades stiff, erect, cylindrical or flattened-cylindrical, gradually acuminate, 0.4–0.6 (0.7) mm diam, with 7–9 vascular bundles, glabrous outside, with (1) 3–5 veins within; densely covered with 0.06–0.15 mm long trichomes; compact cover of sclerenchyma under lower epidermis. Stems more or less pilose or glabrous. Lemma 2.5–3.8 (4.2) mm long; palea glabrous or subglabrous along keels. Plate IV (4).

Sandy steppes and sand of river terraces above meadows. **Cen. Sib.:** KR—Ve. —Europe, Mid. Asia. Described from Lower Volga. Map 171.

16. **F. beckeri** subsp. **polesica** (Zapal.) Tzvelev 1970 in Spisok rast. Gerb. fl. SSSR, 18: 15. —*F. polesica* Zapal. —*F. duriuscula* auct. non L.

Roots covered with sand grains glued with secretions of root hairs. Plant compactly caespitose, with intravaginal shoot regeneration. Sheath laciniate almost to base. Leaf blades stiff, erect, cylindrical or flattened-cylindrical, abruptly acuminate, (0.4) 0.65–0.85 (1.1) mm diam, with (7) 9–11 (15) vascular bundles, scabrous outside, rarely smooth, with 5–7 (9) veins within; densely covered with 0.08–0.15 mm long trichomes; compact sclerenchyma cover under lower epidermis. Stems usually pilose throughout length. Lemma covered with spinules usually along upper third of keels. Plate IV (5).

In sand on river terraces above meadows, steppified pine forests, sandy steppes. **West. Sib.:** TYU—Tb (environs of Tyumen'), KU (Tobol river valley), AL—Ba. —East. Europe, Nor. Kazakhstan. Described from environs of Kiev. Map 172.

Two subspecies of *F. beckeri* s. l. —subspecies *polesica* and more southern subspecies *beckeri* related through transitional forms in the contact zone—are found in southern European USSR, Kazakhstan, and Siberia.

17. **F. borissii** Reverd. 1965 in Sist. zam. Gerb. Tomsk. un-ta 83: 8.

More or less compactly cespitose plant with intravaginal shoot regeneration. —Sheath closed for 2/3 to 3/5 length. Leaf blades green, with 7 vascular bundles, smooth outside, with 3–5 veins within; 3 sclerenchyma strands under lower epidermis. Panicle branches smooth, glabrous or subglabrous. Spikelets brownish. Plate IV (6).

Meadows, rocky slopes, rocks, debris; bald-peak hill belt. **West. Sib.**: AL—Ba (Inya river upper courses), Go. —Mid. Asia, West. China (Junggar). Described from Kazakhstan (Narymsk mountain range). Map 173.

18. **F. brachyphylla** Schultes et Schultes fil. 1827 in Add. ad Mant. 3: 646. —*F. brevifolia* R. Br., non Muhl.

More or less compactly cespitose plant with intravaginal shoot regeneration. Shoots usually (but not always) with few dead leaf sheaths at base. Sheaths closed for 1/2 to 3/4 length. Leaf blades green, with isolated sclerenchyma strands under lower epidermis. Stems glabrous or covered with diffuse spinules only for 1–2 cm under panicle. Latter with branches appressed to rachis, rarely declinate, somewhat branching; branches covered with spinules, sometimes transformed into hairs. Spikelets pinkish-violet, rarely green. Anthers 0.7–1 (1.2) mm long. Ovary glabrous. Yakutia (Tiksi settlement; Shandrin river), $2n = 42$. Plate IV (7).

Grasslands, rocky slopes, rocks, pebble beds, in various types of tundras; usually dominant in the Arctic, less as in bald-peak hill belt. **West. Sib.**: TYU—Yam, AL—Go. **Cen. Sib.**: KR—Ta, Pu, Kha (Kuznets Alatau, sources of Chernyi Iyus river), Ve (East. Sayan), TU (East. Tannu-Ol and Sangilen mountain ranges). **East. Sib.**: IR—An (East. Sayan), ChI—Ka, Shi (Sokhondo bald peak), YAK—Ar, Ol, Yan, Ko. —Europe, Mid. Asia, Far East, West. China, Mongolia, Nor. America. Described from Nor. Canada. Map 176.

19. **F. chionobia** Egor. et Sipl. 1970 in Novosti sist. vyssh. rast. 6: 226. —*F. auriculata* auct. quoad pl. *vivipar* non Drobov.

Compactly cespitose plant with intravaginal shoot regeneration. Sheath closéd for 2/5 to 3/4 length. Leaf blades with 3 weakly developed sclerenchyma strands under lower epidermis. All or at least most spikelets in panicles viviparous. Lemma covered with spinules in upper third on back, with 0.6–1.8 mm long awn in lower florets. Plate V (1).

Grasslands, rocky slopes, rocks; in bald-peak hill belt. **East. Sib.**: BU—Se (source of Muzhinai river on Baikal'sk mountain range. —class. hab.—and others), ChI—Ka. —Far East. Map 174.

Species close to *F. auriculata*, representing its viviparous derivative. Transitional forms between *F. auriculata* and *F. chionobia* with partly nonviviparous spikelets in panicles are extremely rare, which suggests that these taxa can be regarded as independent species.

20. **F. dahurica** (St.-Yves) V. Krecz. et Bobrov 1934 in Fl. SSSR, 2: 517, 771. —*F. beckeri* auct. non Trautv.

Compactly cespitose plant with intravaginal shoot regeneration. Sheath closed for 1/6 to 1/4 length. Leaf blade cross section oval or round, rarely obovate, smooth outside, with solitary midrib and sometimes almost indistinct lateral veins within; densely covered with 0.06–0.1 mm long trichomes; different numbers of frequently connective sclerenchyma strands under lower epidermis. Spikelets green or brownish. Lemma awnless, acuminate. Plate V (2).

Sand steppes, sometimes in pine forests. **East. Sib.**: BU—Yuzh (Ulan-Ude vicinity—class. hab.—and others), ChI—Shi (Onon river valley).—Mongolia, NE China. Map 175.

21. **F. hyperborea** Holmen ex Frederiksen 1977 in Bot. Not. (Lund), 130 (3): 273. —*F. brachyphylla* auct. non Schultes et Schultes fil.

Compactly cespitose plant with intravaginal shoot regeneration. Shoots with many (but not forming covers around rather isolated shoot tufts) dead leaf sheaths at base. Sheath closed for 1/2 to 3/4 length. Leaf blades with rather distinct glaucous bloom, smooth outside, with 3 (5–7) sclerenchyma strands under lower epidermis. Stems glabrous. Panicle branches glabrous or with stray short spinules appressed to rachis, not branched. Spikelets grayish or glaucous violet. Anthers 0.7–0.9 mm long. Ovary glabrous. Plate V (3).

Rubble limestone and loamy-limestone slopes, sometimes sand and pebble beds. **Cen. Sib.**: KR—Ta (Oktyabr'skii Revolyutsii Island, Bikada and Tareya river valleys on Taimyr peninsula). **East. Sib.**: YAK—Ar (Tiksi settlement; Cape Svyatoi Nos; Medvezh'i Island). —Distributed almost throughout Circumpolar region. Described from Nor. Greenland. Map 179.

F. *hyperborea* finds are possible at other sites in East. Siberian Arctic as well. Confused rather often (especially in herbariums) with *F. brachyphylla* because of external similarity; differs from latter not only morphologically but also in very early anthesis when growing together.

22. **F. jacutica** Drobov 1915 in Tr. Bot. muz. Akad. nauk, 14: 163.

Compactly cespitose plant with intravaginal shoot regeneration. Sheath closed for 1/2 length. Leaf blades green, with 5–7 (9) vascular bundles 4–6-angled in cross section, somewhat scabrous outside, covered with rather dense 0.05–0.08 mm long trichomes within; 7–9 (11) sometimes connective sclerenchyma strands under lower epidermis. Panicles not nutant. Spikelets usually pale green. Lemma up to 4.5 mm long, awnless; palea covered with spinules along upper third of keels. Plate VI (1).

149

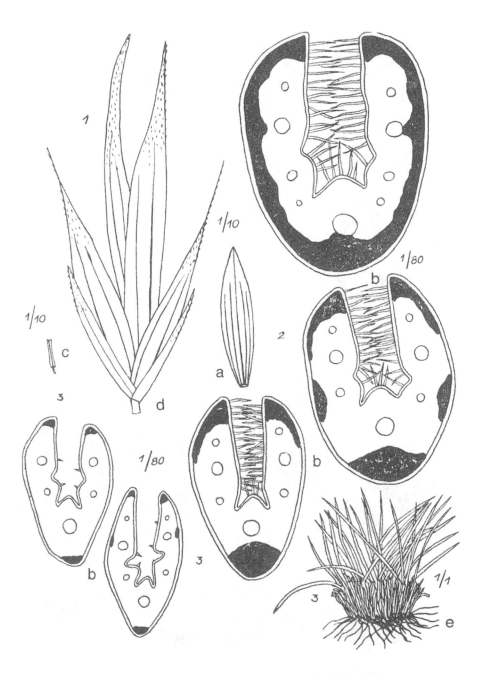

Plate V. 1—*Festuca chionobia*; 2— *F. dahurica*; 3—*F. hyperborea*:
a—lemma, b—leaf blade cross sections, c—stamens, d—spikelet, e—mat
section.

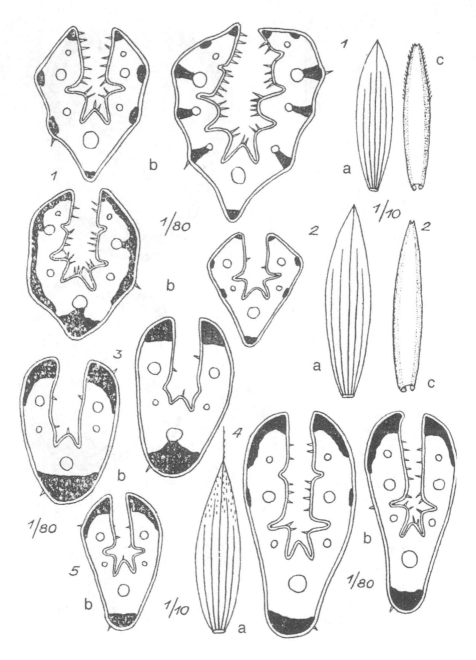

Plate VI: 1—*Festuca jacutica*; 2—*F. karavaevii*; 3—*F. kolymensis*; 4—*F. kryloviana*; 5—*F. kurtschumica*:
a—lemma, b—leaf blade cross sections, c—palea.

Sparse forests (predominantly larch and pine), shrubs, forest glades, arid meadows, rocky slopes, rocks; up to mid-mountain belt. **East. Sib.**: BU—Se (Kudalkan river upper courses), ChI—Shi, YAK—Vi, Al (Amginsk road, Krestyakh river upper courses—class. hab.—and others), Yan, Ko. —Far East, Nor. Mongolia, NE China. Map 177.

23. **F. karavaevii** E. Alexeev 1979 in Byul. Mosk. o-va isp. prir. otd. biol. 84 (5): 122.

Plants with mixed shoot regeneration and short decumbent subsurface shoots. Sheath straw-colored, lustrous, closed up to 3/4 length. Leaf blades green, with 5 vascular bundles, 4–6-angled in cross section, smooth outside, covered with sparse up to 0.03 mm long trichomes within; 7 sclerenchyma strands under lower epidermis. Panicles not nutant. Spikelets green or pinkish-green. Lemma 5.5–5.8 mm long, awn up to 0.7 (1.3) mm long. Palea glabrous or subglabrous along keels. Plate VI (2).

On shifting sand. **East. Sib.**: YAK—Vi (Nidzheli Lake environs in Kobyaisk region—class. hab.—and others). —Endemic.

Species close to *F. jacutica*. Along with several other grasses, endemic in floristically highly typical shifting sand (tukulans) around Nidzheli Lake.

24. **F. kolymensis** Drobov 1915 in Tr. Bot. muz. Akad. nauk. 14: 155.

Compactly cespitose plant with intravaginal shoot regeneration. Sheath laciniate almost to base. Leaf blades with 5 (7) vascular bundles, scabrous outside, with solitary midrib and sometimes 1 or 2 nearly indistinct lateral veins within; 3 very firm sclerenchyma strands under lower epidermis. Spikelets brownish. Plate VI (3).

Rocky and sandy steppified slopes, arid steppified meadows; up to lower mountain belt. **East. Sib.**: YAK—Ar, Ol, Vi, Al, Yan, Ko (Srednekolymsk vicinity—class. hab.—and others). —Far East. Map 178.

25. **F. kryloviana** Reverd. 1927 in Sist. zam. Gerb. Tomsk. un-ta 2: 3, excl. var. *musbelica*.

Compactly cespitose plant with intravaginal shoot regeneration. Sheath closed for 1/3 to 1/2 length. Leaf blades green or with glaucous bloom, (0.4) 0.55–0.85 (1) mm diam, with (5) 7 vascular bundles, scabrous outside, with 3–5 veins within; 3 (5) sclerenchyma strands under lower epidermis. Panicle branches scabrous. Spikelets brownish. Lemma (4.5) 4.8–5.5 (6) mm long. Plate VI (4).

Grasslands, rocky slopes, rocks, pebble beds, rocky steppes; up to upper mountain belt; often dominant in bald-peak as well as steppe mountain belts. **West. Sib.**: AL—Ba, Go (mountain range between Berezovka and Khapsyn rivers—class. hab.—and others). **Cen. Sib.**: all regions. **East. Sib.**: IR—An (Baikal coast), Pr (Kuta river valley), BU—Se, Yuzh, ChI—Shi (Bukukun settlement). —Mid. Asia, West. China (Junggar). Mongolia. Map 180.

Morphological and ecological polymorphism of this species is undoubtedly associated with its hybrid genesis—*F. borissii* × *F. valesiaca* s. l. or *F. kurtschumica* × *F. valesiaca* s. l.

26. **F. kurtschumica** E. Alexeev 1976 in Novosti sist. vyssh. rast. 13: 24, 30.

Compactly cespitose plant with intravaginal shoot regeneration. Sheath closed for 2/3 to 3/4 length. Leaf blades green, 0.4–0.5 (0.55) mm diam, with 5 vascular bundles, scabrous outside, with (1) 3 veins within; 3 sclerenchyma strands under lower epidermis. Panicle branches scabrous. Spikelets brownish. Lemma 3.5–4.5 mm long. Anthers 1.8–2.2 mm long. Plate VI (5).

Rocky meadows and rocks, predominantly in upper hill belt. **West. Sib.**: AL—Go (Archala river valley, Kok-Su tributary). **East. Sib.**: IR—An (Ol'khon Island). —Mid. Asia, West. China (Junggar), Mongolian Altay. Described from Kazakhstan.

Rare species, close to Middle Asian *F. altaica* Drobov. Evidently, "absorbed" as a result of hybridization processes over much of its original, highly extensive distribution range.

27. **F. lenensis** Drobov 1915 in Tr. Bot. muz. Akad. nauk, 14: 158. — *F. albifolia* Reverd. —*F. lenensis* var. *albifolia* (Reverd.) Tzvelev. —*F. lenensis* subsp. *albifolia* (Reverd.) Tzvelev. —*F. kolymensis* auct. non Drobov.

Compactly cespitose plant with intravaginal shoot regeneration. Mats divided into rather isolated shoot tufts enveloped by dead leaf sheaths at base. Sheath closed for 2/5–1/2 length. Leaf blades usually with 7 vascular bundles, rather scabrous outside due to fine spinules or smooth, rarely pilose, with solitary midrib within and sometimes 1 or 2 nearly indistinct lateral veins; 3 firm sclerenchyma strands under lower epidermis. Spikelets brownish, green or pinkish-violet. Plate VII (1).

Rocky slopes, rocks, debris and pebble beds, frequently dominant in rocky tundras and steppes, sometimes on sand; up to upper mountain belt. **West. Sib.**: AL—Go (Severo- and Yuzhno-Chuisk mountain ranges). **Cen. Sib.**: KR—Kha, Ve, TU. **East. Sib.**: IR—An (Baikal coast; Ol'khon island), BU—Se, Yuzh, ChI—Ka, Shi, YAK—Ar (Lena and Kolyma river lower courses), Ol (valleys of Motorchuna and Menkere rivers, Lena river tributary; Anabar river near Saskyllakh settlement), Vi (Lena river near Kyatchinskoe village in Olekminsk region—class. hab.—and others), Yan, Ko (Srednekolymsk town). —Far East, China, Inner Mongolia, North America (western Arctic part). Map 183.

28. **F. litvinovii** (Tzvelev) E. Alexeev 1976 in Novosti sist. vyssh. rast. 13: 31. —*F. pseudosulcata* var. *litvinovii* Tzvelev. —*F. pseudosulcata* auct. non Drobov.

Compactly cespitose plant with intravaginal shoot regeneration. Sheath closed for 1/4–1/3 length. Leaf blades with 5–7 vascular bundles, smooth or almost so outside, with solitary midrib within and

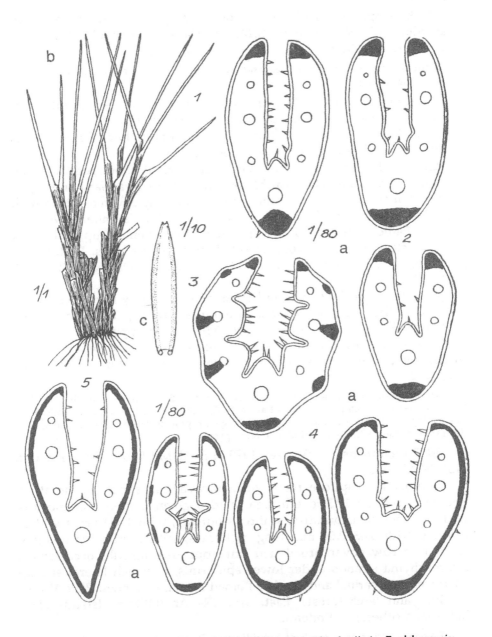

Plate VII: 1—*Festuca lenensis*; 2—*F. litvinovii*; 3—*F. malyschevii*; 4—*F. olchonensis*; 5—*F. ovina* s. str. and *F. ovina* subsp. *sphagnicola*: a—leaf blade cross sections, b—mat section, c—palea.

sometimes 1 or 2 nearly indistinct lateral veins; 3 firm sclerenchyma strands under lower epidermis. Spikelets green, glaucescent, rarely pilose. Plate VII (2).

Rocky slopes and rocks, rarely on sand, up to mid-mountain belt. **East. Sib.**: ChI—Shi. —Far East, Manchuria, East. Mongolia. Described from NE China. Map 181.

Petrophilous derivative of *F. pseudosulcata*, differing well from it morphologically as well as geographically.

29. **F. malyschevii** E. Alexeev nomen et staus nov. —*F. jacutica* var. *nutans* Malyschev 1995 in Vysokogorn. fl. Vost. Sayana: 70, non *F. nutans* Moench, 1794. —*F. jacutica* subsp. *nutans* (Malyschev) Tzvelev. —*F. jacutica* auct. fl. prov. Krasnojarsk, non Drobov.

Compactly cespitose plant with intravaginal shoot regeneration. Sheath closed 1/2 length. Leaf blades green, with 7 vasclar bundles, 4–6-angled in cross section, scabrous outside, densely covered within with 0.05–0.08 mm long trichomes; 9 sclerenchyma strands under lower epidermis. Panicles rather nutant. Spikelets violet. Lemma up to 4.5 mm long, awnless; palea glabrous or subglabrous along keels. East. Sayan (Tunkinsk mountain range), $2n \doteq 14$; Tuva (Kadyr-Os river), $2n = 20$. Plate VII (3).

Grasslands, rocky slopes and rocks predominantly in bald-peak hill belts. **Cen. Sib.**: KR—Ve (Ardanskii mountain range), TU (Kemchikskii and Uyukskii mountain ranges; Sistyg-Khem river upper courses). **East. Sib.**: IR—An (East. Sayan, Doda hill—class. hab.—and others), BU—Yuzh (Tunkinsk bald peaks, Khamar-Daban mountain range). —Endemic. Map 182.

Morphological differences between *F. malyschevii* and *F. jacutica*, which is very close to the former, although poorly distinct are entirely stable. Taking also into consideration the ecological differences and the complete geographic isolation of these taxa, it is desirable to regard the two as independent species.

30. **F. olchonensis** E. Alexeev 1979 in Byul. Mosk. o-va isp. prir., otd. biol. 84 (5): 125.

Compactly cespitose plant with intravaginal shoot regeneration. Sheath closed for 1/3 length. Leaf blades green, flattened-cylindrical, with 5 (7) vascular bundles, scabrous outside, with 1 (3) veins within, covered with 0.03–0.06 mm long trichomes, with continuous or interrupted narrow cover broadened only opposite midrib; occasionally sclerenchyma strands under lower epidermis, of which lateral strands narrower than carinal and nearly as broad as marginal ones. Plate VII (4).

Pine and larch forests. **East. Sib.**: IR—An (Ol'khon Island—class. hab.—and others). —Endemic.

Probably occurs as a result of hybridization *F. ovina* × *F. kurtschumica* imparting transitional characteristics of anatomical structure of leaf blades.

31. **F. ovina** L. s. str. 1753, Sp. Pl.: 73. —*F. supina* var. *elata* Drobov.

—*F. ruprechtii* (Boiss.) V. Krecz. et Bobrov. —*F. ovina* subsp. *ruprechtii* (Boiss.) Tzvelev. —*F. supina* auct. non Schur.

Compactly cespitose plant with intravaginal shoot regeneration. Sheath closed for 1/6 to 1/3 length. Leaf blades usually green, flattened-cylindrical, usually with 7 vascular bundles, scabrous (rarely smooth) outside, with solitary vein within; sparsely covered with 0.02–0.05 mm long trichomes; narrow cover of sclerenchyma under lower epidermis continuous, rarely interrupted, uniformly broad. Spikelets green or pinkish-violet. Lemma up to 4 mm long. Altay (Ongudaisk region) and Yakutia (Nera river basin), $2n = 14$. Plate VII (5).

Meadows, forest glades, sand, rocks and pebble bends, thin forests and shrubs; up to upper mountain belt, quite common. **West. Sib.**: TYU—Yam, Khm, Tb, OM (Krasnoyarka village on Irtysh), TO, NO (?), KE, AL—Go. **Cen. Sib.**: KR—Ta, Pu, Tn, Kha, Ve. **East. Sib.**: IR—An, Pr, BU—Se, Yuzh, ChI—Ka, Shi, YAK—Ar (Chersk settlement of Kolyma river), Ol, Vi, Al, Yan, Ko (Kolyma river valley). —Caucasus, Asia, North America. Described from West. Europe. Map 184.

32. **F. ovina** subsp. **sphagnicola** (B. Keller) Tzvelev 1971 in Bot. zhurn. 56 (9): 1255. —*F. sphagnicola* B. Keller.

Compactly cespitose plant with intravaginal shoot regeneration. Sheath closed for 1/4 to 1/3 length. Leaf blades green, flattened-cylindrical, smooth (rarely, scabrous) outside, with solitary vein within, sparsely covered with 0.02–0.05 mm long trichomes; narrow cover of sclerenchyma under lower epidermis continuous, rarely interrupted, uniformly broad. Spikelets brownish. Lemma 3.8–4.8 mm long. West. Sayan, $2n = 14$. Altay (Aktash settlement), $2n = 28$. Plate VII (5).

Meadows, rocky slopes, rocks and pebble beds, forest glades; often dominant in upper mountain belt, generally very rare. **West. Sib.**: AL—Go (Terektinsk mountain range—class. hab.—and others). **Cen. Sib.**: KR—Kha, Ve (West. Sayan), TU. **East. Sib.**: BU—Yuzh (Munku-Sardyk Mt.). —Kazakhstan and Mongolian Altay. Map 186.

Transitional forms between this subspecies and *F. ovina* subsp. *ovina* with spikelets faintly tinged brown and leaf blades scabrous outside to different extent are frequently found.

33. **F. ovina** subsp. **vylzaniae** E. Alexeev 1979 in Byul. Mosk. o-va isp. prir. otd. biol. 84 (5): 128. —*F. beckeri* auct. fl. Transbaical. non Trautv.

Compactly cespitose plant with intravaginal shoot regeneration. Sheath closed only at very base. Leaf blades with glaucous bloom, flattened-cylindrical, smooth outside, with 1 (3) veins within, densely covered with 0.06–0.08 mm long trichomes; sclerenchyma cover under lower epidermis continuous and uniformly broad. Spikelets green or pinkish-green. Lemma 4.4.–4.8 mm long. Plate VIII (1).

Sand dunes, sometimes rocky slopes and rocks, up to upper mountain belt; predominant on east coast of Baikal. **East. Sib.**: BU—Se

156

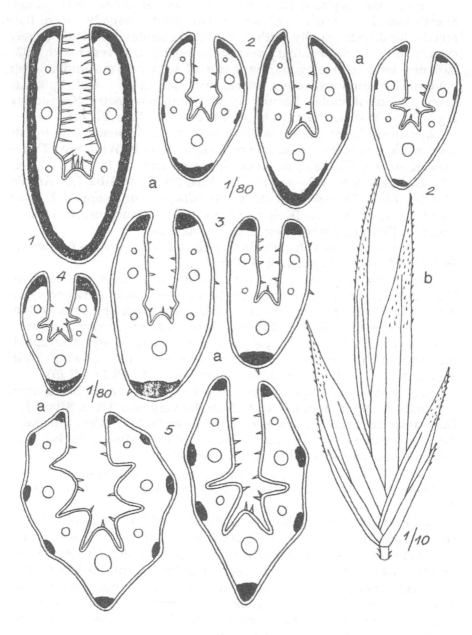

Plate VIII: 1—*Festuca ovina* subsp. *vylzaniae;* 2—*F. viviparoidea;* 3—*F. pseudosulcata;*
4—*F. pseudovina;* 5—*F. rubra* s.l.:
a—leaf blade cross sections; b—spikelet.

(Sosnovka settlement—class. hab.—and others), Yuzh. —Endemic. Map 187.

Generally grows together with subsp. *ovina* and is related to it through various transitional forms.

34. **F. pseudosulcata** Drobov 1915 in Tr. Bot. muz. Akad. nauk, 14: 156, excl. hab. "prope Jacutsk". —*F. jenissejensis* Reverd.—*F. kolymensis* auct. non Drobov.

Compactly cespitose plant with intravaginal shoot regeneration. Sheath closed for 1/4 to 1/3 length. Leaf blades with 5–7 vascular bundles, scabrous outside, with solitary midrib within and sometimes with 1 or 2 nearly indistinct lateral veins; sclerenchyma strands 3, firm. Spikelets green. Plate VIII (3).

Steppes, rocky slopes, among shrubs, arid pine and larch forests, forest glades; up to lower mountain belt. **Cen. Sib.**: KR—Kha, Ve. **East. Sib.**: IR—An, Pr, YAK—Vi (Chona river lower course—class. hab.—and others). —Far East (isolated find on Auri cliff in Amur river valley). Map 185.

35. **F. pseudovina** Hackel ex Wiesb. 1880 in Oesterr. Bot. Zeitschr. 30: 126. —*F. valesiaca* var. *pseudovina* (Wiesb.) Schinz et R. Keller. —*F. valesiaca* subsp. *pseudovina* (Wiesb.) Hegi.

Compactly cespitose plant with intravaginal shoot regeneration. Sheath laciniate almost to base. Leaf blades green, (0.3) 0.4–0.6 (0.75) mm diam, with 5 (7) vascular bundles, scabrous outside, with 3 highly prominent veins within; sclerenchyma strands 3 (5) under lower epidermis. Spikelets green, (4) 4.5–5.5 (6) mm long. Lemma (2.3) 2.8–3.8 (4.2) mm long. Plate VIII (4).

Predominant in meadows, grass steppes, steppified meadows, sparse forests, sand ridges in river valleys, sometimes on limestones, solonchaks and solonetzes; up to lower mountain belt. **West. Sib.**: TYU—Tb, KU, OM, TO, NO, KE, AL—Ba. **Cen. Sib.**: KR—Kha, Ve. **East. Sib.**: IR—An, BU—Yuzh (Tunka settlement). —Europe and Kazakhstan. Described from Austria. Map 191.

36. **F. rubra** L. s. str. 1753, Sp. Pl.: 74. —*F. eriantha* Honda. —*F. egena* V. Krecz. et Bobrov.

Loosely cespitose plant, usually with mixed shoot regeneration. Sheath closed almost to top. Leaf blade 4–6-angled in cross section, with 7–9 (11) vascular bundles, smooth outside, covered with sparse 0.03–0.05 mm long trichomes within; isolated sclerenchyma strands usually only under lower epidermis. Panicle branches scabrous. Lemma 4–7 mm long, from lanceolate to broadly lanceolate, more or less pilose or glabrous, with 0.7–3.5 mm long awn. Palea covered with spinules throughout length of keels. Anthers longer than 2.5 mm. Ovary glabrous. Putoran plateau (Baselak Lake), $2n = 56$, Altay (Kosh-Agach settlement), $2n = 4$. Plates VIII (5), IX (1, 2).

158

Various types of tundras, meadows, forest glades, sparse forests, fallow lands, sand and pebble beds, roadsides and inhabited sites, gardens and parks, various disturbed and secondary habitats, often abundant. In all regions, districts and provinces of Siberia. Distributed in all extratropical countries of the Northern Hemisphere; introduced in South America. Described from West. Europe. Map 192.

Exhibits tendency toward maximum dispersal everywhere.

37. **F. rubra** subsp. **arctica** (Hackel) Govor. 1937 in Fl. Urala: 127. — *F. richardsonii* Hooker. —*F. kirelovii* Steudel. —*F. cryophila* V. Krecz. et Bobrov. —*F. rubra* var. *cryophila* (V. Krecz. et Bobrov) Reverd. —*F. rubra* auct. non L.

Plant usually not forming mats, with extravaginal shoot regeneration and decumbent subsurface shoots. Sheath closed almost to top. Leaf blades 4–6-angled in cross section, with 7–9 (11) vascular bundles, smooth outside, covered with sparse 0.03–0.05 mm long trichomes within; isolated sclerenchyma strands usually only under lower epidermis. Panicle branches pilose or scabrous. Lemma lanceolate-ovate, 4–5.3 mm long, villous, rarely glabrous, with up to 1.5 mm long awn. Palea covered with spinules all along keels. Anthers longer than 2.5 mm. Ovary glabrous. Putoran plateau (Khakoma Lake), $2n = 42$; West. Sayan (Karasulla river), $2n = 14$. Plate IX (4).

Often dominant in various types of tundras; on bald peaks, rocky slopes, rocks, sand and pebble beds. **West. Sib.:** TYU—Yam, AL—Go. **Cen. Sib.:** KR—Ta, Pu. **East. Sib.:** BU—Se (Barguzinsk mountain range), Yuzh (Tunkinsk bald peaks), YAK—Ar, Ol. —Circumpolar subspecies found in hills of East., Mid. and Cen. Asia, and North America. Described from European Arctic. Map 193.

This subspecies and the very common transitional forms between it and F. rubra subsp. rubra can be found in all the hilly regions of South. Siberia.

38. **F. rubra** subsp. **baicalensis** (Griseb.) Tzvelev 1971 in Bot. zhurn. 56 (9): 1254. —*F. rubra* var. *baicalensis* Griseb. —*F. baicalensis* (Griseb.) V. Krecz. et Bobrov.

Plant not forming mats, with extravaginal shoot regeneration, with decumbent subsurface shoots. Sheath closed almost to top. Leaf blades 4–6-angled in cross section, with 7–9 (11) vascular bundles, smooth outside, covered with sparse, 0.03–0.05 mm long trichomes within; isolated sclerenchyma strands usually only under lower epidermis. Panicle branches scabrous. Lemma lanceolate, 6–6.5 mm long, glabrous or puberulent, with 1.5–3.5 mm long awn. Palea covered with spinules all along keels. Anthers 2–3 mm long. Ovary glabrous. Plate IX (3).

Sand and pebble beds, predominantly along Baikal banks. **East. Sib.:** IR—An, BU—Se, Yuzh. —Endemic. Described from Baikal. Map 188.

Often grows along with F. rubra subsp. rubra, is very weakly distinguished from it, and related to it through transitional forms.

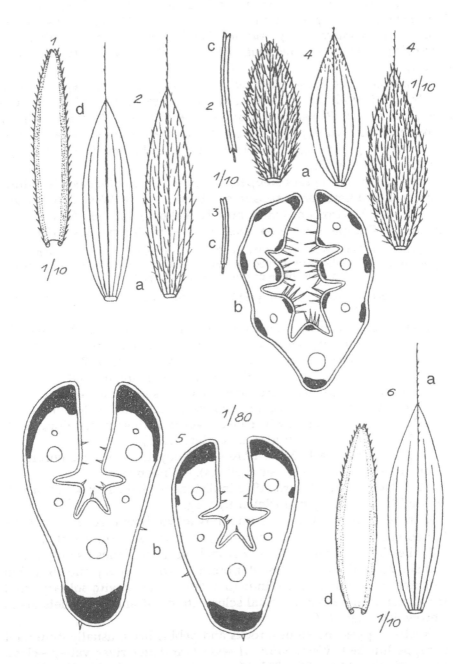

Plate IX. 1—*Festuca rubra* s.l.; 2—*F. rubra* s. str.; 3—*F. rubra* subsp. *baicalenris*;
4—*F. rubra* subsp. *arctica*; 5—*F. rupicola*; 6—*F. skrjabinii*:
a—lemma, b—leaf blade cross sections, c—stamens, d—palea.

39. F. rupicola Heuffel 1858 in Verh. Zool.-Bot. Ges.: Wien, 8: 233. —
F. sulcata (Hackel) Nyman nom. illeg. —*F. taurica* (Hackel) A. Kerner ex
Trautv. —*F. valesiaca* subsp. *sulcata* (Hackel) Schinz et R. Keller. —*F.
ganeschinii* Drobov. —*F. recognita* Reverd. —*F. pseudosulcata* auct. non
Drobov.

Compactly cespitose plant with intravaginal shoot regeneration.
Sheath laciniate almost to base. Leaf blades green, (0.4) 0.55–0.85 (1.1)
mm diam, with 5 (7) vascular bundles, scabrous outside, with 3 highly
prominent veins within; 3 (5) sclerenchyma strands under lower epidermis. Spikelets green, (5.5) 6.5–8.5 (10) mm long. Lemma (4.5) 4.8–5.5 (6)
mm long. Plate IX (5).

Herbaceous-chee grass steppes, steppified sparse forests, old fallow
lands, rock and limestone exposures, solonetzes; rare. **West. Sib.**: TYU—
Tb, KU, OM (environs of Omsk and Tyukalinsk), AL—Ba (environs of
Barnaul). —Europe, Mid. Asia. Described from Europe. Map 189.

Species found in West. Siberia at the northeastern extremity of its distribution range.
Frequently grows together with related species *F. pseudovina* but differs from it in repro-
duction features due to their different diurnal flowering patterns.

40. F. skrjabinii E. Alexeev 1979 in Byul. Mosk. o-va isp. prir. otd.
biol. 84 (5): 123.

Plant predominantly with extravaginal shoot regeneration, loosely
caespitose, with decumbent subsurface shoots. Sheath glabrous, closed
for 2/3 length. Leaf blade cross section polygonal, with 7–11 vascular
bundles, smooth outside, densely covered with up to 0.08 mm long
trichomes within; isolated sclerenchyma strands under lower and upper
epidermis. Spikelets glaucescent. Lemma lanceolate, 5.5–6.5 mm long,
with 0.8–2 (2.5) mm long awn. Palea covered with spinules along keels
for 1/3–1/2 on top. Ovary glabrous. Plate IX (6).

Shifting sand. **East. Sib.**: YAK—Vi (Kobyaisk region; environs of
Nidzheli Lake—class. hab.—and others). —Endemic.

41. F. tschujensis Reverd. 1936 in Sist. zam. Gerb. Tomsk. un-ta 3: 1.
—*F. albifolia* var. *tschujensis* (Reverd.) Serg.

Compactly cespitose plant with intravaginal shoot regeneration. Mats
divided into rather isolated shoot tufts, enclosed by dead leaf sheaths at
base. Sheath closed for 2/5 to 1/2 from base. Leaf blades usually somewhat sinuate at tip, with 7 vascular bundles, with solitary midrib within
and sometimes, 1 or 2 nearly indistinct lateral veins; with solitary stout
carinal and 2 very weak marginal sclerenchyma strands. Spikelets green
or brownish. Plate X (1).

Rocky steppes and slopes, rocks and pebble beds; usually dominant
in steppe hill belt. **West. Sib.**: AL—Go (Tarkhatta river valley—class.
hab.—and others) **Cen. Sib.**: TU (Mongun-Taiginsk region). —Kazakhstan
Altay, Mongolia (north).

Species close to *F. lenensis* differ reliably only in anatomical structure of leaf blades.

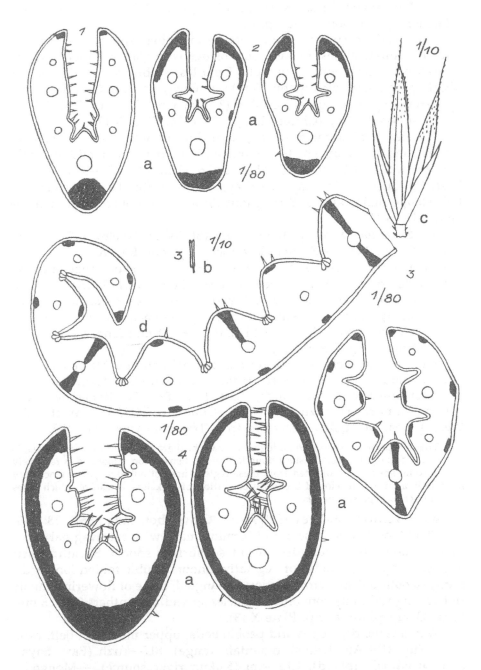

Plate X. 1—*Festuca tschujensis*; 2—*F. valesiaca*; 3—*F. venusta*; 4—*F. wolgensis*:
a—cross sections of leaf blades of vegetative shoots, b—stamens, c—spikelet,
d—cross section of leaf blade of flowering shoot.

42. **F. valesiaca** Gaudin s. str. 1811 in Agrost. Helv. 1: 242. —*F. sulcata* (Hackel) Nyman nom. illeg., p. p.

Compactly cespitose plant with intravaginal shoot regeneration. Sheath laciniate almost to base. Leaf blades with glaucous bloom, (0.35) 0.4–0.6 mm diam, with 5 vascular bundles, scabrous outside, with 3 highly prominent veins within; 3 (5) sclerenchyma strands under lower epidermis. Spikelets green, 4 (5)–6 (7.5) m long. Lemma (2.8) 3.2–4.2 (4.7) mm long. Altay (Chibit village), $2n = 14$. Plate X (2).

Usually dominant in arid steppes, arid meadows, rocky slopes, limestones, sand and pebble beds, steppified pine groves, up to midmountain belt. **West. Sib.**: KU, OM, KE, AL—Ba, Go. **Cen. Sib.**: KR—Kha, Ve, TU. **East. Sib.**: IR—An (in Angara valley), ChI—Shi (Shilka settlement). — Hill steppe plains almost throughout Eurasia. Described from Switzerland. Map 194.

43. **F. valesiaca** subsp. **hypsophila** (St.-Yves) Tzvelev 1971 in Bot. zhurn. 56 (9): 1255. —*F. kryloviana* auct. p. p. non Reverd.

Compactly cespitose plant with intravaginal shoot regeneration. Sheath closed for 1/4 to 1/3 length. Leaf blades with glaucous bloom, (0.3) 0.4–0.6 (0.7) mm diam, with.5 vascular bundles, scabrous outside, with 3 highly prominent veins within; 3 (5) sclerenchyma strands under lower epidermis. Spikelets brownish, (4) 5–6 (7) mm long. Lemma 3.2–4.2 (4.6) mm long. Plate X (2).

Hill steppes, rocky slopes, grasslands, rocks and pebble beds, upper and moderate hill belts, rare. **West. Sib.**: AL—Go. **Cen. Sib.**: TU (Kurtushibinsk mountain range). **East. Sib.**: BU—Se (Romanovka village on Vitim river), Yuzh, ChI—Shi. —Caucasus, Mid. Asia, West. Asia, West. China, Mongolia. Described from Transcaucasus. Map 195.

This subspecies, representing a high-mountain background plant in Middle Asia, is found in Siberia at the northeastern boundary of its distribution range and is relatively rare. Several specimens collected from these regions are not entirely typical since they have only a faintly developed brownish coloration of spikelets, a feature which takes them closer to type subspecies.

44. **F. venusta** St.-Yves 1929 in Izv. Glavn. bot. sada SSSR, 28: 383.

Plant loosely cespitose or not forming mats, with extravaginal shoot regeneration, with short decumbent subsurface shoots. Sheath closed almost to top. Leaf blades of vegetative shoots double folded longitudinally, smooth outside, cross section 4–6-angled; those of flowering shoots flat or longitudinally convolute. Panicles spreading. Anthers 0.5–0.8 mm long. Ovary pilose at tip. Plate X (3).

Wet grasslands, slopes and pebble beds; upper mountain belt, rare. **East. Sib.**: IR—An (Udinsk mountain range), BU—Yuzh (East. Sayan and Dzhidinsk upland), ChI—Shi (Kukun river source). —Mongolia. Described from Khangai. Map 196.

45. **F. viviparoidea** Krajina et Pavlick, 1984 in Canad. Journ. Bot. 62 (11): 2454. —*F. ovina* f. *vivipara* (L.) Reverd. —*F. supina* var. *elata* f. *vivipara*

(L.) Reverd. —*F. brachyphylla* f. *vivipara* A. Skvortsov nom. nud. —*F. vivipara* auct. non Smith.

Compactly cespitose plant with intravaginal shoot regeneration. Sheath closed for 1/2 to 3/4 length. Leaf blades with distinct, isolated and rarely some connective sclerenchyma strands under lower epidermis. All spikelets viviparous. Lemma of lower florets glabrous on back or covered with spinules only in upper third, awnless. Plate VIII (2).

Grasslands, rubble and loamy, usually rather eroded slopes, rocks, talus and pebble beds; different types of tundras and bald-peak hill belt. **West. Sib.**: TYU—Yam. **Cen. Sib.**: KR—Ta, Pu, Ve (East. Sayan). **East. Sib.**: BU—Se (Baikal'sk mountain range), Yuzh (East. Sayan). — Circumpolar distribution. Map 190.

46. **F. wolgensis** P. Smirnov 1945 in Byul. Mosk. o-va isp. prir. otd. biol. 50 (1–2): 100. —*F. wolgensis* subsp. *arietina* (Klokov) Tzvelev.

Compactly cespitose plant with intravaginal shoot regeneration. Sheath rather pilose, laciniate almost to base. Leaf blades cylindrical or flattened-cylindrical, (0.3) 0.5–0.8 mm diam, scabrous outside, with 3–5 veins within, covered with 0.06–0.1 mm long trichomes; continuous sclerenchyma, rarely interrupted only opposite midrib by enlarged cover, under lower epidermis. Lemma (4.2) 4.5–6 mm long, with 1–2 mm long awn; palea covered with spinules throughout length of keels. Plate X (4).

Limestones. **West. Sib.**: KU (Iset' river valley). —Europe. Described from Zhiguli.

Siberian record of this calciphilous East. European species is entirely isolated from its main distribution range.

34. Lolium L.

1. Loosely cespitose plant with few sharply scabrous stems .... 2.
+ Compactly cespitose plant with glabrous stems .... 2. *L. perenne*.
2. Spikelets not declinate from spike rachis .... 3.
+ Spikelets fairly strongly declinate from spike rachis ....
.... 1. *L. multiflorum*.
3. Spikes 10–20 cm long, with rather scabrous rachis. Glumes longer than or as long as spikelets .... 4. *L. temulentum*.
+ Spikes 6–13 cm long, with smooth rachis. Glumes shorter than spikelet .... 3. *L. remotum*.

1. **L. multiflorum** Lam. 1778, Fl. Fr. 3: 621.
Biennial or annual loosely cespitose plant. Stems 25–100 cm high, scabrous under spikes. Leaves flat, 1–1.5 mm broad, densely covered with spinules on upper surface and along edges. Spikes straight, rarely

slightly inclined, up to 20–22 cm long. Spike rachis flexuous, scabrous along ribs. Spikelets with 5–15 florets, declinate from main rachis of inflorescence. Glumes narrowly lanceolate, sparse, usually shorter than or half as long as spikelet. Lemma narrowly lanceolate, with 5 nerves; upper florets with slender, 2–3 mm long awn; lower ones usually awnless. Anthers 2.5–3.2 mm long.

Wastelands, roadsides. **West. Sib.**: NO (Akademgorodka environs). —Europe, Caucasus, Mid. Asia, Mediterranean, West. Asia, Far East. Described from France.

2. **L. perenne** L. 1753, Sp. Pl.: 83.

Compactly cespitose perennial 15–65 cm tall. Leaves linear, glabrous or covered with sparse spinules on upper surface. Spikes straight or slightly inclined. Spike rachis flexuous. Spikelets with 5–10 florets, laterally compressed, arranged singly in recesses of main rachis of inflorescence. Glumes sparse, rather obtuse, covering spikelets only from one side; shorter than spikelet. Lemma glabrous, awnless, 5–7 mm long, stiff, with barely distinct midrib; palea with concave keel, ciliolate along margin. Anthers 3–4 mm long.

Meadows, roadsides, fields. **West. Sib.**: TYU—Tb (Tobol'sk environs), OM (Ekaterinovsk village). —Europe, Caucasus, Mid. Asia, Mediterranean, West. Asia. Described from Europe.

3. **L. remotum** Schrank 1789 in Baier Fl. 1: 382.

Loosely cespitose annual. Stems up to 80 cm tall, few. Spikes erect, 6–13 cm long, with smooth and slender rachis. Spikelets minute, 8–14 mm long, oblong-elliptical, with 5–7 florets. Glumes linear-lanceolate, obtuse, shorter than spikelet, 6–10 mm long. Lemma awnless, rarely with very short awn. Anthers 1.8–2.2 mm long.

In flax, rarely other plantations. **Cen. Sib.**: KR—Ve (Tolstikhinsk village). **East. Sib.**: IR—An (introduced). —Europe, Mediterranean, West. Asia, Far East. Described from Bavaria (FRG).

4. **L. temulentum** L. 1753, Sp. Pl.: 83.

Stems few, occasionally solitary, 40–80 cm tall, sharply scabrous under spikes. Leaves scabrous on upper surface glabrous, on lower surface glassy. Spikes straight, 10–20 cm long, with sharply scabrous rachis. Spikelets with 5–9 florets. Glumes narrowly lanceolate, acuminate, longer than or as long as spikelet. Lemma obtuse, stiff, with 2 sharply developed lateral nerves arising from lower third; midrib extends gradually from upper third into rather scabrous, 7–12 mm long awn. Palea lanceolate, shortly ciliate along margin. Anthers 2–2.5 mm long.

Various types of plantations. **West. Sib.**: TYU—Tb (Tobol'sk environs). **East. Sib.**: KR—Pu (Noril'sk environs). —Europe, Caucasus, Mid. Asia, Mediterranean, West. Asia, the Himalayas, Far East. Described from Europe.

35. Poa L.

1   Plant not viviparous; if so, high-altitude or arctic plant ....
.... 2.

\+   Viviparous steppe plant .... 9. *P. bulbosa.*

2.   Anthers not longer than 1 mm .... 3.

\+   Anthers longer than 1 mm .... 7.

3.   Panicle branches scabrous .... 4.

\+   Panicle branches smooth .... 5.

4.   Anthers 0.4–0.6 mm long .... 31. *P. pseudoabbreviata.*

\+   Anthers 0.6–1 mm long .... 30. *P. abbreviata.*

5.   Callus of lemma glabrous .... 6.

\+   Callus of lemma with tuft of long flexuous hairs ....
.... 26. *P. paucispicula.*

6.   Panicles pyramidal .... 27. *P. annua.*

\+   Panicles narrow, compressed .... 31. *P. pseudoabbreviata.*

7.   Lemma somewhat pubescent at least along keel and marginal
nerves .... 8.

\+   Lemma totally devoid of pubescence along keel and nerves
(sometimes covered with short appressed spinules) ....
.... 25. *P. sibirica.*

8.   Callus of lemma without tuft of long flexuous hairs .... 9.

\+   Callus of lemma with tuft of long flexuous hairs .... 20.

9.   Stems thick, more than 2.5 mm diam in midportion, flattened
at base. Leaf cross section with chain of colorless parenchyma
in 1 or 2 rows; mechanical tissue large-celled .... 10.

\+   Stems less thick, less than 2.5 mm diam, not flattened at base.
Leaf cross section without chain of colorless parenchyma; me-
chanical tissue fine-celled .... 13.

10.   Lemma densely covered with long flexuous hairs along keel
and lateral nerves. Panicles narrow, compressed .... 11.

\+   Lemma glabrous or covered with long sparse hairs only in
lower part. Panicles with long firm branches arising almost at
right angle .... 12.

11.   Lemma usually with 5 nerves; palea with very short spinules
along keels .... 3. *P. tibetica.*

\+   Lemma usually with 7 nerves; palea with long spinules along
keels .... 4. *P. trautvetteri.*

12.   Lemma covered with long sparse hairs along lower part of
keel .... 1. *P. schischkinii.*

\+   Lemma practically without pubescence ...... 2. *P. subfastigiata.*

13 (9).   Gray-green plant with stiff convoluted leaves .... 14.

\+   Green plant with fairly tender flat leaves .... 15.

14.   Lemma glabrous between nerves. High-altitude plant of altay
.... 39. *P. litvinoviana.*

+ Lemma pubescent between nerves. Hill-steppe plant of altay, Khakass and Tuva .... 42. *P. reverdattoi.*
15. Panicle branches wholly smooth .... 16.
+ Panicle branches scabrous due to spinules .... 17.
16. Florets usually dark-colored, lemma densely pubescent along and often between nerves .... 6. *P. alpina.*
+ Florets usually green, lemma pubescent only along keel and intermediate nerves; glabrous between nerves ..... 28. *P. supina.*
17. Upper node toward base (below 1/6 of stem height) .... 18.
+ Upper node between 1/3 and 1/2 of stem height ....
.... 38. *P. krylovii.*
18. Plant with dense spicate panicles; appressed branches not longer than 2 cm. Spikelets minute, many, densely arranged on panicle branches. Leaves stiff, longitudinally folded ..... 19.
+ Plant with lax, elongated panicles; branches long, recurved. Spikelets large, 5-8 mm long; 1–3 (5) on each branch. Leaves soft, long, flat .... 32. *P. altaica.*
19. Lemma glabrous between nerves, spikelets usually minute, not exceeding 4.5 mm .... 34. *P. attenuata.*
+ Lemma between nerves pubescent, spikelets usually longer than 4.5 mm .... 42. *P. reverdattoi.*
20 (8). Sheath of upper leaves closed for more than half; keel in leaf cross section highly elongated .... 21.
+ Sheath of upper leaves closed for less than half; keel in cross-section of leaf weakly elongated .... 22.
21. Stems with lower sheaths orbicular; panicle branches smooth, sometimes with diffuse spinules. East Siberian plant ....
.... 23. *P. ircutica.*
+ Stems with lower sheaths flat, subancipital; panicles broad, branches densely covered with spinules .... 24. *P. remota.*
22. Panicle branches glabrous or covered with sparse spinules. Lemma usually densely pubescent along keel and nerves ....
.... 23.
+ Panicle branches densely covered with spinules. Pubescence of lemma usually sparse .... 38.
23. Panicle branches wholly smooth .... 24.
+ Panicle branches with sparse spinules .... 35.
24. Lemma pubescent between nerves .... 25.
+ Lemma glabrous between nerves .... 30.
25. Spikelets large, 5–8 mm long .... 26.
+ Spikelets smaller, up to 5 mm long .... 28.
26. Rather low, 5–40 cm plant. Pubescence along keel and nerves of lemma appressed; in spread-out form, hairs shorter than lemma breadth .... 27.

+ Tall, 40–60 cm plant. Pubescence along keel and nerves of lemma very long; in spread-out form, hairs longer than lemma breadth .... 19. *P. sublanata.*

27. Plant 5–20 cm tall. Stems usually single .... 11. *P. mariae.*

+ Plant 15–40 cm tall. Stems numerous, aggregated into lax mats .... 17. *P. smirnovii.*

28. Cespitose plant .... 29.

+ Plant with slender, long, decumbent rhizome .... 8. *P. arctica.*

29. Mats lax, panicles dense, compressed ....10. *P. lindbergii.*

+ Mats compact, panicles broad, pyramidal. .... 21. *P. tolmatchewii.*

30. Stems robust, flattened at base, leaves rather thick, gray-green .... 31.

+ Stems of usual thickness, not flattened at base; leaves slender, green .... 33.

31. Plant tall, 25–45 (60) cm .... 32.

+ Plant low, up to 25 cm .... 15. *P. sabulosa.*

32. Leaves with profuse waxy bloom. Stems strongly flattened at base. East. Siberian plant .... 13. *P. pruinosa.*

+ Leaves without waxy bloom, stems stiff, weakly flattened at base. Plants of Atlay, Khakass, Tuva .... 20. *P. tianschanica.*

33. High-altitude plant with compressed ovate panicles .... 34.

+ Meadow-forest or marsh plant with spreading pyramidal panicles .... 35.

34. Lemma glabrous along intermediate nerves, pubescence along keel not more than 1/2 lemma length .... 5. *P. alpigena.*

+ Lemma pubescent along intermediate nerves, pubescence along keel reaching 2/3 lemma length ..... 18. *P. sobolevskiana.*

35 (23). Leaves narrow, convolute, filiform or setaceous .... 36.

+ Leaves broad, flat, sometimes longitudinally folded; vegetative shoots single .... 37.

36. Vegetative shoots aggregated into clusters .... 7. *P. angustifolia.*

+ Vegetative shoots single .... 22. *P. turfosa.*

37. Forest plant with narrow, weakly pubescent lemma and small tuft of long flexuous hairs on callus .... 16. *P. sergievskajae.*

+ Meadow-forest plant with broad, profusely pubescent lemma and dense tuft of flexuous hairs on callus .... 12. *P. pratensis.*

38 (22). Stems terete .... 39.

+ Stems strongly flattened in lower part; whole plant highly scabrous .... 47. *P. compressa.*

39. Lemma with strongly developed nerves .... 40.

+ Lemma with weakly developed nerves .... 42.

40. Ligules of upper cauline leaves obtuse, 2–3 mm long, glumes lanceolate; keel cross section close to an equilateral triangle

.... 41.

+   Ligules of upper cauline leaves acuminate, 3–5 mm long, glumes curved; keel cross section elongated, midrib displaced toward upper leaf surface .... 29. *P. trivialis.*

41.   Sheath of lower cauline leaves scabrous due to short decurved spinules; panicle branches densely covered with spinules .... 14. *P. raduliformis.*

+   Sheath of lower cauline leaves puberulent or scabrous due to nonacuminate tubercular spinules; panicle branches covered with sparse spinules .... 16. *P. sergievskajae.*

42.   Upper node in upper half of stem .... 43.

+   Upper node in lower half of stem .... 44.

43.   Ligules of cauline leaves not longer than 1 mm, rachilla puberulent .... 40. *P. nemoralis.*

+   Ligules of cauline leaves longer than 2 mm, rachilla glabrous .... 41. *P. palustris.*

44.   Upper node between 1/2–1/6 of stem height .... 45.

+   Upper node at base of stem, below 1/6 its height .... 49.

45.   Ligules of cauline leaves longer than 2 mm .... 46.

+   Ligules of cauline leaves shorter than 2 mm .... 48.

46.   Lemma glabrous between nerves .... 47.

+   Lemma pubescent between nerves .... 36. *P. filiculmis.*

47.   Stems 25–60 cm tall, covered with tubercles under panicle .... 44. *P. stepposa.*

+   Stems 25–90 cm tall, covered with sharp, recurved spinules under panicle .... 45. *P. transbaicalica.*

48.   Ligules up to 1 mm long, rachilla pubescent .... 43. *P. skvortzovii.*

+   Ligules up to 1.5 mm long, rachilla glabrous .... 46. *P. urssulensis.*

49 (44).   Lemma glabrous between nerves .... 50.

+   Lemma pubescent between nerves .... 33. *P. argunensis.*

50.   High-altitude plant with somewhat thickened stems and loose, elongated panicles. Spikelets 5–6 mm long, 1–3 (5) each on long branches .... 51.

+   Plant with tender stems and dense spicate panicles. Spikelets up to 4.5 mm long .... 52.

51.   Leaves green, soft .... 32. *P. altaica.*

+   Leaves grayish-green, stiff, longitudinally folded .... 37. *P. glauca.*

52.   High-altitude plant, stems less than 25 cm long, predominantly smooth or covered with tubercles .... 34. *P. attenuata.*

+   Plant of steppes and lower hill belt; stems more than 25–30 cm long, predominantly covered with spinules .... 35. *P. botryoides.*

SUBGENUS **Arctopoa** (Griseb.) Probat.

Plant with strong decumbent subsurface shoots and fleshy leaf blades. Mesophyll of leaf blades with stellate lobed cells. Vascular bundles contain hypoderma.

Section A p h y d r i s (Griseb.) Tzvelev

Callus of lemma covered with hairs only at base of keel and marginal nerves or totally glabrous; sheath of cauline leaves closed for 1/4 to 1/6 from base.

1. **P. schischkinii** Tzvelev 1974 in Novosti sist. vyssh. rast. 11: 32. — *P. subfastigiata* var. *hirsutiflora* Krylov.

Robust plant 25–40 (60) cm high, with thick decumbent rhizome and smooth stems. Leaves fleshy, grayish, 2.5–5 mm broad. Ligules of upper leaves 1.5–2 mm long. Panicles broadly spreading, with long thick branches, frequently arising perpendicular to rachis. Spikelets 5–7 mm long, sometimes dark-colored. Lemma with poorly distinct nerves, keel and marginal nerves covered with long and slender hairs at base, callus glabrous. Plate XI (1).

Saline wet meadows. **West. Sib.**: AL—Go (Chuisk steppe—class. hab.—and others). **Cen. Sib.**: KR—Kha, TU. **East. Sib.**: BU—Yuzh (Borgoisk steppe, Borgoi river valley). —Endemic. Map 211.

2. **P. subfastigiata** Trin. 1829 in Ledeb., Fl. Alt. 1: 96.

Robust plant with thick rhizome and smooth stems 30–80 cm high. Leaves 2.5–5 mm broad, stiff, fleshy gray-green, sometimes longitudinally folded. Ligules of upper cauline leaves 2–2.5 mm broad. Panicles broadly spreading, with long thick branches, frequently arising perpendicular to rachis. Spikelets 5–10 mm long. Lemma without pubescence along nerves and keel. Callus glabrous. Plate XI (2).

Saline wet meadows. **Cen. Sib.**: TU. **East. Sib.** IR—An, Pr, BU—Se, Yuzh (Uda river—class. hab.—and others), ChI—Ka, Shi, YAK—Ol, Vi. —Far East, Mongolia, NE China. Map 216.

3. **P. tibetica** Munro ex Stapf 1896 in Hooker fil., Fl. Brit. Ind. 7: 399.

Plant with thick decumbent rhizome forming subsurface shoots. Stems stout, smooth, 20–90 cm tall. Leaves stiff, grayish-green, more often longitudinally folded, 2–5 mm broad. Ligules of upper leaves 0.5–2 mm long. Panicles usually narrow, dense. Spikelets with 3–6 florets, 4–8 mm long. Lemma covered with long hairs along keel and marginal nerves, frequently running onto callus. Palea with very short spinules along keels. Altay (southeastern part), $2n = 42$. Plate XI (3).

Desert-steppe valleys of hill rivers, solonetz meadows, fringes of saline marshes. **West. Sib.**: AL—Go. **Cen. Sib.**: KR—Kha, Ve, TU. — Mid. Asia, the Himalayas, West. China, Tibet, Mongolia. Described from Tibet. Map 218.

170

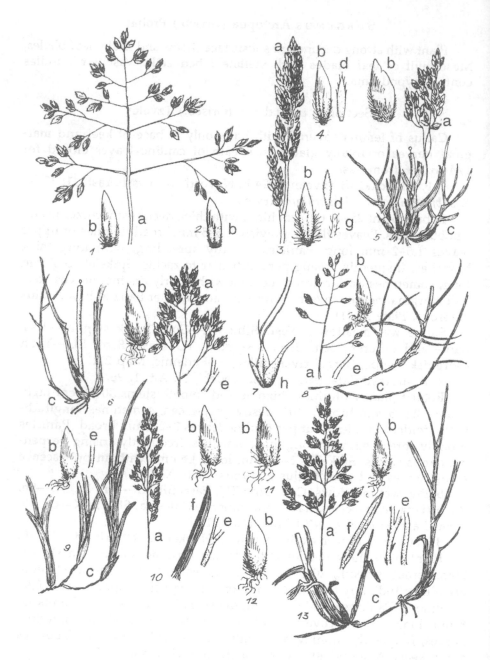

Plate XI. 1—*Poa schischkinii*; 2—*P. subfastigiata*; 3—*P. tibetica*; 4—*P. trautvetteri*; 5—*P. alpina*; 6—*P. mariae*; 7—*P. bulbosa*; 8—*P. arctica*; 9—*P. pruinosa*; 10—*P. alpigena*; 11—*P. sobolevskiana*; 12—*P. sabulosa*; 13—*P. pratensis*: a—panicles, b—lemma, c—vegetative shoots, d—palea, e—surface of panicle branches, f—leaf, g—rachilla, h—viviparous spikelet.

4. **P. trautvetteri** Tzvelev 1964 in Novosti sist. vyssh. rast.: 122.

Plant with thick decumbent rhizome. Stems thick, firm, 35–60 cm tall. Leaves 2–5 mm broad, flat or longitudinally folded, stiff. Ligules of upper leaves 0.5–2 mm long. Panicles weakly spreading, narrow. Spikelets 4–8 mm long. Lemma covered with long hairs along keel and lateral nerves, running onto callus. Palea covered with fairly long spinules along keels. Arctic Siberia (Lena river lower course), $2n = 42$. Plate XI (4).

Meadows and shoals along riverbeds. **East. Sib.**: YAK—Ar (Lena river lower course—class. hab.).—Endemic.

S U B G E N U S **Poa**

Plant cespitose or with decumbent subsurface shoots. Mesophyll of orbicular cells; hypoderma lacking in vascular bundles.

Section P o a

Perennial, with decumbent subsurface shoots, occasionally forming mat. Sheath of cauline leaves usually closed for 1/4–1/3 from base; viviparous forms known. Lemma often pilose, with tuft of long flexuous hairs on callus.

5. **P. alpigena** (Fries) Lindman 1918, Svensk. Fanerogamfl.: 91. —*P. pratensis* var. *alpigena* Fries. —*P. pruinosa* auct. non Korotky. —*P. pratensis* subsp. *alpigena* (Blytt) Hiit.

Low, 10–45 cm plant with short arcuately ascending shoots. Stems smooth. Leaves 0.6–2 mm broad, usually longitudinally folded. Ligules of upper leaves up to 2 mm long. Panicles oblong, somewhat compressed; branches smooth. Spikelets up to 5 mm long, often dark-colored. Lemma profusely pubescent along keel and marginal nerves but for not more than half length. Callus with well-developed tuft of long flexuous hairs. Viviparous forms known. Yakutia, $2n = 56$ (Shandrin river), 60 (Pokhodsk settlement). Plate XI (10).

Tundra and meadow habitats, banks of brooks and rubble slopes. **West. Sib.**: TYU—Yam, Khm, KE, AL—Go. **Cen. Sib.**: KR—Ta, Pu, TU. **East. Sib.**: YAK—Ar, Ol, Vi, Al, Yan, Ko. —Holarctic region. Described from Scandinavian hills.

6. **P. alpina** L. 1753, Sp. Pl.: 67.

Stems 7–35 cm tall, aggregated into mat. Leaves 2–5 mm broad, flat, soft, green. Ligules of upper leaves up to 3.5 mm long. Panicles dense, ovate, with smooth branches. Spikelets 5–8 (10) mm long, often speckled. Lemma puberulent along keel and nerves, usually pubescent between nerves in lower part. Callus glabrous. Viviparous form known. East. Sayan (Tubota river upper course), $2n = 42$. Plate XI (5).

Alpine grasslands, rubble and rocky slopes. **West. Sib.**: TYU—Yam, KE, AL—Go. **Cen. Sib.**: KR—Ta, Pu, TU. **East. Sib.**: BU—Se (Barguzinsk mountain range), Yuzh. —Holarctic region. Described from Europe.

7. **P. angustifolia** L. 1753, Sp. Pl.: 67. —*P. pratensis* var. *angustifolia* (L.) Sm. — *P. pratensis* subsp. *angustifolia* (L.) Arcang.

Plant with slender rhizome; vegetative shoots aggregated into small clusters and enveloped by sheaths of year-old dead leaves. Stems 30–50 (70) cm high, smooth. Leaves narrow, 1.5 mm broad, radical, sometimes even cauline leaves convolute, stiff. Ligules of upper leaves 0.6–2 mm long. Panicles usually slightly narrowed, with weakly scabrous branches. Spikelets 4–5 mm long. Lemma profusely pubescent along keel and marginal nerves. Tufts of long, flexuous hairs on callus well developed. Plate XII (2).

Arid exposed sites, steppe and dry valley meadows, steppified slopes. **West. Sib.**: TYU—Khm, Tb, KU, OM, TO, NO, KE, AL—Ba, Go. **Cen. Sib.**: KR—Kha, Ve, TU. **East. Sib.**: IR—An, Pr, BU—Se, Yuzh, ChI—Ka, Shi, YAK—Ar, Ol, Vi, Al. —Eurasia. Described from Europe.

8. **P. arctica** R. Br. 1824 in Parry, Journ. Voy. N.W Pass. Suppl. App.: 288.

Plant with slender decumbent rhizome, vegetative shoots arcuate. Stems 10–25 (40) cm tall, smooth. Leaves tender, 1–2 (3) mm broad, flat or longitudinally folded. Ligules 1–1.5 mm long. Panicles pyramidal, spreading, with slender smooth branches. Spikelets 4–5 mm long, often dark-colored. Lemma pubescent with soft hairs along nerves, and usually between them. Tuft of long flexuous hairs on callus poorly developed. Viviparous form known. Putoran plateau (Bogatyr' Lake) and Stanovoi upland (Severo-Muisk mountain range), $2n = 56$; Putoran plateau (Baselak Lake) and Yakutia (Ulakhan-Tas mountain range), $2n = 70$. Plate XI (8).

Rubble banks of brooks in tundras. **West. Sib.**: TYU—Yam. **Cen. Sib.**: KR—Ta, Pu, TU (Mongun-Taiga mountain range). **East. Sib.**: BU—Se (Barguzinsk mountain range, Levyi Mal. Sarankhur and Tompuda rivers upper courses), Yuzh (Khamar-Daban mountain range, foot of Margasansk mud volcano; sources of Stolbik river; Khan-Ula hill), ChI—Shi (Sokhondo bald peak, Agutsakan river upper course), YAK—Ar, Ol, Vi, Al, Yan, Ko. —Circumpolar distribution. Described from Canada (Melville Island).

9. **P. bulbosa** L. 1753, Sp. Pl.: 70. —*P. bulbosa* subsp. *vivipara* (Koeler) Arcang.

Stems 10–30 cm tall, slender, smooth, leaf sheaths of shoots thickened and enlarged at base, forming bulbous thickenings. Leaves narrow, filiform, grayish-green. Ligules 2–3.5 mm long. Panicles dense, ovate, with short scabrous branches. Spikelets most often viviparous. Least altered lemma along keel and nerves pubescent. Callus with small tuft of long flexuous hairs. Plate XI (7).

Steppes, sand-pebble riverbeds, rocky and rubble slopes. **West. Sib.**: AL—Ba. —Europe, Caucasus, Mid. Asia, Mediterranean, West. Asia, West. China (Junggar). Described from France.

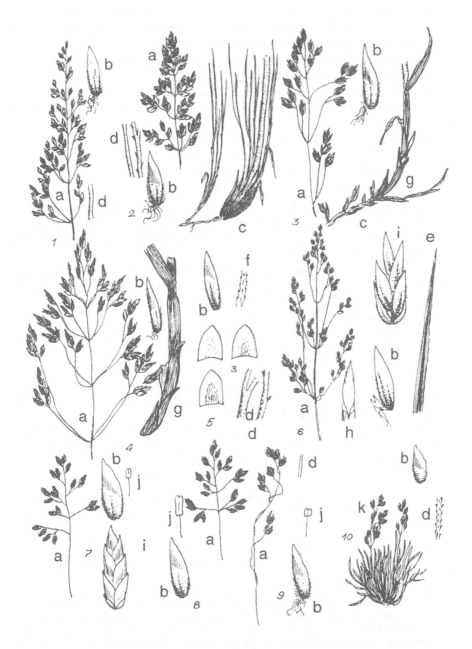

Plate XII. 1—*Poa sergievskajae;* 2—*P. angustifolia;* 3—*P. ircutica;* 4—*P. remota;* 5—*P. sibirica;* 6—*P. trivialis;* 7—*P. annua;* 8—*P. supina;* 9—*P. paucispicula;* 10—*P. pseudoabbreviata:* a—panicles, b—lemma, c—vegetative shoots, d—surface of panicle branches, e—shape of leaf tip, f—rachilla, g—shoot base, h—pubescence of ligule, i—spikelets, j—stamens, k—general view.

10. **P. lindbergii** Tzvelev 1974 in Novosti sist. vyssh. rast.: 27.

Plant with slender smooth stems 10–35 cm tall, forming fairly dense mats. Leaf blades usually longitudinally folded, 1–2 mm broad. Ligules of upper leaves 0.5–2 mm long. Panicle ovate, fairly dense, branches smooth. Spikelets usually viviparous.

Rocky, rubble and clayey slopes. **East. Sib.**: YAK—Ar (Lena river lower course, Tuora-Sis mountain range). —Norway. Described from Scandinavian hills.

11. **P. mariae** Reverd. 1933 in Sist. zam. Gerb. Tomsk. un-ta 2–3: 2. —*P. smirnovii* subsp. *mariae* (Roshev.) Tzvelev.

Rhizomatous plant 10–25 cm tall, stems smooth. Leaves soft, 2–4 mm broad, longitudinally folded or flat. Ligules of upper leaves 2–4 mm long. Panicles pyramidal, 2–7 cm long, with smooth branches. Spikelets 5–8 mm long. Lemma with distinctly developed nerves, pubescent in lower part along and between nerves. Callus with tuft of long flexuous hairs. Plate XI (6).

Alpine grasslands, rubble and rocky slopes. **West. Sib.**. AL—Go. **Cen. Sib.**: KR—Kha (Surla river basin in Abakan river basin—class. hab.—and others), TU. —Endemic. Map 203.

12. **P. pratensis** L. 1753, Sp. Pl.: 67.

Plant with decumbent rhizome. Stems smooth, 20–100 (120) cm tall. Leaves 1.5–4 mm broad, green, usually flat. Ligules of upper leaves 1.5–3 mm long. Panicles pyramidal, branches usually with sparse spinules or smooth. Spikelets 3–5 mm long, sometimes dark-colored. Lemma pubescent along keel and marginal nerves. Tuft of long flexuous hairs on callus well developed. Plate XI (13).

Meadows, pebbled riverbeds, forest glades, sparse forests and scrubs. **West. Sib.**: all regions. **Cen. Sib.**: KR—Pu, Tn, Kha, Ve, TU. **East. Sib.**: IR—An, Pr, BU—Se, Yuzh, ChI—Ka, YAK—Al, Yan, Ko. —Eurasia and North America. Described from Europe.

P. pratensis is one of the most complex and polymorphous species of genus *Poa*. In Siberia, some tendencies toward morphological changes have been noted. Throughout the Siberian territory, leaf blades and sheaths of many specimens of *P. pratensis* are densely pubescent, more frequently in lower leaves. This feature correlates with no other features; found in extremely diverse forms of plants throughout the region under study and independent of habitat. Density of pubescence of lemma varies and is most profuse in plants of exposed habitats.

In West. Siberia, comparatively short specimens, 20–45 cm tall, predominate, with flat leaves up to 4 mm broad, scabrous branches of fairly narrow panicle, spikelets 4.5–5 mm long and profuse pubescence of lemma. In Baikal and Transbaikal regions, this form is almost absent. There, *P. pratensis* is more often characteristic of herbaceous cover. Plants are very large, up to 120 cm tall, length and breadth of leaves increase, panicles become very lax; their slender and long branches, frequently wholly smooth and slightly nutant, bear only 2–4 spikelets each. Spikelets larger than in West. Siberian form, with lemma usually very narrow and nerves more prominent. Pubescence of lemma very scant. Similar plants are found in hill forests of Altay and Kuznets Alatau. These specimens are similar and some come very close to *P. raduliformis* described by N.S. Probatova from the Far East which, according to her, is a hybrid of *P. pratensis* and *P. remota*.

13. **P. pruinosa** Korotky 1915 in Feddes Repert. (Beihn.) 13: 291. —
P. tianschanica auct.

Rhizomatous plant, covered with waxy bloom. Stems 20–80 cm tall, smooth. Leaves 1–2 mm broad, usually longitudinally folded; arcuate, stiff in vegetative shoots. Ligules of upper leaves 2–3 mm long. Panicles oblong, rarely pyramidal, branches usually smooth. Spikelets 3–4.5 (5) mm long, lemma pubescent along keel and marginal nerves. Tuft of long flexuous hairs on callus well developed. Plate XI (9).

Banks of rivers and lakes, sand-pebble riverbeds, occasionally on coastal grasslands. **East. Sib.**: IR—Pr, BU—Se, Yuzh (Konstantinovka village, environs of Eravninsk lakes—class. hab.—and others). —Endemic. Map 209.

14. **P. raduliformis** Probat. 1971 in Novosti sist. vyssh. rast. 8: 25.

Rhizomatous plant 50–120 cm tall. Stems highly foliated, scabrous due to several minute spinules. Leaves green, flat, 4–6 mm broad, also scabrous. Ligules of upper leaves 2–3 mm long. Panicles somewhat compressed, branches sharply scabrous. Spikelets 4–5 mm long. Lemma narrow, lanceolate, weakly pubescent along keel and marginal nerves. Tuft of long flexuous hairs on callus poorly developed.

Among shrubs, forest glades, grasslands, banks of water reservoirs. **East. Sib.**: BU—Yuzh, YAK—Ko. —Far East, NE china. Described from Amur region.

15. **P. sabulosa** (Roshev.) Roshev. 1934 in Fl. SSSR 2: 394. —*P. pratensis* subsp. *sabulosa* (Roshev.) Tzvelev.

Rhizomatous plant, stems smooth, 10–25 cm tall. Leaves rather thick, grayish-green, radical leaves arcuate; 1.5–2 mm broad. Ligules 1.5–2 mm long. Panicles pyramidal, branches smooth. Spikelets 3.5–4.5 mm long. Lemma pubescent in lower part along keel and marginal nerves. Tuft of long flexuous hairs on callus well developed . Plate XI (12).

Banks of water reservoirs, saline short-grass meadows. **West. Sib.**: Al—Go (Chuisk steppe, Kosh-Agach settlement; Kosh-Agach region, Tashtu-Khol' river midcourse). **Cen. Sib.**: KR—Kha (Shira Lake), TU. **East. Sib.**: BU—Se (NE bank of Lake Baikal, Sosnovka river), Yuzh (Posol'sk village on Baikal—class. hab.—and others), YAK—Vi (environs of Vilyuisk). —Endemic. Map 208.

16. **P. sergievskajae** Probat. 1971 in Novosti sist. vyssh. rast. 8: 27. —*P. pratensis* subsp. *sergievskajae* (Probat.) Tzvelev.

Rhizomatous plant. Stems 45–120 cm tall, scabrous in lower part due to minute spinules. Leaves green, soft, usually flat, 2–4 mm broad. Ligules of upper leaves 2–2.5 mm long. Panicles narrow, oblong, with smooth or somewhat scabrous branches. Spikelets 3–5 mm long. Lemma lanceolate, with well-developed nerves. Pubescence poor along keel and marginal nerves. Tuft of long flexuous hairs on callus barely developed. Plate XII (1).

Mixed and birch forests, forest glades, scrubs. **East. Sib.**: IR—An, BU—Se (northern Bank of Baikal, Kholodnoe village, Yuzhno-Muisk mountain range, Muya village), ChI—Shi. —Far East. Described from Amur region.

17. **P. smirnovii** Roshev. 1929 in Izv. Glavn. bot. sada SSSR, 28: 424.

Rhizomatous plant, usually forming several decumbent shoots, aggregated into loose mat. Stems smooth, 25–40 cm tall. Leaves tender, flat or longitudinally folded, 3–4 mm broad. Panicles pyramidal, 5–8 cm long, with smooth branches. Lemma with well-developed nerves; pubescent along nerves and in lower part between nerves. Tuft of long flexuous hairs on callus well developed. East. Sayan (Sagan-Shuluta river), $2n = 42$.

Alpine grasslands, rocky and rubble slopes. **West. Sib.**: KE (Kuznets Alatau, Tigiri-Tish town). **Cen. Sib.**: TU. **East. Sib.**: IR—An, Pr, BU—Se, Yuzh (Irkut river in Tunkinsk bald peaks—class. hab.—and others), ChI—Ka, Shi. —Mongolia. Map 213.

18. **P. sobolevskiana** Gudoschn. 1963 in Izv. SO AN SSSR 4, ser. biol.-med. 1: 73. —*P. pratensis* subsp. *sobolevskiana* (Gudoschn.) Tzvelev.

Rhizomatous plant 10–45 cm tall, with arcuately ascending vegetative shoots. Leaves 1–3 mm broad, usually longitudinally folded. Ligules of upper leaves 2–3 mm long. Panicles oblong, with smooth branches. Spikelets up to 5–6 mm long; lemma pubescent not only along marginal, but also along intermediate nerves; pubescence along keel extending 2/3 lemma length. Tuft of long flexuous hairs on callus well developed. Plate XI (11).

Wet coastal meadows, sand-pebble beds. **Cen. Sib.**: TU (Ierikhol' Lake in Bai-Taiginsk region—class. hab.—and others). **East. Sib.**: BU—Se (Baikal'sk mountain range, Khibelen river sources). —Endemic. Map 212.

19. **P. sublanata** Reverd. 1934 in Sist. zam. Gerb. Tomsk. un-ta 2–3: 1.

Plant 40–70 cm tall, with long decumbent rhizome. Leaves tender, flat, sometimes longitudinally folded, 2–3 mm broad. Ligules of upper leaves up to 5 mm long, acuminate. Panicles lax, pyramidal, 10–17 cm long, with slender smooth branches. Spikelets 6–8 mm long. Lemma pubescent up to 2/3 along midrib and up to 1/2 along marginal nerves; pubescence between nerves short. Tuft of long flexuous hairs on callus very well developed.

Coastal sand-pebble beds. **West. Sib.**: TYU—Yam. **Cen. Sib.**: KR—Ta (Leont'evsk Island in Yenisey lower course—class. hab.—and others), Pu. **East. Sib.**: YAK—Ar, Vi, Ko. —Far East, North America (Alaska). Map 214.

20. **P. tianschanica** (Regel) Hackel ex O. Fedtsch. 1903 in Tr. Peterb. bot. sada 21: 441.

Plant with thick decumbent rhizome and strong smooth stems 20–80 (90) cm tall. Leaves somewhat fleshy, gray-green, 2–5 mm broad, usually longitudinally folded, radical leaves arcuate. Panicles pyramidal, 5–15 cm long, with smooth branches. Spikelets 2–5 mm long. Lemma with distinctly developed nerves; pubescent along keel and marginal nerves. Tuft of long flexuous hairs on callus well developed.

Banks of water reservoirs, coastal pebble beds, weakly saline and steppified meadows. **West. Sib.**: AL—Go. **Cen. Sib.**: KR—Kha (Charkovsk nomad village), TU. —Mid. and Central Asia. Described from Tien Shan. Map 217.

21. **P. tolmatchewii** Roshev. 1932 in Izv. Bot. sada AN SSSR 30: 299.

Compactly cespitose plant with smooth stems 10–25 cm tall. Leaves green, tender, 1–2 mm broad, flat or longitudinally folded. Ligules up to 1.5 mm long. Panicles pyramidal, with smooth branches. Spikelets up to 5 mm long; lemma pubescent along nerves and in lower part between nerves. Tuft of long flexuous hairs on callus poorly developed. Plate XIII (4).

Rocky and rubble slopes. **Cen. Sib.**: KR—Ta (Yamu-Tarida river lower course—class. hab.—and others), Pu. **East. Sib.**: BU—Se (Barguzinsk mountain range, Tompuda river upper course), YAK—Ar. —European Arctic, Far East. Map 215.

22. **P. turfosa** Litv. 1922 in Spisok rast. Gerb. russk. fl. 8: 135.

Plant with slender decumbent rhizome, vegetative shoots single. Stems smooth, 30–60 (80) cm tall. Leaves narrow, longitudinally folded, 1–2 mm broad. Ligules of upper leaves 2–4 mm long. Panicles pyramidal, 4–10 cm long, branches usually smooth, sometimes with sparse spinules. Spikelets up to 5.5 mm long; lemma profusely pubescent along keel and marginal nerves. Tuft of long flexuous hairs on callus well developed.

Peat swamps. **West Sib.**: TYU—Tb, OM, TO. **Cen. Sib.**: KR—Tn. **East. Sib.**: ChI—Shi (Gazimura river valley, Yamkun settlement). —European USSR. Described from Vladimir region.

Section H o m a l o p o a Dum.

Perennial plant, usually with short decumbent subsurface shoots forming loose mats. Stems often scabrous due to sharp spinules. Sheath of cauline leaves closed for 2/3 to 5/6 from base. Palea with fairly long spinules along keels, frequently transformed into hairs in lower part.

23. **P. ircutica** Roshev. 1922 in Bot. mat. (Leningrad) 3: 91.

Plant with thick decumbent rhizome. Stems 20–70 cm tall. Leaves scaled at base, with underdeveloped blades; upper bright green, 3–7 mm broad, flat, long. Panicles lax, spreading, with slender branches. Spikelets 5–8 mm long, lemma with distinctly developed nerves,

Plate XIII. 1–3—*Poa glauca;* 4—*P. tolmatchewii;* 5—*P. sibirica:*
a—lemma.

pubescent with short hairs. Tuft of long flexuous hairs on callus poorly developed or lacking. East. Sayan (Tubota river), $2n = 28$. Plate XII (3).

Grasslands, rocky and rubble slopes, sparse forests. **East. Sib.**: IR—An, BU—Yuzh (Khamar-Daban mountain range, Bol. Bystraya and Slyudyanka river sources—class. hab.—and others). —Endemic. Map 201.

24. **P. remota** Forsell. 1807 Skr. Linn. Inst. Upsal. 1: 1.

Plant with decumbent rhizome. Stems up to 100–120 cm tall, scabrous, highly flattened in lower part. Leaves soft, flat, 5–10 mm broad. Ligules of upper leaves 2–3 mm long. Panicles lax, pyramidal, with slender branches, densely covered with minute spinules. Spikelets 5–6 (7) mm long, lanceolate. Lemma narrow, with distinctly developed nerves; pubescence poor along keel and marginal nerves. Tuft of long flexuous hairs on callus poorly developed. Plate XII (4).

Forests, forest glades, scrubs. **West. Sib.**: TYU—Khm, Tb, TO, KE, AL—Ba, Go. **Cen. Sib.**: KR—Kha, Ve. **East. Sib.**: IR—An (Taishetak region, Patrikha settlement), BU—Yuzh (Khamar-Daban mountain range, Bryansk settlement; Mishikhi river bank). —Europe, Caucasus, Mid. Asia, West. China. Described from Sweden.

Section M a c r o p o a F. Hermann ex Tzvelev

Perennial, usually loosely cespitose plant, frequently forming decumbent subsurface shoots. Sheath of cauline leaves closed for 2/3 to 5/6 length from base. Spikelets not viviparous. Lemma glabrous, with glabrous callus; palea with spinules along keels.

25. **P. sibirica** Roshev. 1912 in Izv. Peterb. bot. sada 12: 121.

Plant with decumbent rhizome. Stems tall, smooth, 40–120 cm high. Leaves bright green, flat, 2–4 mm broad. Ligules of upper leaves 2–2.5 mm long. Panicles elongated-pyramidal, with scabrous branches. Spikelets 4–6 mm long, frequently dark-colored; lemma with distinctly developed nerves, totally without pubescence, sometimes with short-appressed spinules. Callus invariably glabrous. Putoran plateau (Talnakh settlement) and Nor. Yakutia (Dvaibalakh river), $2n = 14$. Plate XII (5).

Forests, meadows, dwarf arctic birch groves. **West. Sib.**: TYU—Yam, Khm, Tb, TO, NO, KE, AL—Ba, Go. **Cen. Sib.**: KR—Pu, Tn, Kha, Ve (Balai village—class. hab.—and others), TU. **East. Sib.**: IR—An, Pr, YAK—Ar, Ol, Al, Yan. —Ural region, Mid. Asia, Far East, China, Mongolia, Korean peninsula.

Among Siberian specimens, length of spikelets, width of leaf blade, and thickness of stem at lower nodes vary greatly and these characteristics are manifest in highly diverse combinations. Among all Siberian specimens, ligules of leaves are pubescent to various extent; based on the degree of their pubescence, all the plants studied can be classed into three groups with transitions.

In the first group, pubescence of ligules is very weakly developed and often uneven, with very short hairs of length only slightly more than thickness; some rather long hairs, not longer than 0.06 mm, are seen only in the midportion. Members of this group are common throughout Siberia.

The second group is characterized by uniform and dense pubescence of ligules with 0.03–0.06 mm long hairs. Members of this group are found throughout Siberia but very rarely in Transbaikal.

The third group covers plants with ligules uniformly pubescent with hairs longer than 0.06 mm, sometimes reaching 0.25 mm. Such plants are found only in Altay and Kuznets Alatau.

Section O r e i n o s (Ascherson et Graebner) Jirásek

Perennial, forming dense mats, usually with short subsurface shoots. Sheath of cauline leaves closed for 1/3 to 1/2 length from base. Panicle branches smooth or almost so. Lemma pubescent in lower part; tuft of long flexuous hairs on callus poorly developed. Anthers 0.6–1.3 mm long.

26. **P. paucispicula** Schribner et Merr. 1910 Grass Alaska: 69.

Plant with slender filiform rhizome. Stems smooth, 10–30 (40) cm tall, frequently forming loose mats. Leaves narrow, 1–2 mm broad, soft, usually longitudinally folded. Ligules 1–1.5 mm long. Panicles 3–8 cm long, with few spikelets, lax, with slender smooth trichoid branches. Spikelets 3–5 mm long. Lemma with short pubescence along keel and marginal nerves. Tuft of long flexuous hairs on callus weakly developed. Anthers not longer than 1 mm. Putoran plateau (Talnakh settlement), $2n = 28$. Plate XII (9).

Moss-covered tundras and turf-covered spots among debris. **Cen. Sib.**: KR—Ta, Pu. **East. Sib.**: IR—Pr, BU—Se, ChI—Ka (Kodar mountain range, Apsat river), YAK—Ar. —Far East, North America (Alaska, Cordilleras). Described from Alaska. Map 205.

Section O c h l o p o a (Ascherson et Graebner) Jirásek

Perennial or annual cespitose plant without decumbent subsurface shoots; sheath of cauline leaves closed for 1/2 to 2/3 length from base; spikelets not viviparous, lemma rather pubescent, usually with glabrous callus.

27. **P. annua** L. 1753, Sp. Pl.: 68.

Cespitose plant. Stems smooth, 10–30 cm tall. Leaves soft, green, usually flat, 2–3 (4) mm broad. Ligules 1.5–2 mm long. Panicles 2.5–6 cm long, pyramidal, with smooth branches. Lemma with well-developed nerves; nerves and keel poorly pubescent or altogether glabrous in lower part. Callus invariably glabrous. Anthers not longer than 1 mm. Plate XII (7).

Roadsides, grasslands, coastal pebble beds. **West. Sib.**: all regions. **Cen. Sib.**: KR—Tn, Kha, Ve, TU. —Nearly cosmopolitan. Described from Europe.

28. **P. supina** Schrader 1806, Fl. Germ. 1: 289.

Biennial cespitose plant. Stems smooth, 10–30 cm tall. Leaves flat, 2–4 mm broad. Ligules of upper leaves up to 2 mm long. Panicles 3–6 cm long, pyramidal, with smooth branches. Lemma with distinctly developed nerves; keel and marginal nerves pubescent with short appressed hairs in lower part. Callus invariably glabrous. Anthers longer than 1 mm. East. Siberia (northeastern Baikal coast), $2n = 14$. Plate XII (8).

Roadsides, inhabited sites, grasslands and pebble beds, river banks. **West. Sib.**: TYU—Tb, OM, TO, NO, KE. **Cen. Sib.**: KR—Kha, Ve. **East. Sib.**: IR—An, Pr, BU—Se, Yuzh, ChI—Ka, Shi, YAK—Vi, Al. —Atlantic and Mid Europe, Scandinavia, Urals, Mid. Asia, Nor. Iran, Mediterranean, West. Himalayas, West. China, Mongolia. Described from Austria.

Section C o e n o p o a Hyl.

Perennial scabrous plant with or without short decumbent subsurface shoots, usually forming loose mats. Sheath of cauline leaves closed for 1/4 to 1/3 length from base, lemma pubescent along nerves, with tuft of long flexuous hairs on callus.

29. **P. trivialis** L. 1753, Sp. Pl.: 67.

Plant with decumbent rhizome. Stems often scabrous, sometimes smooth, up to 90 cm tall. Leaves soft, flat, 2–4 mm broad. Ligules of upper leaves acuminate, 3–5 mm long. Panicles elongated-pyramidal, spreading, with highly scabrous branches. Spikelets 3–4.5 mm long. Lemma with well-developed nerves, pubescent with short appressed hairs along keel and marginal nerves in lower part. Tuft of long flexuous hairs on callus poorly developed. Plate XII (6).

Fringes of dark coniferous and mixed forests, less often in birch groves. **West. Sib.**: TYU—Tb, OM, TO, NO, KE, AL—Ba, Go. **Cen. Sib.**: KR—Ve, TU. **East. Sib.**: BU—Yuzh (Kabansk region, Tankhoi station). — Nearly cosmopolitan. Described from Europe.

Section A b b r e v i a t a e Nannf. ex Tzvelev

Perennial compactly cespitose plant without decumbent subsurface shoots. Sheath of cauline leaves closed for 1/6 to 1/4 length from base. Lemma pubescent in lower part. Tuft of long flexuous hairs on callus poorly developed, sometimes totally lacking. Palea with spinules along keels, sometimes transformed below into short hairs.

30. **P. abbreviata** R. Br. 1824 in Parry, Jour. Voy. N.W. Pass. Suppl. App.: 287.

Compactly cespitose plant. Stems smooth, 5–25 cm long. Leaves narrow, up to 3 mm broad, longitudinally folded. Ligules of upper leaves 1–2 mm long. Panicles narrow and dense, 1.5–6 cm long, branches short, smooth or weakly scabrous. Spikelets 3–4 mm long. Lemma with weakly developed nerves; keel and marginal nerves covered with short appressed hairs in lower part. Callus glabrous. Anthers 0.6–1 mm long.

Rocky and rubble tundras, coastal pebble beds. **Cen. Sib.**: KR—Ta, Pu. **East. Sib.**: YAK—Ar (Lena river lower courses, Tuora-Sis mountain range, Sokuidakh hill). —Circumpolar distribution. Described from Canada. Map 197.

31. **P. pseudoabbreviata** Roshev. 1922 in Bot. mat. (Leningrad) 3: 91.

Compactly cespitose plant. Stems glabrous, smooth, 5–25 (30) cm tall. Leaves 1–2 mm broad, longitudinally folded. Ligules of upper leaves 1–2 mm long. Panicles 2–5 cm long, spreading, branches usually slender and long, scabrous due to sharp spinules. Spikelets 3–4 mm long. Lemma with poorly developed nerves, puberulent along keel and marginal nerves in lower part, callus glabrous. Anthers 0.4–0.6 mm long. Plate XII (10).

Rocky tundras and pebble beds. **Cen. Sib.**: KR—Pu. **East. Sib.**: BU— Se (Yuzhno-Muisk mountain range, Ukuolkit river, Mui tributary), Yuzh (Tunkinsk mountain range, Gargansk pass—class. hab.), ChI—Ka (Kodar mountain range, Srednii Sakukan river, Mramornaya hill; Verkh. Sakukan river upper course), YAK—Ar. —Novaya Zemlya, Far East (Chukchi), North America (Alaska). Map 206.

Section S t e n o p o a Dum.

Perennial, usually compactly cespitose plant, without decumbent subsurface shoots. Sheath of cauline leaves closed for less than 1/5 length from base. Panicle branches invariably covered with minute spinules. Spikelets never viviparous. Lemma pubescent in lower part, rarely glabrous. Tuft of long flexuous hairs on callus poorly developed, sometimes totally lacking.

32. **P. altaica** Trin. 1829 in Ledeb. Fl. Alt. 1: 97.

Compactly cespitose plant 10–40 cm tall, stems usually smooth. Leaves green, soft, 1–3 mm broad, flat or longitudinally folded. Ligules of upper leaves 0.5–3 mm long. Panicles elongated, branches sharply scabrous. Spikelets 4–8 mm long, frequently dark-colored. Rachilla glabrous or pubescent. Lemma with barely distinct nerves; appressed-pilose along keel and marginal nerves and sometimes even between nerves in lower part. Tuft of long flexuous hairs on callus poorly developed or totally lacking. East. Sayan (Tubota river upper course), $2n = 42$. Plate XIV.

Rocky and rubble slopes. **West. Sib.**: AL—Go. **Cen. Sib.**: KR—Ve, TU. **East. Sib.**: IR—An, Pr, BU—Se, Yuzh, ChI—Ka, Shi, YAK—Al. — Endemic. Described from Altay.

33. **P. argunensis** Roshev. 1934 in Fl. SSSR 2: 404.

Compactly cespitose plant. Stems 10–45 cm tall, gray-green, stiff, scabrous. Upper node at stem base, below 1/6 of stem. Leaves 0.5–1.5 mm broad, short, longitudinally folded, stiff. Ligules of upper leaves (1) 2–3 mm long. Panicles short, compressed, dense, branches densely covered with sharp spinules. Spikelets 3–4.5 mm long. Lemma pubescent along keel and nerves in lower part; pubescence between nerves very short, appressed. Tuft of long flexuous hairs on callus poorly developed, sometimes totally lacking. Plate XV (10).

Steppes, steppified rocky and rubble slopes. **West. Sib.**: AL—Go. **Cen. Sib.**: KR—Kha, Ve, TU. **East. Sib.**: BU—Se (Lake Baikal, Bol. Ushkanii Island), Yuzh, ChI—Shi (Lake Orabuduk, valley of Sukhoi Urulyungui river—class. hab.—and others). —Mongolia. Map 198.

34. **P. attenuata** Trin. 1835 in Mém. Sav. Étr. Pétersb. 2: 527.

Compactly cespitose gray-green rigid plant 5–25 cm tall. Stems usually smooth or covered with tubercles. Upper node at base below 1/6 of stem. Leaves 0.5–1.5 mm broad, short, stiff, usually longitudinally folded. Ligules 1.5–2 mm long. Panicles narrow, dense, branches densely covered with spinules. Spikelets 3–4.5 mm long. Lemma pubescent along keel and marginal nerves in lower part, glabrous between nerves. Tuft of long flexuous hairs on callus poorly developed, totally lacking in *P. attenuata* var. *dahurica* (Trin.) Griseb. Gornyi Altay (Kosh-Agach settlement), $2n = 28$. Plate XV (9).

High-altitude steppes and rocky slopes. **West. Sib.**: AL—Go. **Cen. Sib.**: KR—Kha, Ve, TU. **East. Sib.**: IR—An, Pr, BU—Se, Yuzh, ChI—Ka, Shi, YAK—Al. —Mid. and Cen. Asia, Mongolia. Described from Altay. Map 199.

35. **P. botryoides** (Trin. ex Griseb.) Roshev. 1934 in Fl. SSSR 2: 403. —*P. attenuata* var. *botryoides* Trin. ex Griseb.

Compactly cespitose gray-green rigid plant 25–60 cm tall. Stems usually covered with spinules, upper node at base of stem below 1/6, occasionally between 1/3 and 1/6 stem length. Leaves 0.5–1.5 mm broad, short, stiff, usually longitudinally folded. Ligules 1.5–2 mm long. Panicles dense, spicate, rarely elongated-pyramidal, with sharply scabrous branches. Spikelets 3–5 mm long, pubescent along keel and marginal nerves in lower part. Tuft of long flexuous hairs on callus poorly developed, sometimes totally lacking. Yakutia (Byttangai river), $2n = 28$. Plate XV (8).

Steppes of lower mountain belt. **West. Sib.**: AL—Go. **Cen. Sib.**: KR—Kha, Ve, TU. **East. Sib.**: IR—An, Pr, BU—Se, Yuzh, ChI—Ka, Shi, YAK—Ar, Al, Yan. —Far East, Mongolia, NE China. Described from Transbaikal.

184

Plate XIV. *Poa altaica*:
a—lemma.

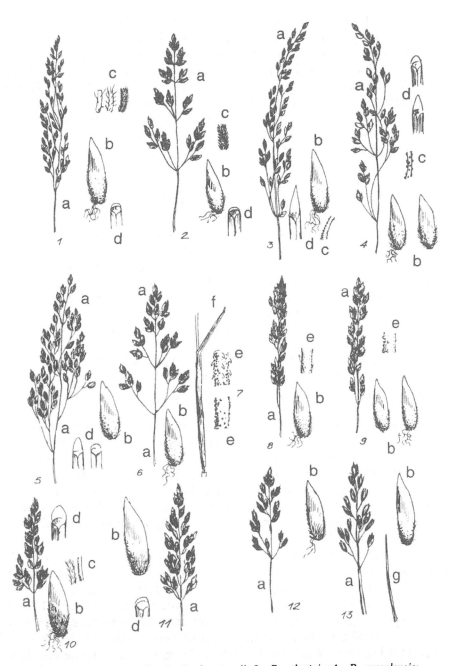

Plate XV. 1—*Poa nemoralis*; 2—*P. skvortzovii*; 3—*P. palustris*; 4—*P. urssulensis*;
5—*P. krylovii*; 6—*P. stepposa*; 7—*P. transbaicalica*; 8—*P. botryoides*; 9—*P. attenuata*;
10—*P. argunensis*; 11—*P. reverdattoi*; 12—*P. filiculmis*; 13—*P. litvinoviana*:
a—panicles, b—lemma, c—surface of rachilla, d—ligules, e—surface of stem under
panicle, f—upper leaf with sheath, g—leaf.

36. **P. filiculmis** Roshev. 1949 in Bot. mat. (Leningrad) 6: 29.

Compactly cespitose grayish-green plant 20–40 (50) cm tall. Stems slender, scabrous under panicle, upper node at stem base. Leaves short, stiff, 0.5–1.5 mm broad. Ligules of uppermost leaves 1.5–2.5 mm long. Panicles elongated-pyramidal, rarely dense, spicate, with sharply scabrous branches. Spikelets 3.5–5 mm long. Lemma appressed-pilose in lower part along and between nerves. Tuft of long flexuous hairs on callus poorly developed, occasionally lacking. Plate XV (12).

Sand-pebble riverbeds, rocks and rocky slopes. **Cen. Sib.**: KR—Ta (northern part of Central Siberian plateau, Khatanga river basin). **East Sib.**: YAK—Ar, Yan (Indigirka river midcourse). —Far East. Described from Far East (Anadyr' river valley). Map 200.

37. **P. glauca** Vahl 1790, Fl. Dan. 17: 3.

Compactly cespitose glaucous green plant 5-40 (60) cm tall. Leaves short, stiff, 1–2 mm broad, usually longitudinally folded. Ligules 0.5–2 mm long. Panicles with few spikelets, spreading, usually with ascending sharply scabrous branches, rarely compressed. Spikelets 3–8 mm long, rachilla glabrous or pubescent. Lemma puberulent along keel and marginal nerves, sometimes between them. Tuft of long flexuous hairs on callus well developed. Yakutia (Chersk settlement), $2n = 56$. Plate XIII (1–3).

Rocky and rubble slopes, in tundra. **West. Sib.**: TYU—Yam, AL—Go. **Cen. Sib.**: KR—Ta, Pu, Ve, TU. **East. Sib.**: IR—Pr, BU—Se, Yuzh, ChI—Ka, YAK—Ar, Vi, Yan. —Circumpolar distribution. Described from Norway.

38. **P. krylovii** Reverd. 1936 in Sist. zam. Gerb. Tomsk. un-ta 8: 3.

Cespitose plant 35–60 cm tall, stems scabrous under inflorescence. Upper node between 1/3 and 1/2 of stem. Leaves 1–2 mm broad, flat or longitudinally folded. Ligules of upper leaves 0.5–1.5 mm long. Panicles 7–15 cm long, lax, elongated-pyramidal, with long scabrous branches. Spikelets 3–5 mm long, rachilla glabrous or covered with tubercles. Lemma with indistinct nerves; pubescent along keel and marginal nerves. Callus glabrous. Plate XV (5).

Steppified slopes up to midmountain belt. **West. Sib.**: KE, AL—Go. **Cen. Sib.**: KR—Kha (Oznachennaya village—class. hab.—and others), Ve, TU. **East. Sib.**: IR—An, BU—Yuzh. —Endemic. Map 202.

39. **P. litvinoviana** Ovcz. 1932 in Izv. Tadzh. bazy AN SSSR 1 (1): 22.

Cespitose gray-green plant 5–25 cm tall; stems thickened in lower part. Leaves short, stiff, usually longitudinally folded, 1–2 mm broad. Ligules 0.5–2 mm long. Panicles 2–8 cm long, with few spikelets, dense, or rarely pyramidal, with sharply scabrous branches. Spikelets 3–8 mm long. Lemma pubescent along keel and marginal nerves, sometimes between them. Callus glabrous. Plate XV (13).

Rocky and rubble slopes of high hills. **West. Sib.**: AL—Go. **Cen. Sib.**: TU. —Mid. Asia. Described from Mid. Asia (Zeravshan mountain range). Map 204.

40. **P. nemoralis** L. 1753, Sp. Pl.: 69.

Loosely cespitose plant with smooth stems 25–80 (90) cm tall. Upper node in upper half of stem. Leaves flat, soft, 2–4 mm broad, leaf blades usually longer than sheaths. Ligules of upper leaves in typical specimens not longer than 1 mm. Panicles lax, elongated-pyramidal, with long scabrous branches. Spikelets 3–5 mm long, rachilla usually pubescent. Lemma pubescent along keel and marginal nerves. Tuft of long flexuous hairs on callus poorly developed. Plate XV (1).

Larch and mixed forests, their fringes, scrubs. **West. Sib.**: TYU—Tb, KU, OM, TO, NO, KE, AL—Ba, Go. **Cen. Sib.**: KR—Tn, Ve, TU. **East. Sib.**: BU—Yuzh. —Holarctic. Described from Europe.

Typical specimens of this species from Siberia vary in a broader range than those from the midpart of the European USSR. In typical specimens of *P. nemoralis* from the European broad-leaved forests, ligules are barely visible and may generally be absent. Among Siberian specimens, however, ligules are generally longer, more often up to 1 mm long, sometimes even slightly more. Extent of pubescence and length of hairs on rachilla vary greatly. At the same time, plants are found in Siberia which differ from typical forms and combine in themselves the characteristics of *P. nemoralis* and *P. pratensis*. These are mesomorphous plants with glabrous rachilla and short ligule and plants with pubescent rachilla and long ligule.

41. **P. palustris** L. 1759, Syst. Pl. 10: 874.

Loosely cespitose plant with smooth stems 25–90 (120) cm tall. Upper node in upper half of stem. Leaves soft, flat, 1–4 mm broad, leaf blades usually longer than sheaths. Ligules of upper leaves 2–3 mm long. Panicles spreading, with long scabrous branches. Spikelets 3–5 mm long. Rachilla glabrous. Lemma pubescent along keel and marginal nerves. Tuft of long flexuous hairs on callus poorly developed. Plate XV (3).

Meadows, forests, and scrubs. **West. Sib., Cen. Sib., East. Sib.**: all regions. —Holarctic. Described from Europe.

42. **P. reverdattoi** Roshev. 1934 in Fl. SSSR 2: 407.

Stiff, cespitose plant with thick stout scabrous stems 20–50 cm tall. Upper nodes at stem base. Leaves gray-green, stiff, usually setaceous, 1–2 mm broad. Ligules of upper leaves 1–2 mm long. Panicles narrow, compressed, usually not more than 2 cm broad. Spikelets large, 5–6 mm long, rachilla pubescent. Lemma pubescent along keel and intermediate nerves; pubescence very short between nerves. Callus glabrous. Plate XV (11).

Low and moderate altitude steppes, rubble slopes. **West. Sib.**: AL—Go (Ininsk steppe). **Cen. Sib.**: KR—Kha (Abakansk steppe—class. hab.— and others), TU. —Endemic. Map 210.

43. **P. skvortzovii** Probat. 1972 in Novosti sist. vyssh. rast. 9: 72.

Compactly cespitose stiff plant 40–60 (80) cm tall. Stems scabrous, thick; upper node in lower third of stem. Leaves short, stiff, 1–2 mm broad, more often longitudinally folded. Sheath of upper leaves longer than leaf blades. Ligules not longer than 1 mm. Panicles elongated-pyramidal, rarely narrow, compressed, with sharply scabrous branches. Spikelets 3.5–5 mm long. Rachilla pubescent. Lemma pubescent along keel and marginal nerves. Tuft of long flexuous hairs on callus well developed. Plate XV (2).

Steppified rocky and rubble slopes, in steppes. **East. Sib.**: IR—An, BU—Se, Yuzh, ChI—Shi, YAK—Al. —Far East, NE China, Korean peninsula. Described from NE China. Map 207.

44. **P. stepposa** (Krylov) Roshev. 1934 in Fl. SSSR 2: 401. —*P. attenuata* var. *stepposa* Krylov. —*P. versicolor* subsp. *stepposa* (Krylov) Tzvelev.

Compactly cespitose stiff plant 25–60 cm tall. Stems glabrous, almost smooth under panicle, without sharp spinules. Upper node in lower third of stem. Leaves short, stiff, 1–2 mm broad, more often longitudinally folded. Sheath of upper leaf longer than blade. Ligules 1.5–2 mm long. Panicles elongated-pyramidal with ascending sharply scabrous branches, occasionally narrow, spicate. Spikelets 3.5–5 mm long. Rachilla glabrous. Lemma pubescent along keel and marginal nerves. Tuft of long flexuous hairs on callus well developed. Plate XV (6).

Steppified rocky and rubble slopes, in steppes. Throughout Siberia. —European USSR, Caucasus, Mid. Asia, West. China (Junggar), Mongolia. Described from Altay (between Bortuldag and Kair).

45. **P. transbaicalica** Roshev. 1929 in Izv. Glavn. bot. sada SSSR 2: 382.

Compactly cespitose stiff plant 50–90 cm tall. Stems glabrous, sharply scabrous, densely covered with sharp recurved spinules under panicle. Leaves short, stiff, longitudinally folded or flat, 1.5 mm broad. Panicles elongated-pyramidal, with sharply scabrous branches. Spikelets up to 5–6 mm long, 3–5-flowered. Rachilla glabrous. Lemma pubescent along keel and marginal nerves in lower part. Tuft of long flexuous hairs on callus well developed. Plae XV (7).

Steppified rocky and rubble slopes. **East. Sib.**: BU—Se, Yuzh, ChI—Ka, Shi (Nercha river, 200 versts (1 verst = 1.067 km) above Nerchinsk—class. hab.—and others). —Endemic.

46. **P. urssulensis** Trin. 1835, Mém. Sav. Étr. Pétersb. 2: 527. —*P. nemoralis* var. *urssulensis* Krylov.

Compactly cespitose plant 25–30 cm tall. Stems glabrous, smooth or covered with tubercles. Upper node in lower half of stem. Leaves 1–2 mm broad, flat or longitudinally folded, leaf blades usually shorter than sheath. Ligules of upper leaves not longer than 1.5 mm. Panicles elongated-pyramidal, with sharply scabrous branches turned upward. Spikelets 3.5–5 mm long. Rachilla glabrous. Lemma pubescent along

Spikelets 3.5–5 mm long. Rachilla glabrous. Lemma pubescent along keel and marginal nerves. Tuft of long flexuous hairs on callus well developed. Plate XV (4).

Steppified rocky and rubble slopes. **West. Sib.**: KE, AL—Go (Ursul river—class. hab.—and others). **Cen. Sib.**: KR—Kha, Ve, TU. **East. Sib.**: BU—Se, ChI—Shi. —Nor. Urals, Mid. Asia, West. China, Mongolia, the Himalayas. Map 221.

Section T i c h o p o a Ascherson et Graebner

Perennial, often with decumbent subsurface shoots. Sheath of cauline leaves closed for less than 1/5 length from base. Stems flattened throughout length; whole plant scabrous due to minute spinules.

47. **P. compressa** L. 1753, Sp. Pl.: 69.

Plant with long decumbent rhizome. Stems 20–80 cm tall, strongly flattened together with sheaths. Leaves gray-green, flat, 1.5–3 mm broad. Ligules of upper leaves 0.5–3 mm long. Panicles rather oblong, 3–10 cm long, with scabrous branches. Spikelets 4–5 mm long. Lemma with indistinctly visible nerves, pubescent along keel and marginal nerves in lower part. Tuft of long flexuous hairs on callus poorly developed.

Only as introduced species east of Urals. **East. Sib.**: IR—An (environs of Tulunsk experimental station). —Europe, Caucasus, North America. Described from Europe.

36. Eremopoa Roshev.

1. **E. altaica** (Trin.) Roshev. 1934 in Fl. SSSR 2: 431. —*Nephelochloa persica* (Trin.) Griseb.

Rather low, pale green plant 2–8 (10) cm tall. Stems single, slender, glabrous, with fibrous trichoid roots. Leaves narrowly linear, up to 1 mm broad, longitudinally folded, glabrous, with few hairs only at tip; sheaths smooth, split almost to base. Panicles spreading, up to 3–5 cm long, with few spikelets, with slender scabrous obliquely erect branches. Spikelets 2–4-flowered, 4–5 mm long, lanceolate, slightly laterally compressed. Glumes shorter than spikelet, lanceolate, glabrous, keeled, strongly unequal, upper acuminate, with 3 lateral nerves, scarious along margin, 3.7–3.8 mm long, nearly 3–4 times longer than lower. Lemma lanceolate, with awnlike terminal cusp. Anthers about 0.5 mm long.

Swampy meadows. **West. Sib.**: AL—Go (Chuisk steppe—class. hab.; Kokorya and Kosh-Agach villages). —Mid. Asia, West. Asia.

37. Hyalopoa (Tzvelev) Tzvelev

1. Lemma with or without short (0.2–0.3 mm long) terminal cusp, usually glabrous between nerves .... 2.

+ Lemma with up to 0.8 mm long terminal cusp, densely pilose along and between nerves .... 3. *H. lanatiflora* subsp. *momica*.
2. Sheath of cauline leaves closed for 2/3 length from base; styles 0.3–0.7 mm long. Caryopsis ellipsoidal .... 1. *H. lanatiflora* s. str.
+ Sheath of cauline leaves closed for 3/4 length from base; styles 0.8–1 mm long. Caryopsis globose-ellipsoidal ....
.... 2. *H. lanatiflora* subsp. *ivanoviae*.

1. **H. lanatiflora** (Roshev.) Tzvelev s. str. 1966 in Novosti sist. vyssh. rast.: 32. —*Poa lanatiflora* Roshev. —*Colpodium lanatiflorum* (Roshev.) Tzvelev.

Perennial, 20–30 cm tall, with decumbent rhizome. Stems straight, smooth, with several glabrous green leaves at base. Sheath of cauline leaves closed for 1/3 to 2/3 length. Panicles 4–8 cm long, pyramidal, with long, slender branches turned upward. Spikelets 5–7 mm long, 3–4-flowered. Glumes violet, shorter than lemma, unequal; lower acuminate, with solitary nerve, upper obtuse, with 3 nerves. Lemma rather membranous, keeled, with 3 nerves, obtuse, without cusp, densely pilose along keel and lateral nerves, with corona of straight, sparse, long hairs at base. Anthers 2.7 mm long. Caryopsis ellipsoidal or fusiform, with short 0.3–0.7 mm long style. Yakutia (environs of Tiksi), $2n = 42$.

Rocky slopes, debris and pebble beds of bald peaks. **East. Sib.:** YAK— Ar, Ol (Verkhoyansk region—class. hab.—and others), Vi (Verkhoyansk mountain range), Al, Yan. —Endemic. Map 158.

2. **H. lanatiflora** subsp. **ivanoviae** (Malyschev) Tzvelev 1976, Zlaki SSSR: 485. —*Colpodium ivanoviae* Malyschev.

Perennial, 15–45 cm tall, with decumbent rhizome. Stems straight, smooth. Sheath of cauline leaves closed for 3/4 length. Panicles 4–6 cm long, secund or pyramidal. Spikelets 5–7 mm long, 3–4-flowered. Glumes violet-green, shorter than lemma, unequal; lower acuminate, with solitary nerve; upper obtuse, with 3 nerves and irregularly toothed edges. Lemma rather membranous, keeled, with nerves, without cusp at tip; lanate-pilose for 1/3 length along keel and lateral nerves in lower part. Caryopsis with long style, 0.8–1 mm long. Anthers 2.5–2.7 mm long.

Bald peaks—coastal sand and pebble beds, alpine grasslands, rocky slopes and debris. **East. Sib.:** BU—Se (Yuzhno-Muisk mountain range, Ankundakan river—class. hab.—and others), ChI—Ka (Leprindo Lake). —Endemic. Map 219.

3. **H. lanatiflora** subsp. **momica** (Tzvelev) Tzvelev 1966 in Novosti sist. vyssh. rast.: 33. —*Colpodium lanatiflorum* subsp. *momicum* Tzvelev.

Perennial, 15–30 cm tall. Stems straight, smooth, with many glabrous green leaves at base. Sheaths of cauline leaves closed for 1/3 to 2/3 from base. Panicles 4–8 cm long, pyramidal, with slender branches turned upward. Spikelets 5–6 mm long, 3–4-flowered. Glumes violet, shorter than lemma, unequal; lower acuminate; upper obtuse with 3 nerves and toothed edges. Lemma membranous, keeled, with 0.5–0.8 mm long terminal cusp, lanate-pilose along keel, lateral nerves, and between them. Anthers 2.5–2.7 mm long.

Bald peaks—pebble beds and rock debris. **East. Sib.**: YAK—Yan (Momsk mountain range—class. hab.).

More field investigations are needed for a more accurate understanding of the taxonomic status of subsp. *ivanoviae* and subsp. *momica*.

38. Arctophila (Rupr.) Andersson

1. **A. fulva** (Trin.) Andersson 1852 in Gram. Scand.: 49. —*A. fulva* subsp. *similis* (Rupr.) Tzvelev. —*Colpodium pendulinum* (Laest.) Griseb.

Stems 15–70 cm long, straight, smooth, thick. Rhizome long, funicular, shoots covered with long root fibrils at nodes. Leaves flat, glabrous, with short sheaths, acuminate. Panicles 8–15 cm long, 4–10 cm broad, lax, spreading, with smooth, rather nutant branches. Spikelets 2–7-flowered, 4–7 mm long, at branch ends. Glumes coriaceous-membranous, with rather reddish edges, acute or rather obtuse; upper insignificantly longer than lower. Lemma and palea truncate-terete or straight, with irregularly dentate edges; lemma 3–4 mm long, with membranous edges, obtuse, sometimes with up to 0.8 mm long cusp; palea narrower and shorter than lemma. Callus with corona of straight, sparse hairs. Putoran plateau (Talnakh settlement) and Yakutia (Pokhodsk settlement), $2n = 42$.

Banks of water reservoirs, swamps and swampy meadows. **West. Sib.**: TYU—Yam. **Cen. Sib.**: KR—Ta, Pu. **East. Sib.**: YAK—Ar, Ol, Vi (Segen-Kyuel' settlement), Al (Khandyga settlement), Europe (north), Far East, North America. Described from Alaska. Map 222.

39. Dupontia R. Br.

1. Lemma pubescent. Panicles narrow, uninterrupted, with branches turned upward or appressed to rachis of inflorescence. Spikelets 2–(3)-flowered .... 2.
+ Lemma glabrous. Panicles pyramidal, interrupted, with branches declinate at right angle from rachis of inflorescence. Spikelets 1–(2)-flowered .... 3. *D. psilosantha.*
2. Panicles 3–5 cm long, compressed and dense, spicate, with branches shorter than 5 mm. Plant 6–18 cm tall .... 1. *D. fisheri* s. str.
+ Panicles 4–18 cm long, compressed or rather spreading, often quite dense but lower branches usually more than 5 mm long. Plant 10–50 cm tall .... 2. *D. fisheri* subsp. *pelligera.*

1. **D. fisheri** R. Br. 1824 in Parry, Jour. Voy. N.W. Pass. Suppl. App.: 291.

Perennial, with decumbent subsurface shoots. Stems glabrous, 10–40 cm high. Leaves linear, short, longitudinally folded, stems violet at base. Panicles narrow, compressed, with short (shorter than 5 mm) obliquely erect branches. Spikelets 5–7 mm long, lustrous, golden-violet, 2–4-flowered; upper floret usually underdeveloped. Glumes glabrous, lanceolate, membranous, rather obtuse, rarely acuminate, only slightly longer than spikelet. Lemma and palea equal; lemma broadly lanceolate, acuminate, slightly shorter than glumes, pubescent for 1/3 to 3/4, occasionally completely or just along nerves, with very stiff and long hairs on callus; palea narrowly lanceolate, glabrous, with 2 keels.

Arctic tundras on marshy meadows and along banks of water reservoirs. **West. Sib.**: TYU—Yam. **Cen. Sib.**: KR—Ta, Pu (Syndasko river). **East. Sib.**: YAK. —Arctic Europe (north), North America. Described from Canada. Map 223.

2. **D. fisheri** subsp. **pelligera** (Rupr.) Tzvelev 1973 in Novosti sist. vyssh. rast. 10: 91. —*D. fisheri* auct. non R. Br.

Perennial, with decumbent subsurface shoots. Stems glabrous, 20–50 cm tall. Panicles 4–18 cm long, dense, branches obliquely erect; lowermost longer than 5 mm. Spikelets 5–7 mm long, 2–3-flowered, bright golden. Glumes glabrous, lanceolate, rather obtuse. Lemma broadly lanceolate, rather pilose along nerves, sometimes only in lower part; palea narrowly lanceolate, glabrous. Callus diffusely pilose.

Arctic tundras—on marshy meadows, marshes, banks of water reservoirs. **West. Sib.**: TYU—Yam. **Cen. Sib.**: KR—Ta. **East. Sib.**: YAK—Ar. —Europe (north), North America. Described from Europe (Kanin peninsula). Map 224.

3. **D. psilosantha** Rupr. 1846 in Beítr. Pfl. Russ. Reich. 2: 64. —*D. fisheri* var. *psilosantha* (Rupr.) Trautv.

Perennial, with decumbent subsurface shoots. Stems 10–40 cm tall, glabrous, foliated. Leaves narrowly linear, longitudinally folded, acuminate. Panicles 5–18 cm long, spreading, pyramidal, with straight or weakly flexuous, long, horizontally declinate lateral branches. Spikelets 5–7 mm long, 1–2-flowered. Glumes membranous, narrowly lanceolate, lustrous. Lemma 3–4 (4.5) mm long, glabrous, lustrous, acuminate, considerably (almost by half) shorter than glumes; palea glabrous, acuminate or obtuse-acuminate. Callus shortly bristled, with few sparse hairs.

Swampy meadows, banks of tundra rivers and lakes, near sea coast. **West. Sib.**: TYU—Yam. **Cen. Sib.**: KR—Ta. **East. Sib.**: YAK—Ar. —North. Europe, North America. Described from Europe (Kolgujev Island). Map 225.

40. Catabrosa Beauv.

1. **C. aquatica** (L.) Beauv. 1812, Ess. Agrost.: 97.

Stems 20–60 cm tall, single, geniculate at base, rooting at nodes, with long decumbent rhizome. Leaves glabrous, with sheath split in upper half and blade obtuse. Ligules 3–4 mm long, rounded apically. Panicles 10–20 cm long, 3–7 mm broad, brownish-violet, pyramidal, with horizontal or decurved, weakly scabrous branches. Spikelets brown-violet, 1-, rarely 2-flowered. Glumes whitish-violet, unequal; upper truncate-straight, flabellate, with toothed edges, equal to half spikelet length; lower oval, half size of upper. Lemma and palea identical, with scarious terminal fringe, glabrous, cinnamon, with 3 distinctly developed thick nerves. Anthers 0.7–1.5 mm long.

Banks of water reservoirs, marshy meadows. **West. Sib.:** AL—Go. **Cen. Sib.:** KR—Kha, TU. **East. Sib.:** BU—Yuzh (environs of Kyakhta, Engorboi and Torei villages). —Europe, Urals, Caucasus, Mid. Asia, Mediterranean, West. Asia, West. China, the Himalayas, Mongolia, North America. Described from Europe. Map 220.

41. Paracolpodium (Tzvelev) Tzvelev

1. **P. altaicum** (Trin.) Tzvelev in Bot. zhurn. 1965, 50 (9): 1320. — *Colpodium altaicum* Trin.

Stems 10–40 cm tall, smooth, somewhat flattened, aggregated into loose mats. Leaves green, glabrous, acinaciform, longitudinally folded. Ligules elongated, conical, 2–4 mm long. Panicles 3–11 cm long, 0.7–3 cm broad, rather lax or weakly compressed, with smooth, upward, sometimes slightly nutant branches. Spikelets 1-, rarely 2-flowered, 3.5–4.5 mm long, violet or greenish-violet. Glumes glabrous, lanceolate, acuminate, unequal, with white scarious edges. Lemma broad and white scarious in upper part, oblong, obtuse, nearly as long as glumes, with indistinct lateral nerves, covered with long and soft hairs along keel and nerves. Callus glabrous. Stamens 2, with linear, dark violet anthers up to 2.5 mm long.

Alpine belt on rubble meadow slopes, banks of brooks, near neve basins. **West. Sib.:** AL—Go. **Cen. Sib.:** KR—Kha, Ve, TU. **East. Sib.:** IR—An, BU—Se, Yuzh. —Mid. Asia. Described from NE Kazakhstan (Ul'binsk mountain range). Map 226.

42. Phippsia (Trin.) R. Br.

1. Panicles compressed, up to 3 cm long, branches short, turned upward. Glumes present. Lemma and palea glabrous .... 1. *P. algida.*
+ Panicles rather spreading, 3–8 cm long, branches long, horizontal or inclined downward. Glumes absent. Lemma and palea profusely pubescent along nerves .... 2. *P. concinna.*

1. **P. algida** (Solànder) R. Br. 1824 in Parry, Jour. Voy. N.W. Pass. Suppl. App.: 295.

Extremely small, dwarf cespitose plant 2–7 (10) cm tall. Leaves gla-brous, narrow, usually longitudinally folded, obtuse. Panicles 0.7–3 cm long, compressed, subcylindrical, with short glabrous branches. Spikelets small, about 1 mm long, single-flowered, greenish-yellow or reddish-violet. Glumes very small, sometimes rudimentary, in form of 2 mem-branes. Lemma and palea equal, obtuse, glabrous or weakly pilose along nerves in lower part, with reddish scarious margin along edge. Stamens usually 2. Anthers 0.4 mm long.

River alluvium, pebble beds, banks of water reservoirs, silt bars on sea coast. **West. Sib.**: TYU—Yam. **Cen. Sib.**: KR—Ta, Pu (Kutaramakan Lake). **East. Sib.**: YAK—Ar. —Nor. Europe, Far East, North America (north). Described from Spitsbergen. Map 227.

2. **P. concinna** (Th. Fries) Lindeb. 1898 in Bot. Not. (Lund): 155.

Dwarf loosely cespitose plant 15 (20) cm tall. Leaves flat, smooth, obtuse. Panicles 3–8 cm long, rather spreading, with horizontally declinate, glabrous, verticillate branches. Spikelets single-flowered, about 1.5 mm long, pale- or yellowish-green. Glumes lacking. Lemma oblong, profusely pubescent with white hairs along nerves and between them; palea densely ciliate along keel. Stamens very often solitary. Anthers up to 0.5 mm long. Putoran plateau (Bogatyr' Lake), $2n = 28$.

Arctic region, on exposed clayey slopes, near rivers and lakes, wet spots of loam and rubble, in rocky tundra. **West. Sib.**: TYU—Yam. **Cen. Sib.**: KR—Ta, Pu. **East. Sib.**: YAK—Ar. —Nor. Europe, Urals, Far East. Described from Spitsbergen. Map 228.

43. Puccinellia Parl.

1. Panicle branches smooth or with few diffuse spinules. Plant of northern Siberia .... 2.
+ Panicle branches densely covered with spinules throughout length or almost so; rarely smooth (only in southern regions of Siberia) .... 9.
2. Anthers 0.4–1.0 mm long .... 3.
+ Anthers 1.1–2.0 mm long .... 6.
3. Lemma 2.3–4.2 mm long, with indistinct nerves, densely pu-bescent with long hairs in lower third, ciliate or smooth along margin. Palea with spinules along keels in upper part and hairs in lower part. Panicle branches smooth or with spinules in upper part. Plant of sand hills (baidzharakhs) and coastal swells (located inland) .... 4.
+ Lemma 2.25–2.6 mm long, with distinctly prominent nerves, glabrous or with sparse hairs at base; smooth, without cilia along margin. Palea smooth along keels. Panicle branches wholly smooth. Plants of marine shoals and rocks ....
.... 25. *P. tenella.*

4.   Panicles 7–12 cm long, rather spreading, with smooth branches (only occasionally with isolated spinules under spikelets). Spikelets 5–8-flowered. Lemma 2.3–3.2 mm long .... 5.

+   Panicles 2–8 cm long, compressed, with weakly scabrous branches. Spikelets 3–5-flowered. Lemma 3.2–4.2 mm long, ciliolate along margin. Anthers 0.6–1.0 mm long ....
.... 2. *P. angustata.*

5.   Lemma 2.7–3.2 mm long, truncate-terete, with membranous smooth unequal (but without cilia!) toothed margin. Anthers 0.5–0.8 mm long .... 18. *P. lenensis.*

+   Lemma 2.3–2.5 mm long, obtuse, with lacerated triangular margin covered with cilia. Anthers 0.4–0.5 mm long ....
.... 21. *P. neglecta.*

6.   Plant without procumbent shoots, forming mat. Panicles invariably formed. Lemma densely pilose at base. Panicle branches smooth or weakly scabrous. Anthers 1.1–1.9 mm long
.... 7.

+   Plant with many procumbent surface shoots rooting at nodes. Panicles rarely formed. Lemma glabrous. Panicle branches smooth. Anthers 1.6–2.0 mm long .... 22. *P. phryganodes.*

7.   Panicle branches with spinules under spikelets. Spikelets 4–7-flowered. Lemma with narrow membranous margin .... 8.

+   Panicle branches totally smooth. Spikelets 2–3-flowered. Lemma with broad white membranous margin; palea with sparse hairs beneath along keels. Anthers 1.2–1.7 mm long .....
.... 4. *P. byrrangensis.*

8.   Panicles with erect branches. Spikelets 2–4-flowered. Lemma terete, subacute, puberulent at base, 2.4–3.0 mm long. Palea weakly scabrous above along keels. Anthers 1.1.–1.5 mm long
.... 8. *P. gorodkovii.*

+   Panicles with pendent, branches. Spikelets with 5–7 florets. Lemma oblong-plicate, lanceolate-acuminate, with long hairs beneath along nerves, 3.4–3.8 mm long. Palea with spinules above along keels, transformed beneath into hairs. Anthers 1.3–1.9 mm long .... 13. *P. jenisseiensis.*

9 (2).   Anthers 0.3–0.8 mm long .... 10.

+   Anthers 0.8–2.0 mm long .... 15.

10.   Panicles 10–25 cm long, pyramidal, broadly spreading. Plants of coastal sand and terraces (located inland) .... 11.

+   Panicles 8–12 cm long, oval-oblong, compressed. Plants of coastal shoals .... 29. *P. vaginata.*

11.   Anthers 0.3–0.5 mm long. Lemma 1.5–2.25 mm long, glabrous or with isolated hairs at base. Plant of temperate latitudes ......
.... 12.

+   Anthers 0.6–0.8 mm long. Lemma 2.25–3.4 mm long, with few or abundant hairs at base. Plants of northern Siberia .... 13.

12.   Lemma 1.5–1.8 (1.9) mm long. Anthers 0.3–0.5 mm long. Spikelets 5–8-flowered .... 10. *P. hauptiana.*

+   Lemma 2.0–2.25 mm long. Anthers 0.5 mm long. Spikelets 3–4-flowered .... 11. *P. interior.*

13.   Plant large, spreading, 15–60 cm tall. Palea as long as lemma, with small spinules and hairs along margin .... 14.

+   Plant small, loosely bushy, 10–30 cm tall. Palea distinctly longer than lemma, with long sharp spinules along keels in upper part .... 15. *P. kamtschatica.*

14.   Panicle branches with diffuse spinules. Lemma 2.25–2.5 (2.8) mm long, with few hairs at base. Palea with spinules above along keels. Anthers 0.6–0.7 mm long .... 3. *P. borealis.*

+   Panicle branches densely covered with spinules throughout length. Lemma 2.6–3.4 mm long, with long hairs beneath along ribs. Palea with spinules in upper part of keels and hairs below. Anthers 0.7–0.8 mm long .... 24. *P. sibirica.*

15 (9).   Anthers 0.8–1.3 mm long. Lemma glabrous at base or with sparse hairs .... 16.

+   Anthers (0.9) 1.1–2.0 mm long. Lemma profusely pilose in lower third or totally glabrous; then anthers 1.2–1.8 mm long .... 22.

16.   Stems geniculate at base, greenish glaucescent .... 17.

+   Stems arcuate or straight, green .... 19.

17.   Radical leaves short, narrow, convolute. Plants of high-altitude saline meadows and pebble beds .... 18.

+   Radical leaves long, broad, flat or weakly convolute. Plants of meadows on plains, often semiweedy .... 5. *P. distans.*

18.   Lemma 2.5–2.8 mm long. Anthers 0.7–0.9 mm long .... 9. *P. hackeliana.*

+   Lemma 2.0–2.25 mm long. Anthers 1.2–1.4 mm long .... 14. *P. kalininiae.*

19.   Plant supple, with straight stems. Panicles broadly spreading, pyramidal, short or long. Lemma 1.5–2.25 mm long. Anthers 0.9–1.3 mm long .... 20.

+   Plant rigid, with ascending, arcuate stems. Panicles narrow, secund and long, usually reaching half stem length. Lemma 2.7–3.4 mm long. Anthers 0.8–1.1 mm long .... 23. *P. schischkinii.*

20.   Plant tall, 40–80 cm, with few broad, long radical leaves and large, 11–18 cm long, panicles with highly scabrous slender branches throughout length .... 21.

+   Plant low, 10–40 cm, with many narrow short radical leaves and small, 4–10 cm long panicles with interrupted branches

glabrous, rarely scabrous under spikelets, slender ....
.... 16. *P. kreczetoviczii.*
21.	Lemma 2.0–2.1 mm long, with few hairs at base. Plant of Western Siberia .... 17. *P. kulundensis.*
+	Lemma 1.6–1.9 mm long, glabrous at base. Plant of Eastern Siberia .... 26. *P. tenuiflora.*
22 (15).	Stems thick, 2–3 mm diam, geniculate at base. Leaves broad, 2–4 mm, flat .... 23.
+	Stems thick or very slender, 0.5–1.0 mm diam, straight. Leaves narrow, 1–2 mm, convolute .... 24.
23.	Lemma 2.3–3.0 mm long, profusely pilose in lower third. Palea with spinules in upper part along keels and entangled hairs below .... 19. *P. macranthera.*
+	Lemma 2.0–2.3 mm long, glabrous. Palea smooth along keels .... 20. *P. mongolica.*
24.	Plant 40–80 cm tall, with straight thick stems and large interrupted panicles with dense branches and spikelets. Lemma 2.0–2.5 mm long .... 25.
+	Plant 20–50 cm tall, with straight, very slender stems and small panicles with loosely arranged branches and spikelets. Lemma 2.0–3.5 mm long .... 26.
25.	Lemma green, triangularly truncate, with prominent rostellum along midrib and some hairs at base. Anthers 1.1–1.4 mm long .... 7. *P. gigantea.*
+	Lemma greenish-violet, obtuse-ovate, densely pilose at base. Anthers 1.3–1.6 mm long .... 28. *P. waginiae.*
26.	Lemma profusely pilose in lower half. Palea with spinules along keels in upper part and hairs below. Panicle branches densely covered with spinules .... 27.
+	Lemma totally glabrous. Palea with sparse spinules upward along keels. Panicle branches smooth .... 1. *P. altaica.*
27.	Lemma 2.0–2.75 mm long. Anthers 0.9–1.65 mm long. palea 1.4–2.0 mm long .... 28.
+	Lemma 2.8–3.5 mm long. Anthers 1.4–2.0 mm long. Palea 2.0–2.5 mm long .... 6. *P. dolicholepis.*
28.	Palea with spinules along keels upward for 2/3 length. Anthers 1.2–1.65 mm long .... 12. *P. jacutica.*
+	Palea with spinules along keels in upper part and hairs below. Anthers 0.9–1.3 mm long .... 27. *P. tenuissima.*

1. **P. altaica** Tzvelev 1968 in Rast. Tsentr. Azii 4: 152. —*P. dolicholepis* var. *paradoxa* Serg.

Perennial, 20–40 cm tall, cespitose, with slender stems and several extravaginal shoots covered with coriaceous 0.6–2 mm broad scale leaves. Ligules 0.6–1.5 mm long. Panicles 5–15 cm long, compact, compressed, with barely scabrous or, more often, smooth short branches of varying length; some spikelets subsessile. Spikelets 5–9-flowered, golden-violet. Glumes lanceolate; lower 1.1–1.5 mm long, upper 2–2.5 mm long. Lemma sharply lanceolate, smooth, 2.8–3.5 (3.9) mm long. Palea slightly aculeate upward along keels; glabrous at base. Anthers 1.3–1.75 mm long. Altay (Kosh-Agach settlement), $2n = 14$. Plate XVI (1).

Solonetz meadows. **West. Sib.**: AL—Go (Chuisk steppe, Dzhenishke-Tal—class. hab.—and others). —Junggar. Map 231.

2. **P. angustata** (R. Br.) Rand et Redf. 1894 in Fl. Mount Desert Is. Maine: 181. —*Atropis angustata* (R. Br.) Krecz.

Perennial, 10–30 cm tall, forming very dense mats. Leaf sheaths glabrous, smooth, often enveloping inflorescence. Leaves linear, acuminate, flat, 2–4 mm broad, glabrous on lower surface, scabrous toward top. Ligules scarious, up to 2 mm long, laciniate. Panicles 2–8 cm long, oblong-compressed, with branches smooth or diffusely scabrous in upper part. Spikelets 3–4-(5)-flowered, oblong, violet. Glumes lanceolate, rather acute, uneven along margin; lower 1.6-2.4 mm long, upper 1.9–3.25 mm long. Lemma 3.2–4.2 mm long; pubescent with long entangled hairs in lower third along nerves and on callus and usually even between nerves; poorly visible cilia like terminal teeth. Palea scabrous due to spinules along keels in upper part, with hairs below. Anthers 0.6-1 mm long. Plate XVII (3).

Clayey and sandy soils on coastal slopes, landslides, tundras and inland sites (away from sea coast). **West. Sib.**: TYU—Yam (Venuieuo river; Drovyanoi settlement). **Cen. Sib.**: KR—Ta. **East. Sib.**: YAK—Ar. —Circumpolar distribution; high-arctic species. Described from nor. Canada, Melville Island. Map 229.

3. **P. borealis** Swallen 1944 in Journ. Washington Acad. Sci. 34: 19.

Compactly cespitose plant 50–70 cm tall. Leaves 1–2 (3) mm broad, smooth, spinescent on upper surface. Ligules 2–3.5 mm long, membranous, with rounded, sometimes lacerated margin. Panicles 15–25 cm long. Pyramidal, spreading, with slender diffusely prickly branches arranged 2–4 each at internodes. Spikelets 4–6-flowered, green, rarely purple. Lower glume 0.9–1.2 (1.4) mm long, upper 1.25–1.8 mm long, sometimes with barely distinguishable spinules on back. Lemma (2.0) 2.25–2.5 (2.8) mm long, slightly obtuse, with eroded-ciliate margin and few long hairs at base; palea sharply prickly upward along keels. Anthers 0.5–0.7 mm long. Yakutia (Shandrin river), $2n = 42$. Plate XVI (5).

Sandy floodplains and dumps, river banks. **East. Sib.**: YAK—Ar, Yan (Indigirka river 45 km below Belaya Gora settlement), Ko. —Far East (north), North America (Arctic Region). Described from Alaska (Port Clarence town). Map 249.

4. **P. byrrangensis** Tzvelev 1971 in Novosti sist. vyssh. rast. 8: 80. — *P. vahliana* auct.

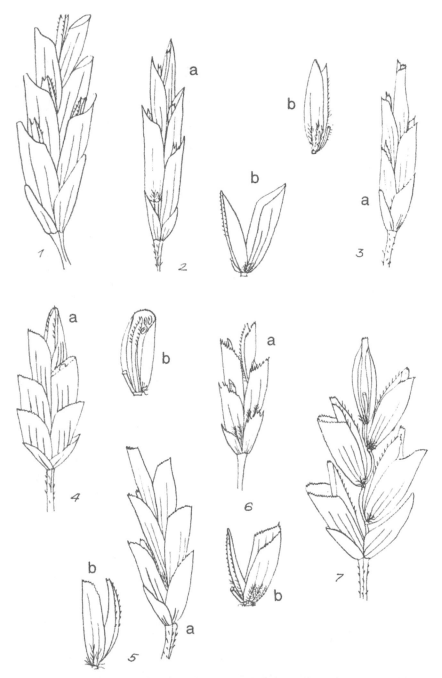

Plate XVI. 1—*Puccinellia altaica;* 2—*P. schischkinii;* 3—*P. hackeliana;* 4—*P. hauptiana;* 5—
P. borealis; 6—*P. neglecta;* 7—*P. kamtschatica:*
a—spikelets, b—lemma and palea with rachilla.

200

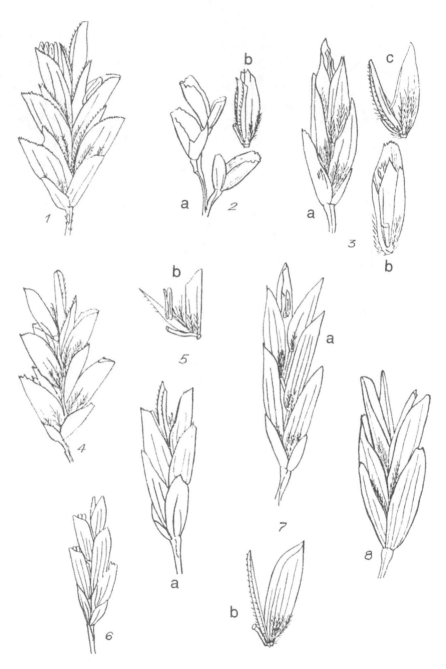

Plate XVII. 1—*Puccinellia sibirica*; 2—*P. lenensis* (a—glumes); 3—*P. angustata*;
4—*P. vaginata*; 5—*P. gorodkovii*; 6—*P. tenella*; 7—*P. jenisseiensis*;
8—*P. byrrangensis*:
a—spikelets, b and c—lemma and palea with rachilla.

Perennial, compactly cespitose, 6–25 cm tall with longitudinally folded glabrous, smooth narrow leaves up to 2 mm broad, reaching 1/3 stem height. Ligules up to 2.5 mm long, scarious, subacute, toothed. Panicles 2–6 cm long, lax, with few spikelets, branches compressed and totally smooth. Spikelets 3.5–6 mm long, 2–5-flowered, pinkish-violet. Glumes subacute, with 1–3 nerves; lower 1.5–2.75 mm long, upper 2.1–3.25 mm long. Lemma 3.0–4.0 mm long, with 5 nerves, subacute, broadly membranous along margin, densely puberulent in lower third along and between nerves, with oblong folds. Palea with 2 teeth at tip, glabrous or with sparse hairs beneath along keels. Anthers 1.2–1.7 mm long. Plate XVII (8).

Patchy and rocky tundras. **Cen. Sib.:** KR—Ta (Byrranga mountain range western spurs, Tareya river—class. hab.—and others).—Endemic. Map 230.

5. **P. distans** (Jacq.) Parl. 1848 in Fl. Ital. 1: 367. —*Atropis distans* (L.) Griseb. —*A. distans* var. *typica* Trautv. —*P. distans* (L.) Parl.

Perennial, spreading-bushy green plant with geniculate stems in lower part, 15–50 cm tall and up to 2 mm thick. Leaves 2–5 mm broad, stiff, flat, with enlarged sheaths; radical leaves reaching half stem length. Panicles 10–15 cm long, pyramidal, with spreading, later decurved scabrous unequal branches, some shortened. Spikelets 4–5 mm long, 3–6-flowered, green or somewhat violet. Glumes ovate, ciliate along margin; lower up to 1 mm long and upper 1.5–2 mm long. Lemma 2.0–2.2 mm long, obovate, broadly rounded above, subobtuse, ciliate along margin, glabrous at base. Palea bristly along keels. Anthers 0.8–1 mm long.

Solonetz meadows, riverine sand, weed along roadsides and in inhabited sites. **West. Sib.:** TYU—Yam (Salekhard town), Khm (Khanty-Mansiisk town, Kondinskoe village), Tb, KU, OM, TO (Kargasok village), NO, KE, AL—Ba. **Cen. Sib.:** KR—Pu (Igarka town), Tn (Uchami settlement), Kha (Shira Lake; Kolmakovo nomad village), Ve (Krasnoyarsk town, Astaf'evka village), TU. **East. Sib.:** IR—An (Porog and Taloe villages), BU—Se (Lake Baikal, Kotel'nikovskii cape), YAK—Ar (Tiksi bay). —Europe, Caucasus, Mid. Asia, Far East, Mongolia, China, North America; introduced in many countries of both hemispheres. Described from Austria (Vienna). Map 233.

European semiweed species with its natural distribution range falling in Europe, Urals, and West. Siberia; elsewhere found only as a rare introduced plant.

6. **P. dolicholepis** Krecz. 1934 in Fl. SSSR 2: 488, 764, nom. altern. — *Atropis dolicholepis* Krecz.

Gray-green, compactly cespitose plant with many slender (1–1.2 mm thick) stems up to 50 cm tall. Stems densely covered with brown sheaths at base. Leaves convolute like bristles, smooth. Panicles 6–10 cm long, sparse, small, compressed, with slender short scabrous branches.

Spikelets up to 7 mm long, 4–6-flowered, greenish-violet. Glumes oblong, subacute upward; lower 1.1–1.5 mm long, upper 2–2.5 mm long. Lemma 2.8–3.5 mm long, lanceolate, subacute, profusely pilose at base, keels of palea sharply prickly on top, softly pilose below. Anthers 1.4–2.0 mm long. Plate XVIII (4).

Solonetz in sheep's fescue-wormwood steppes along southern rubble slopes. **West. Sib.**: KU, NO. —Europe (south), Caucasus, Mid. Asia. Described from Kazakhstan. Map 232.

7. **P. gigantea** (Grossh.) Grossh. 1928 in Fl. Kavkaza 1: 114. —*Atropis gigantea* Grossh. —*P. sclerodes* Krecz. p. p.

Large, loosely cespitose plant 35–55 cm tall. Stems totally straight, smooth. Leaves 4–5 mm broad, flat or semiconvolute, sharply scabrous on upper surface. Sheath free, straw-yellow, violet-reddish below. Panicles 10–17 cm long, straight, lanceolate, initially compressed, later with obliquely erect branches. Branches scabrous, 7–10 each arising from single node, varying in length (2–3, up to 6–8 cm), rest shortened, some with subsessile spikelets. Spikelets lanceolate, 5–6-flowered, dense, greenish. Glumes oblong-ovate, obtuse; lower 1.1–1.75 mm long, upper 1.75–2.25 mm long. Lemma 2.25–2.5 mm long, obovate, triangularly truncate, rarely pilose at base. Palea sharply prickly along keels in upper half. Anthers 1.1–1.4 mm long, oblong. Plate XVIII (2).

Solonchak and solonetz along banks of lakes. **West. Sib.**: TYU—Tb (Krivinskoe Lake), KU, NO. —Europe (south), Caucasus, Mid. Asia, West. Asia. Described from Caucasus. Map 237.

8. **P. gorodkovii** Tzvelev 1964 in Arkt. fl. SSSR 2: 199.

Perennial compactly cespitose plant 10–40 (50) cm tall, with several 6–10 cm long radical leaves. Leaf blades flat, scabrous on upper surface. Panicles 3–8 cm long, with erect slender branches scabrous under spikelets; branches varying in length, 1.5–2 cm long, arising 2 each from node. Spikelets 4–7 mm long, lustrous, 2–4-flowered. Glumes smooth, even along margin; lower 1.0–1.55 mm long, upper 1.3–2.0 mm long, subacute. Lemma 2.4–3.0 (3.2) mm long, subacute, pinkish-violet, with broad, even membranous margin, puberulent near base along nerves. Palea weakly scabrous upward along keels, glabrous below. Anthers large, 1.1–1.5 mm long. Plate XVII (5).

Landslides and hill slopes, baidzharakhs, river estuaries, rocky seashores. **Cen. Sib.**: KR—Ta (Nizh. Taimyr river estuary—class. hab.—and others). —Endemic. Map 236.

Known formerly only from Nizh. Taimyr river.

9. **P. hackeliana** Krecz. 1934 in Flora SSSR 2: 484, 762 nom. altern. —*Atropis hackeliana* Krecz.

Perennial glaucescent green plant, 10–30 cm tall, forming dense brushlike mats. Stems geniculately ascending from base, rigid. Leaf sheaths broad, glabrous. Ligules up to 2.2 mm long, orbicular, smooth.

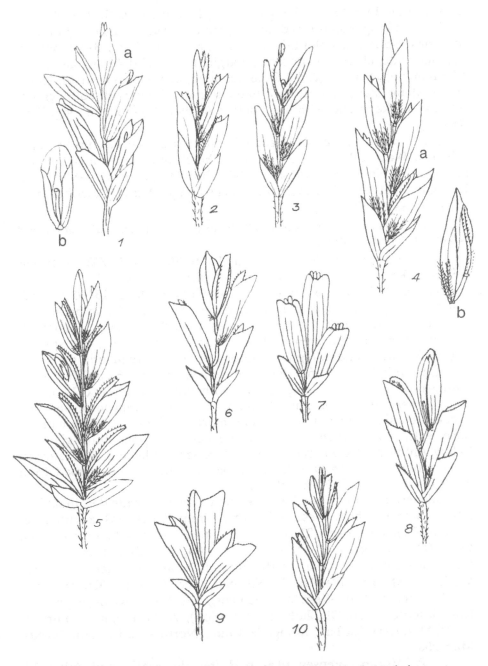

Plate XVIII. 1—*Puccinellia phryganodes;* 2—*P. gigantea;* 3—*P. tenuissima;*
4—*P. dolicholepis;* 5—*P. macranthera;* 6—*P. jacutica;* 7—*P. tenuiflora;* 8—*P. kalininiae;*
9—*P. kreczetoviczii;* 10—*P. mongolica:*
a—spikelets, b—lemma and palea with rachilla.

204

Leaves convolute, bristlelike, glaucous, glabrous, smooth. Panicles 5–7 cm long, oblong, compressed, later spreading, rachis smooth, branches scabrous. Spikelets oblong-lanceolate, green or violet, 3–5-flowered. Glumes ovate, obtuse, ciliate along margin; lower 1.2–1.4 mm long, upper 1.75 mm long. Lemma (2.0) 2.5–2.8 mm long, obovate, deltoid, orbicular, ciliate along margin, with long hairs at base. Palea with spinules upward along keels and with hairs below. Anthers 0.7–0.9 mm long. Plate XVI (3).

Saline meadows, pebble beds with salt efflorescences, up to upper hill belt. **West. Sib.**: AL—Go (Severo-Chuisk mountain range, Kyzky-Nor river). **Cen. Sib.**: TU (Mongun-Taiga hill, Mugur river). —Mid. Asia, Kashgar, Tibet, the Himalayas, Mongolia (Mongol Altay). Described from Pamir (Sassykkul' Lake).

Central Asian species with genetic affinity to *P. distans* (Jacq.) Parl. Not reported before from Siberian territory.

10. **P. hauptiana** Krecz. 1934 in Fl. SSSR 2: 485, 763. —*Atropis hauptiana* Krecz. —*A. iliensis* Krecz. —*P. filiformis* V. Vassil.

Slender light or gray-green cespitose plant. Stems 15–60 cm tall, slender, straight, geniculate at lower nodes. Leaves 1–3 mm broad, flat or semiconvolute, soft, smooth. Ligules 1.4–2.0 mm long, subobtuse, lacerated along margin. Panicles 15–20 cm long, compressed, later spreading, with long, slender, highly scabrous branches, frequently decurved in fruit. Spikelets 4–5 mm long, lanceolate, green, (5)–6–8-flowered. Glumes ovate, uneven along margin, with tiny cilia; lower 0.7–1.0 mm long, upper 1.0–1.5 mm long. Lemma 1.5–1.8 (1.9) mm long, obovate, broad orbicular-obtuse, uneven along margin, ciliate, with few hairs at base. Palea with minute spinules along keels in upper third, glabrous below. Anthers 0.3–0.5 mm long. Altay (Aktash settlement, Bol. Yaloman river estuary, Ust'-Koksa settlement, 365 km along Chuisk road from Biisk town), 2n = 28. Plate XVI (4).

Solonetz meadows, sand and pebble riverbeds, roadsides, inhabited sites, lower hill belt. **West. Sib.**: TYU—Yam (Salekhard town), Khm (Khanty-Mansiisk town), Tb, KU (Kurgan town—typ. hab.—and others), OM, TO (Tomsk town), NO, KE, AL—Ba, Go. **Cen. Sib.**: KR—Ta (Khatanga Settlement), Pu, Tn (Baikit and Osharovo settlements), Kha, Ve, TU. **East. Sib.**: IR—An, Pr, BU—Se, Yuzh, ChI—Ka (Chara settlement environs; Kyus'-Kamda settlement), Shi, YAK—Ar (Kyusyur and Chersk settlements), Ol (Shologontsy station), Vi, Al, Yan, Ko. —Europe, Urals, Mid. Asia, Far East, Mongolia, China, North America (introduced). Map 250.

11. **P. interior** Sørensen 1950 in Hultén, Fl. Alaska and Yukon 10: 1713.

Loosely cespitose plant 40–70 cm tall. Stems yellowish-green, lustrous. Leaves few, reaching half stem length. Ligules up to 3 mm long, membranous, with sparse spinules on back. Panicles (10) 15–25 cm long, broadly spreading, with slender diffusely prickly branches. Spikelets whitish-golden, purple-mottled, lustrous, 3–4-flowered. Lower glume (0.8) 0.9–1.1 (1.25) mm long, upper 1.25–1.46 mm long. Lemma 2.0–2.25 mm long, subacute, ciliate along margin, with barely visible hairs at base. Palea with sparse spinules along keels. Anthers 0.5 mm long.

Saline meadows, banks of lakes, roadsides, inhabited sites. **East. Sib.**: YAK—Ol (Zhigansk settlement), Al (Tattinsk and Churapchinsk regions), Vi. —Far East, Alaska. Described from Alaska. Map 235.

Reported for the first time from USSR and Siberian territories.

12. **P. jacutica** Bubnova 1988 in Bot. zhurn. 73 (9): 1331.

Perennial, with relatively slender, straight, up to 60 cm tall stems, forming small mats. Sheath green, compactly enclosing stem, with few straw-colored sheath remnants of dead leaves. Leaf blades narrow, convolute, subacute, green, reaching half stem length, scabrous along veins. Ligules 1.75–2.75 mm long, transparent, hastate, with few spinules on back. Panicles 8–15 cm long, lax, relatively narrow, with slender branches finely prickly throughout length. Branches arising 5–6 each from rachis, erect, short, with few 6–7.5 mm long spikelets, lax, 4–6-flowered. Glumes subacute, lanceolate; lower (0.7) 0.9–1.4 mm long, upper (1.2) 1.4–2.0 mm long. Lemma 2.0–2.5 (2.75) mm long, sharply lanceolate, greenish-violet, golden along margin, with long hairs in lower part along nerves. Palea obtuse, bidentate, densely covered with fine spinules for 2/3 upward along keels. Anthers 1.2–1.65 mm long, violet. Plate XVIII (6).

Solonchaks, saline, grassy steppified meadows. East. Sib.: YAK—Vi, Al, Yan (Dulgalakh river valley—class. hab.—and others). —Endemic. Map 234.

13. **P. jenisseiensis** (Roshev.) Tzvelev 1964 in Arkt. fl. SSSR 2: 195. —*P. jenisseiensis* (Roshev.) Krecz. —*Atropis jenisseiensis* Roshev.

Bushy cespitose plant 25–45 cm high. Stems more or less geniculate at base, weak. Leaves 2–2.5 mm broad, flat, green, reaching half stem length. Panicles 8–15 cm long, oval-pyramidal, lax, spreading, with slender pendent smooth or faintly scabrous 2–3 cm long branches, arising 2 each from each node. Spikelets oblong, 0.8–1 cm long, with 5–7 florets, violet. Glumes oblong-ovate, subacute, varying greatly in length; lower 1.0–2.2 mm long, upper 2.0–3.0 mm long. Lemma 3.4–3.8 (4.6) mm long, oblong-plicate, lanceolate-acuminate, light violet, golden membranous along margin, profusely hairy along nerves at base. Palea with spinules upward along keels, transformed below into hairs. Anthers 1.35–1.9 mm long. Plate XVII (7).

Crumbling clayey and sandy slopes, landslides. **Cen. Sib.**: KR—Ta (Yenisey lower course, Zverevsk sand, 71° 43′ n.l—class. hab.; Uboinaya river estuary; Efremov Kamen' bay). —Endemic.

Formerly known only from type specimens with anthers not fully developed. In new finds, plants differ from type specimens in stronger pubescence of palea—keels with spinules on top and profuse hairs below—features very closely resembling florets of *P. angustata* but, unlike them, new finds have very large, well-developed anthers and distinctly plicate lemma.

14. **P. kalininiae** Bubnova 1988 in Bot. zhurn. 73 (9): 1332.

Perennial, loosely cespitose. Stems 20–40 cm tall, slightly curved below and straight above, greenish glaucescent. Radical leaves 5–7 cm long, few, narrow, covered beneath like stems with straw-yellow scale sheaths and on upper surface with free green sheaths. Cauline leaves reaching panicles, narrow, convolute. Ligules 0.7–1.2 mm long, membranous, orbicular-obtuse. Panicles 5–12 cm long, initially compressed, narrow, later spreading, pyramidal, with scabrous slender branches arising 2 (3) each from each node. Spikelets 3–4–(5)-flowered, 3.7–4.2 mm long, greenish-violet. Glumes oblong, obtuse, with ciliate margin; lower 0.9–1.0 mm long, upper 1.2–1.6 mm long. Lemma 2.05–2.25 mm long, lanceolate, obtuse, with ciliate golden margin, glabrous, rarely with sparse hairs at base. Palea glabrous along keels or rarely with sparse spinules above. Anthers (1.05) 1.2–1.4 mm long. Plate XVIII (8).

Herbaceous and saline sedge meadows on river banks, up to upper mountain belt. **West. Sib.:** AL—Go (Chuisk steppe, Tarkhatta river midcourse—class. hab.—and others). —Endemic. Map 238.

15. **P. kamtschatica** Holmb. 1927 in Bot. Not. (Lund): 208 s. str. (quoad var. *asperula*). —*Atropis kamtschatica* (Holmb.) Krecz.

Green, loosely bushy plant 10–30 cm tall. Stems straight, slender, few, smooth, weak. Leaves 1–2 mm broad, soft, flat, rarely convolute, slightly scabrous on upper surface, as long or slightly shorter than stem. Ligules 1.5–2 mm long, orbicular, obtuse. Panicles 7–10 cm long, oblong, with ascending and scabrous branches weakly declinate from rachis. Spikelets oblong (4)–5–9-flowered, green, lax. Glumes scarious, ovate, truncate, with extremely fine cilia; lower 1.3–1.5 mm long, upper 2–2.5 mm long. Lemma 2.3–2.5 mm long, obovate, membranous, semitransparent, with indistinct nerves, with cilia on top and long hairs at base. Palea somewhat longer than lemma; with long spinules above along keel, glabrous below. Anthers 0.7–0.8 mm long. Plate XVI (7).

Wet meadows, near thermal springs. **East. Sib.:** YAK—Vi (Olekminsk region—Ebe Lake, Kenemkan winter station, Vilyuisk town). —Far East (Kamchatka), Alaska. Described from Kamchatka (Kashkan river).

16. **P. kreczetoviczii** Bubnova 1988 in Bot. zhurn. 73 (9): 1334. —*P. tenuiflora* var. *capillifolia* Reverd. —*P. filifolia* auct. non Tzvelev.

Perennial, with several radical leaves forming compact mats. Stems 10–40 cm high, straight, green. Radical leaves 3–11 cm long, reaching 1/4 stem length, very narrow, convolute, setaceous, covered beneath with dense greenish sheaths. Ligules 0.6–0.9 (1.1) mm long, membranous, obtuse, transparent, with cilia along margin. Panicles 4–10 cm

long, small, pyramidal, spreading, with few interrupted branches glabrous below and scabrous above under spikelets. Branches short, 2–4.5
cm long, arising 2 each from node, with few (4–12) spikelets. Latter violet,
3–4-flowered. Glumes unevenly obtuse, with cilia along margin; lower 0.6–
0.8 (1.0) mm long, upper 0.9–1.3 (1.5) mm long. Lemma 1.5–2.25 mm long,
subacute, glabrous. Palea glabrous or very rarely with sparse spinules
upward along keels. Anthers 0.9–1.3 (1.4) mm long. Plate XVIII (9).

Solonetz, solonetz meadows, sandy-clayey shoals along banks of salt
lakes. **Cen. Sib.:** KR—Kha (Shira Lake—class. hab.—and others), Ve,
TU. —Mongolia. Map 239.

17. **P. kulundensis** Serg. 1961 in Sist. zam. Gerb. Tomsk. univ. 82: 5.
—*Atropis distans* var. *convoluta* auct. non Trautv. p. p.

Perennial, forming small loose mats. Stems 50–80 cm tall, not slender, 1–2 mm thick, densely foliate almost to base of panicles. Leaves
flat, 2–3 mm broad, smooth beneath, aculeolate on upper surface along
veins. Ligules 1.75–2.75 mm long, rounded on top. Panicles 10–25 (32)
cm long, more or less spreading, with scabrous branches, arising (5) 7–
9 each from node. Longest branches in lower node (5) 7–12 cm. Spikelets
4–5-flowered. Lemma 2–2.1 mm long, orbicular-obtuse on top, with short
hairs at base. Palea with spinules upward along keels. Anthers 0.95–1.2
mm long.

Solonetz and saline meadows. **West. Sib.:** No (Veselovsk region,
Belaya village—class. hab.—and others), AL—Ba. —Endemic. Map 240.

N.N. Tzvelev (Zlaki SSSR (Grasses of the USSR), Nauka, Leningrad division, Leningrad, 1976, 788 pp.) cites an extensive distribution range for the species, obviously
including in it several hybrid forms between *P. macranthera* and *P. tenuiflora*. Plants from
Yakutia treated as *P. kulundensis* by Tzvelev have been described by us as an independent species, *P. jacutica*. They differ from *P. kulundensis* in several morphological features.
Number of chromosomes, $2n = 28$, as determined for plants from environs of Yakutsk
town, pertains evidently either to *P. hauptiana* or to the new species *P. jacutica*.

18. **P. lenensis** (Holmb.) Tzvelev 1971 in Novosti sist. vyssh. rast. 8:
80. —*P. fragiliflora* Sørensen p. p. excl. typo.

Perennial, forming loose mats 10–40 cm tall. Leaves 2 mm broad,
flat, soft, reaching middle of stem. Ligules membranous, coarsely dentate along margin, up to 4.5 mm long. Panicles 5–12 cm long, spreading,
with laterally declinate branches; latter smooth or very weakly scabrous
in upper part. Spikelets greenish-violet, with 5–8 flowers. Glumes del-
toid-ovate, unevenly lobed along margin; lower 0.85–1.6 mm long,
upper 1.5–2.1 mm long. Lemma (2.5) 2.7–3.2 (3.4) mm long, truncate-
orbicular, narrowed toward top, with membranous smooth, uneven (but
without cilia!) toothed margin; profusely pilose downward along nerves.
Palea with spinules along keels, transformed into short hairs below.
Anthers 0.52–0.85 mm long. Plate XVII (2).

Along slopes of sand hills—baidzharakhs, coastal embankments.
Cen. Sib.: KR—Ta (southern foothills of Byrranga mountain range; Yamu-

Nera river lower courses). **East. Sib.**: YAK—Ar (Lena lower course, Krestyakh settlement—class. hab.—and others). —Far East. Map 241.

Reported by authors only from Siberian territory. In the herbarium of the Botanical Institute (Leningrad), we detected specimens very similar to type *P. lenensis* from Chaunsk bay region (floor of dried-up lake near Chaun settlement, 6 VIII 1938, K.F. Yakovlev).

19. **P. macranthera** Krecz. 1934 in Fl. SSSR 2: 171, 759 nom. altern. —*P. macranthera* (Krecz.) T. Norlindh. —*Atropis macranthera* Krecz.

Perennial, 25–50 cm tall, forming loose mats. Stems several, geniculately decurved and covered with withering sheaths of lower leaves. Leaves aggregated in lower third of stem; usually solitary leaf in upper part of stem. All leaves flat, 2–4 mm broad, grayish-green, stiff, highly scabrous on upper surface, transformed into subacute tip. Ligules 2–3 mm long, obtuse, covered with spinules on outer surface. Panicles lax, 5–15 cm long. Panicle branches subhorizontal, highly scabrous, 2 (3) each from node. Spikelets up to 7 mm long, greenish or pale violet, 4–6-flowered. Glumes obtuse; lower 0.8–1.5 mm long, upper 1.2–2.5 mm long. Lemma 2.3–3.0 mm long, broadly deltoid-acuminate, scarious along margin, profusely pilose beneath along nerves. Palea prickly upward along keels, pilose below. Anthers 1.25–1.8 mm long. Plate XVIII (5).

Solonchaks, salt steppes, chee grass groves, banks of salt lakes. **Cen. Sib.**: KR—Kha (Lake Shira—class. hab.—and others), Ve, TU. **East. Sib.**: BU—Yuzh, ChI—Shi. —Mongolia, China. Map 242.

20. **P. mongolica** (T. Norlindh) Bubnova 1988 in Bot. zhurn. 73 (9): 1336. —*P. tenuiflora* var. *mongolica* T. Norlindh.

Perennial, with relatively thick stems, straight or geniculately decurved, 60–80 cm tall, forming loose mats. Sheath broad, high, green, later straw-yellow. Leaves few at base of stems, leaf blades of cauline leaves broad, 2–4 mm, flat, subacute, glabrous, scabrous only along margin. Ligules 1–3 mm long, subobtuse, glabrous. Panicles 10–17 cm long, lax, pyramidal, with long (up to 7 cm), scabrous, ascending branches, arising 4–7 each from rachis. Spikelets 3.5–4 (5.5) mm long, lax, 4–5-flowered. Glumes subacute, lanceolate; lower 0.7–1.1 mm long, upper 1.1–1.5 mm long. Lemma 2.0–2.25 mm long, ovate-obtuse, greenish-violet, glabrous; palea glabrous along keels. Anthers 1.2–1.5 mm long. Plate XVIII (10).

Wet marshy solonetz meadows, banks of rivers and lakes, in pea shrubs, forest hill belt. **East. Sib.**: IR—An, Pr, BU—Se, Yuzh. —Mongolia. Described from Mongolia (Doien settlement). Map 252.

Evidently a hybrid species resulting from cross between *P. macranthera* and *P. tenuiflora*. Its stable characteristic features, not typical of parent species, are: totally glabrous lemma and palea and few glabrous soft broad leaves.

21. **P. neglecta** (Tzvelev) Bubnova 1988 in Spiske rast. Gerb. fl. SSSR 26: 40. —*P. borealis* subsp. *neglecta* Tzvelev.

Loosely cespitose plant with slender stems 20–35 cm tall. Leaves 1.5–2 mm broad, soft, smooth, slightly prickly on upper surface. Ligules up to 2 mm long, membranous, with deltoid margin. Panicles 7–10 cm long, compact, compressed, with slender smooth branches; occasionally prickly only under spikelets. Latter 5–6-flowered, greenish or purple. Glumes subacute, ciliate along margin; lower 1.1–1.3 mm long, upper 1.7–1.8 mm long. Lemma with lacerated deltoid margin, bearing 2.3–2.5 (3.0) mm long cilia, with dense long hairs in lower third along and between nerves. Palea finely prickly in upper part along keels, entangled-pilose in lower part. Anthers 0.4–0.5 mm long. Plate XVI (6).

Sand hills (baidzharakhs) along steep sand banks of rivers. **Cen. Sib.**: KR—Ta (northern bank of Lake Taimyr, Sabler cape—class hab.—and others). **East. Sib.**: YAK—Ar. —Far East (West. Chukchi, Kargyk peninsula). Map 243.

First report from the Far East.

22. **P. phryganodes** (Trin.) Scribner et Merr. 1910 in Contr. US Nat. Herb. 13 (3): 78. —*P. phryganodes* subsp. *asiatica* (Hadač et A. Löve) Tzvelev. —*Atropis phryganodes* (Trin.) Krecz.

Plant 8–30 cm tall, with long stoloniform decumbent shoots. Leaves longitudinally folded, up to 2 mm broad. Panicles 3–8 cm long, with smooth branches, initially appressed, later declinate; developing relatively rarely. Spikelets with 2–6 florets. Glumes slender, flattened, with 3 distinct nerves; lower 1.2–1.9 mm long, upper 2.05–2.6 mm long. Lemma 3–4 mm long, with scarious fringe and distinctly manifest nerves. Keels of palea with extremely fine papillae. Anthers 1.6–2 mm long. Epidermis of upper surface of cauline leaves regularly alveolate, with profuse, distinct papillae. Yakutia (Tiksi bay), $2n = 28$. Plate XVIII (1).

Silt shoals, meadows near sea in zone of high tides. **Cen. Sib.**: KR—Ta. **East. Sib.**: YAK—Ar. —Circumpolar distribution. Described from Alaska (Kotzebue Sound). Map 244.

23. **P. schischkinii** Tzvelev 1955 in Bot. mat. (Leningrad) 17: 57. —*Atropis roshevitsiana* auct. non Schischkin.

Perennial, forming loose mats. Stems 20–40 cm tall, erect, fairly thick; base somewhat arcuate and covered with brownish sheaths. Leaf blades 1–3 mm broad, from compactly convolute to nearly flat, fairly stiff, gray-green. Ligules up to 3 mm long. Panicles 8-22 cm long, narrow, secund and long, usually reaching half stem length, with scabrous branches. Spikelets 5-10 mm long, 3–7-flowered, pale green. Glumes from lanceolate to ovate-lanceolate, acute; lower 1.1–1.5 mm long, upper 2.1–2.6 mm long. Lemma 2.7–3.4 mm long, acuminate, ciliolate at base, with few hairs along callus. Palea prickly above along keels and with few hairs in lower part. Anthers 0.8–1.1 mm long, pale yellow. Plate XVI (2).

210

Solonetz meadows, banks of highly saline lakes and springs. **West. Sib.**: AL—Ba (Zarya settlement, Klyuchevsk region). **Cen. Sib.**: TU (Ak-Chyra settlement; Dus-Khol' Lake). —Mid. Asia, West. China, Mongolia. Described from East. Kazakhstan (Lake Zaisan).

24. **P. sibirica** Holmb. 1927 in Bot. Not. (Lund): 206. —*P. arctica* auct. —*Atropis sibirica* (Holmb.) Krecz.

Perennial, large bushy green plant 50–80 cm tall. Leaves up to 3 mm broad, profuse, flat, scabrous on upper surface and along margin. Panicles 10–25 cm long, compressed like racemes, later branched. Branches slender, scabrous. Spikelets about 5 mm long, with 4–7 florets, green. Glumes ovate, ciliate on top; lower 1.1–1.5 mm long, upper 1.7–2.2 mm long. Lemma obovate, 2.6–3.4 (3.7) mm long, roundly obtuse, ciliate, with distinct crinite nerves below. Palea with spinules up to middle of keels, with long convolute hairs below. Anthers (0.6) 0.7–0.8 mm long. Plate XVII (1).

Coastal sand, sandy slopes of hills, river terraces. **West. Sib.**: TYU—Yam, **Cen. Sib.**: KR—Ta (Yenisey lower course, Tolstyi Nos settlement—class. hab.—and others), Pu. —Europe (north), North America. Map 245.

Not reported before from the Arctic region outside the USSR. In the herbarium of BIN, 2 sheets of this species from northern Canada are preserved under the name *P. borealis* Swallen: from environs of Inuvik, Northwest Canadian Territory, and from Yukon along the Porcupine river.

25. **P. tenella** (Lange) Holmb. 1926 in M. Porsild, Meddel. om Groenl. 63: 45. —*P. capillaries* auct. non Jansen. —*P. langeana* subsp. *asiatica* Sørensen. —*Atropis laeviuscula* Krecz.

Perennial cespitose plant 10–20 cm tall, with ascending stems, without decumbent shoots, with several scarious dead leaf sheaths at base. Leaves up to 2 mm broad, arcuate, narrow, longitudinally folded throughout length, half as long as stem. Panicles up to 8 cm long, compressed, with smooth, recurved branches. Spikelets 4–5-flowered, oblong, sometimes subsessile on thick short stalks. Glumes obtuse, 3-nerved; lower 1.1–1.8 mm long, upper 1.6–1.7 (2.25) mm long. Lemma 2.25–2.6 mm long, more or less keeled, oblong-lanceolate, subacute, diffusely pilose, greenish-purple along nerves in lower third. Palea smooth along keels. Anthers 0.45–0.6 mm long. Plate XVII (6).

Clayey and sandy shoals on sea coast, in tidal zone and rock crevices. **Cen. Sib.**: KR—Ta. **East. Sib.**: YAK—Ar. —Europe (north), Far East (north). Described from Vaigach island. Map 246.

26. **P. tenuiflora** (Griseb.) Scribner et Merr. 1910 in Contr. US Nat. Herb. 13 (3): 78 quoad. nom. —*Atropis tenuiflora* Griseb.

Perennial, forming loose mats. Stems 40–80 cm tall, 1–2 mm thick, straight or curved at lower nodes. Sheath initially green, later straw-yellow, glabrous, compactly enclosing stem bases. Radical leaves few, cauline leaves narrow, 1–2 mm broad, convolute, smooth outside, oxyacanthous on inner surface along nerves. Ligules 1.0–1.25 mm long, transparent, orbicular-obtuse, with very fine spinules on top. Panicles 11–18 cm long, initially compressed, narrow, later broad, spreading. Branches 6–11 cm long, several, slender, scabrous, arising 3–6 each from rachis. Spikelets 2.5–4 mm long, 3–4-flowered. Glumes lanceolate, subobtuse, with extremely fine cilia along margin; lower (0.4) 0.6–0.9 mm long, upper 0.8–1.3 mm long. Lemma 1.6–1.9 (2.0) mm long, orbicular-ovate, glabrous, with cilia along margin. Palea glabrous along keels, rarely with sparse spinules on top. Anthers 0.9–1.2 mm long, yellow. Plate XVIII (7).

Solonetzes, solonchaks, wet solenetz meadows, along banks of salt lakes, on kuzhiras. **Cen. Sib.**: KR—Kha, Ve, TU. **East. Sib.**: IR—An, Pr, BU—Se, Yuzh (Novoselenginsk settlement—class. hab.—and others), ChI—Shi. —Mongolia, China. Map 247.

Available data on chromosome numbers for this species from Gornyi Altay, $2n = 14$, evidently pertain to another species since *P. tenuiflora* s. str. does not grow in Altay.

27. **P. tenuissima** Litv. ex Krecz. 1934 in Fl. SSSR 2: 489, 765 nom. altern. —*Atropis tenuissima* Litv. ex Krecz. —*A. distans* var. *convoluta* auct. non Trautv.

Finely cespitose plant with very slender, stiff, erect stems 20–70 cm tall. Leaves convolute, slender, smooth, sheaths compactly enclosing stem. Panicles 3–12 cm long, with few spikelets, compressed, with slender branches of varying length projecting upward, smooth or rather scabrous in upper part; branches arising 2–7 each from each node; longest branches in lower node (2) 3–7 cm long. Spikelets 3–5-flowered, golden-violet, entirely golden in fruit. Glumes ovate; lower 0.8–1.25 mm long, upper 1.4–2.0 mm long. Lemma 2.0–2.4 mm long, with smooth margin, with short hairs on callus and sometimes along nerves in lower third. Palea finely prickly along keels toward top, puberulent below. Anthers 0.9–1.3 mm long. Altay (Borovskoe village), $2n = 14$. Plate XVIII (3).

Wormwood-grass steppes, solonetzes and solonetz meadows. **West. Sib.**: TYU—Tb, KU, OM, NO, KE (Ur-Bedari Village), AL—Ba, Go (Uzuntal area). **Cen. Sib.**: KR—Kha, Ve, TU. **East. Sib.**: IR—An (Kultuk settlement). —Europe (south), Mid. Asia. —Described from Kazakhstan. Map 248.

28. **P. waginiae** Bubnova sp. nov.

Plantae laxe fruticosae atro-virides 60–70 cm altae. Caules molles 1.5–2 mm in diam., pauci recti, basi plus minusve curvati. Vaginae virides breves, caules arcte ambientes. Folia plana vel semiconvoluta 1.5–2 mm lata, levia, caulibus duplo breviora. Ligulae 1.5–2.75 mm longae, triangulari-acutatae, margine saepe dilaceratae. Paniculae pyramidales, magnae, 15–21 cm longae, plus minusve diffusae, ramulis tenuibus aculeolatis, adscendentibus, ab axi per 5–6 oriundis, inaequilongis, longissimis (in nodo infimo sitis) 7.5–10 cm longis, brevissimis 2–3.5 cm longis. Spiculae lanceolatae, 5–7-florae, 6.5–7.75 mm longae, viridulo-violaceae. Glumae obtusiuscule rotundatae, inferiores 1.0–1.25 mm longae, superiores 1.5–2.1 mm longae. Lemmata (2.2) 2.25–2.75 mm longa, obtusiuscule rotundata, margine minutissime ciliolata, basi longe pilosa. Paleae superiores carinatae, superne aculeolatae, inferne parce pilosae. Antherae (1.0) 1.25–1.6 (1.7) mm longae, polline fertili.

T y p u s: regio Novosibirsk, distr. Czany, praedium auxiliare Karaczi, stationarium sectionis Sibiricae Acad. Sci. URSS, pratum salsuginosum 7 VII 1968, T. Vagina, V. Perevertova (NS, isotypi LE, NS).

A f f i n i t a s: A *P. limosa* (Schur) Holmb., cui affinis est, panicula majore, ramorum dispositione ac longitudine, necnon lemmatibus parce pilosis, a *P. bilykiana* Klok. panicula majore et lemmatum magnitudine et forma differt.

Dark green loosely bushy plant 60–70 cm tall. Stems supple, 1.5–2 mm diam, few, straight, somewhat curved at base. Sheaths green, short, compactly enclosing stem. Leaves flat or semiconvolute, 1.5–2 mm broad, smooth, half as long as stems. Ligules 1.5–2.75 mm long, deltoid-acuminate, often with lacerated margin. Panicles pyramidal, large, 15–21 cm long, rather spreading, with slender finely prickled branches. Branches arising 5–6 each from top of rachis, variable in length, longest in lower node 7.5–10 cm long, shortest 2–3.5 cm long. Spikelets lanceolate, 5–7-flowered, 6.5–7.75 cm long, greenish-violet. Glumes obtusely rounded; lower 1.0–1.25 mm long, upper 1.5–2.1 mm long. Lemma (2.2) 2.25–2.75 mm long, obtusely rounded, finely ciliate along margin, with long hairs at base. Palea with spinules upward along keels and sparse hairs below. Anthers (1.0) 1.25–1.6 (1.7) mm long, with well-developed pollen.

Solonchaks, saline meadows, banks of brackish water reservoirs. **West. Sib.:** NO (Karachi village—class. hab., Gadatskii settlement, Kargan and Chernyi Mys villages; nor. bank of Ubinsk Lake). —Endemic.

29. **P. vaginata** (Lange) Fern. et Weath. 1916 in Rhodora 18: 14.

Plant 15–20 cm tall, grayish-green. Stems geniculate below. Leaves 2–3 mm broad, supple, smooth. Ligules up to 1 mm long. Panicles 8–12 cm long, branches slender, scabrous. Spikelets 5–8 mm long, with 4–6 florets, linear-oblong. Glumes slender, somewhat obtuse, with cilia along margin; lower (1.1) 1.4–1.8 mm long, upper 2.0–2.7 mm long. Lemma 2.8–3.7 mm long, broadly ovate-truncate, with few hairs at base along nerves, distinct cilia like teeth at tip. Palea with spinules along keels in upper part but without hairs. Anthers 0.6–0.8 mm long. Plate XVII (4).

Wet sections on sea coasts. **East. Sib.:** YAK—Ar (Stannakh-Khocho settlement; Yana river estuary). —Far East, North America (north). — Described from Greenland.

While drawing up the description, herbarium data for the entire distribution range of the species was studied.

In the interval between manuscript submission for publication and receipt of printer proofs, the following species new for Siberia were reported.

Puccinellia beringensis Tzvelev 1973 in Novosti sist. vyssh. rast. 10: 86.

Perennial, glaucescent or reddish plant 10–25 cm tall. Stems slender, slightly curved, smooth at base. Sheaths smooth. Leaves 1.5–2 mm broad, slightly convoluted or longitudinally folded, smooth. Ligules 1.5–2 mm long, with large teeth toward tip. Panicles 4–8 cm long, compressed, later with squarrose branches. Branches smooth, 2–3 cm long, with 2–5 spikelets. Spikelets 6–7 mm long, 4–5-flowered, pinkish-lilac. Glumes lanceolate, subacute; lower 1.5 mm long and upper 2.25–2.5 mm long. Lemma 3.25–3.5 mm long, lanceolate, subacute, triangularly incised at top, uneven along margin, but without cilia, with few hairs along nerves in lower part. Palea with spinules at top along keels. Anthers 1.2–1.3 mm long.

Rocky and sandy sites near sea coast. **Cen. Sib.**: KR—Ta (Efremov Kamen' bay). —Chukchi peninsula. Described from Chukchi peninsula.

Regarded as endemic in Far East. In 1981, collected in taimyr by N.V. Matveeva and L.L. Zanokha on a rocky caps in Efremov Kamen' bay. Close to North American species *P. andersonii* Swallen, differing only in much larger anthers.

44. Dactylis L.

1. Leaf sheaths smooth, blades glabrous, covered with sparse spinules only in lower part. Lemma glabrous, palea with very short spinules along keels .... 1. *D. altaica.*
+ Leaf sheaths and blades highly scabrous. Lemma ciliate along keel, palea with slender long spinules .... 2. *D. glomerata.*

1. **D. altaica** Besser 1827 in Schultes et Schultes fil. Add. ad. Mantissa 3: 626.

Rhizome short, not thickened. Stems up to 60 cm high, straight, glabrous, lustrous, smooth. Radical and cauline leaves green, short, smooth, and glabrous. Ligules 3–4 mm long, obtuse, laciniate in middle. Panicles dense, green-violet, branches short, glabrous, lustrous. Spikelets 3–5-flowered, strongly laterally compressed, up to 10 mm long. Glumes glabrous, somewhat keeled, acuminate; lower shorter than upper. Lemma with short terminal cusp, covered with soft long hairs along keel and lateral nerves; palea flattened, with almost indistinct slender spinules along keels. Anthers 3.5–4 mm long.

Middle and upper hill-forest belt. **West. Sib.**: AL—Ba (Zmeinogorsk region), Go. —West. China, Mongolia. Described from Altay.

2. **D. glomerata** L. 1753, Sp. Pl.: 71.

Rhizome short, thickened. Stems straight, up to 150 cm tall, with long internodes. Leaves grayish-green, scabrous along veins, oxyacan-thous along margin. Ligules 4–6 (10) mm long, obtuse, laciniate. Panicles secund, dense, grayish-green; branches scabrous, terminate in capitate, densely aggregated spikelet clusters. Spikelets up to 10 mm long, 3–5-(rarely 6)-flowered, strongly laterally compressed. Glumes keeled, lanceolate-oblong, acute, shorter than spikelet. Lemma scabrous, acuminate awnlike, sharply keeled, prickly or coarsely ciliate along keel; palea flattened, with slender short cilia along keels. Anthers up to 5 mm long. Novosibirsk province (Kotorovo village), $2n = 28$.

Floodplain meadows, forest fringes. **West. Sib.**: TYU—Tb, KU, OM, TO, NO, KE, AL—Ba, Go. **Cen. Sib.**: KR—Kha, Ve, TU. **East. Sib.**: IR—An, Pr (Ust'-Kutsk region, Zvezdnyi settlement), BU—Yuzh. —Europe, Caucasus, Mid. Asia, Mediterranean, West. Asia, West. China, Far East, Mongolia. Described from Europe. Map 256.

45. Cynosurus L.

1. **C. cristatus** L. 1753, Sp. Pl.: 72.
Plant 20–60 cm tall, forming mats. Stems erect, glabrous, smooth. Leaves narrowly linear, about 2 mm broad, flat. Panicles spicate, linear, secund. Fertile spikelets 3–5-flowered, about 3 mm long, with sterile spikelets gathered at base, pectinate-pinnate. Glumes of sterile spikelets with solitary nerve, often transformed into straight up to 10 mm long awn. Glumes of fertile spikelets equal, scabrous along keel. Lemma with 5 nerves, acuminate or with short awn.

Meadows, forest glades, roadsides. Introduced. **East. Sib.**: IR—An (Ust'-Ordynsk region). —Europe, Caucasus. Described from Europe.

46. Briza L.

1. **B. media** L. 1753, Sp. Pl.: 70.
Loosely cespitose plant with short rhizome. Stems 30–70 cm tall, erect, glabrous. Radical leaves long, cauline short, sheaths smooth. Leaf blades 2–4 (5) mm broad, narrowly linear, flat, smooth on upper surface; usually scabrous along margin and lower surface. Ligules short and obtuse. Panicles pyramidal, up to 15 cm long, initially compressed, later spreading, with slender, smooth, sometimes scabrous, horizontally declinate branches and pendent spikelets. Latter 4–6 mm long, orbicular-ovate, 5–9-flowered, strongly laterally compressed, pale green or slightly lilac. Glumes 3–3.5 mm long, 1.5 mm broad, broadly ovate, shorter than spikelets, membranous along margin, without keel. Lemma 3.5–4 mm long, emarginate-cordate at base, with broad white scarious fringe, lustrous. Palea lanceolate, concave, scarious.

Meadows. **East. Sib.**: IR—An (environs of Irkutsk, 38 km on road to Kultuk, introduced). —Europe, Mediterranean. Described from Europe.

47. Cinna L.

1. C. latifolia (Trev.) Griseb. 1852 in Ledeb., Fl. Ross. 4: 435.

Rhizome procumbent, with shoots. Stems up to 2 m tall, erect, smooth. Leaves up to 1.5 cm broad, flat, with projecting white midrib, scabrous along margins and veins. Panicles 15–30 cm long, 5–10 cm broad, spreading, with nutant branches. Spikelets light green, 3–4 mm long, flat, laterally compressed. Glumes linear-lanceolate, acuminate, slightly longer than lemma. Lemma ovate, weakly scabrous, with short hairs along nerves, bidentate at tip, with short (about 0.5 mm) straight cusp in incision between teeth.

Shaded coniferous and mixed forests, river banks. **West. Sib.:** TYU—Tb, OM, TO, KE, AL—Ba (Pikhtovka river). **Cen. Sib.:** KR—Kha, Ve. **East. Sib.:** IR—An, Pr, BU—Se, Yuzh, ChI—Shi (Kalga settlement), YAK—Al (Tagil and Uchur rivers). —Europe, Caucasus, Far East, Japan, China, North America. Described from specimens originating in America and cultivated in Berlin. Map 261.

48. Arctagrostis Griseb.

1. Blade of uppermost cauline leaf nearly equaling sheath. Panicle branches somewhat long, bearing more than 15 spikelets. Latter 2.5–3.5 (3.8) mm long. Anthers 1.5–2 (2.2) mm long ....
.... 1. *A. arundinacea.*
+ Blade of uppermost cauline leaf many times shorter than sheath. Panicle branches short, bearing 6–10 spikelets. Latter 4–6 mm long. Anthers 3–4.5 mm long .... 2. *A. latifolia.*

1. A. arundinacea (Trin.) Beal 1896 in Grass. North Amer. 2: 317. — *A. latifolia* var. *arundinacea* (Trin.) Griseb.

Rhizome decumbent, with many shoots. Stems (30) 40–100 cm tall, with brownish leaves at base. Leaves 5–10 mm broad, flat, acute, weakly scabrous. Leaf blade of uppermost cauline leaf long, nearly as long as sheath, latter usually smooth, rarely weakly scabrous. Ligules 4–6 mm long, obtuse, lacerated. Panicles somewhat spreading, 4–10 (15) cm long, pale violet, with ascending branches, sometimes declinate from rachis of inflorescence. Branches relatively long, bearing more than 15 spikelets. Latter single-flowered, 2.5–3.5 mm long. Glumes narrow, acuminate, scarious, unequal. Lower markedly shorter than upper, with 5 indistinct nerves; upper almost as long as spikelet. Lemma and palea scarious along margins, acuminate. Anthers 1.5–2 mm long.

Meadows, banks of water reservoirs, forests, scrubs. **West. Sib.:** TYU—Yam. **Cen. Sib.:** KR—Ta, Pu, Tn. **East. Sib.:** YAK—Ar, Ol, Vi, Yan, Ko. —Far East, North America (Alaska). Described from Alaska. Map 259.

2. A. latifolia (R. Br.) Griseb. 1852 in Ledeb., Fl. Ross. 4: 434.

Stems 20–60 (80) cm tall, ascending, geniculate at base, with long, decumbent rhizomes. Leaves 3–8 mm broad, flat, occasionally longitudinally folded, scabrous; leaf blades of upper cauline leaves few times shorter than weakly scabrous sheaths. Ligules obtuse, lacerated, 4–6 mm long. Panicles compressed, up to 20 cm long, with short scabrous branches bearing 6–10 (12) spikelets. Latter single-flowered, 4–6 mm long. Glumes glabrous, ovate, obtuse, somewhat unequal. Lemma and palea 1–2 mm longer than glumes, scabrous, equal; lemma with 3 nerves, of which 1 sharply prominent; palea with 2 proximate nerves. Anthers 3–4.5 mm long. Putoran plateau (Talnakh settlement, Bogatyr' Lake, Baselak), East. Sayan (Tunkinsk mountain range) and Yakutia (Shandrin river), $2n = 56$.

Marshes, banks of rivers and lakes, forests, dwarf cedar shrubs. **West. Sib.**: TYU—Yam. **Cen. Sib.**: KR—Ta, Pu, Tn, TU. **East. Sib.**: IR—An, Pr, BU—Se, Yuzh, ChI—Ka, YAK—Ar, Ol, Vi, Al, Yan, Ko. —Nor. Europe, Urals, Mongolia, Far East, North America. Described from nor. Canada. Map 260.

49. Coleanthus Seidel

1. **C. subtilis** (Tratt.) Seidel 1817 in Roemer et Schultes, Syst. Veg. 2: 276.

Annual, with slender fibrous roots, branching from base and forming extremely small bushes 1–6 cm tall. Leaves narrowly linear, longitudinally folded. Sheaths of upper leaves highly inflated, enclosing inflorescence. Spikelets gathered 10–20 each in globose clusters of 2–4 mm diam on weakly geniculate rachis of inflorescence. Glumes absent. Lemma ovate, drawn into terminal cusp, weakly keeled. Palea 1.5 times shorter than lemma.

Floodplain meadows, silty banks. **West. Sib.**: TYU—Khm (Vakh river), Tb (Tura river), TO (environs of Narym village; Inkino village; Parabel' river). —Europe, Far East, North America. Described from Czechoslovakia.

50. Glyceria R. Br.

1. Keels of palea winged. Spikelets 8–20 mm long, orbicular in cross section. Leaf sheaths laterally flattened, keeled .... 2.
+ Keels of palea wingless. Spikelets 4–9 mm long, laterally compressed. Leaf sheaths not flattened, without keel .... 3.
2. Lemma 5.5–7 mm long; anthers 1.5–2 mm long. Sheaths of cauline leaves rather tuberculate along veins but without spinules ....
.... 1. *G. fluitans.*
+ Lemma 3.5–5 mm long; anthers 0.8–1.4 mm long. Leaf sheaths scabrous due to short spinules in upper part along veins ..... 5. *G. plicata.*

3. Panicles somewhat spreading, with many (more than 30) spikelets. Rhizomes thick, funicular. Leaf blades usually 3–10 (12) mm broad; if narrower, stiff and convolute. Stems 50–150 cm or more tall .... 4.

+ Panicles compressed, narrow, with few (not more than 25) spikelets. Rhizomes slender, filiform. Leaf blades 1.5–3 mm broad, slender, delicate, flat. Stems 20–40 cm high .... 2. *G. leptorhiza*.

4. Upper glume up to 3 (3.5) mm long, distinctly shorter than adjoining lemma, retuse or obtuse. Leaves flat, largely green, not stiff .... 5.

+ Upper glume 3.5–4.5 mm long, nearly as long as adjoining lemma, gradually acuminate. Leaves often convolute, grayish-green, stiff, acuminate .... 6. *G. spiculosa*.

5. Rachilla as well as nerves of lemma glabrous and smooth, only sometimes with diffuse, very short and thick spinules. Stamens 3. Anthers 0.9–1.6 mm long .... 6.

+ Rachilla and nerves of lemma bearing dense, very short and slender spinules. Stamens 2. Anthers 0.5–0.8 mm long .... 3. *G. lithuanica*.

6. Lower glume membranous, rather stiff, acuminate, almost entirely enclosing adjoining lemma, latter 3–4 mm long, with relatively large but short and thick spinules along nerves. Leaf blades light green on both surfaces, glabrous above or with diffuse very short spinules but without fine papillae .... 4. *G. maxima*.

+ Lower glume scarious, often lacerated apically, rarely more than half as long as adjoining lemma, latter 2–3 (3.5) mm long, with very fine tubercular spinules along nerves. Leaf blades grayish on upper surface and covered with extremely fine papillae, frequently interspersed with very short spinules .... 7. *G. triflora*.

1. **G. fluitans** (L.) R. Br. 1810, Prodr. Fl. Nov. Holl. 1: 179.

Rhizome decumbent, funicular, with subsurface shoots. Stems 30–100 cm tall, ascending. Sheaths laterally flattened, slightly keeled; those of cauline leaves somewhat tuberculate, but without spinules. Leaf blades 2–7 mm broad, slightly scabrous along margin and beneath. Ligules up to 5–7 mm long, usually laciniate almost to base. Panicles 10–35 cm long, secund, narrow, branches slightly declinate and, appressed to main rachis after anthesis; branches usually arising 1–2 (3) each in whorl and bearing 1 or 2 spikelets; only longest ones with 3 or 4 spikelets. Latter 10–20 mm long, cylindrical-linear, pale green. Glumes up to 3–3.5 mm long, ovate. Lemma 5.5–7 mm long, oblong, acute, with 7 sharp nerves, slightly scabrous, white scarious apically. Stamens 3. Anthers 1.5–2 mm long.

Banks of water reservoirs. **West. Sib.**: TYU—Tb (environs of Tobol'sk) **East. Sib.**: IR (near Irkutsk, introduced). —Europe, Caucasus, North America (NE part). Described from Europe.

218

2. **G. leptorhiza** (Maxim.) Kom. 1901 in Tr. Peterb. bot. sada 20: 307.
Rhizome slender, nearly funicular, decumbent. Stems 20–30 cm tall,
very slender, ascending. Sheath glabrous, smooth. Leaf blades 1.5–3 mm
broad, flat, slender, smooth. Ligules 2–3 mm long, slightly lacerate
apically. Panicles 5–8 cm long, narrow, compressed, with few (less than
25) spikelets on slender smooth branches, appressed to main rachis.
Spikelets 5–12 mm long, oblong, greenish. Glumes 2–3.7 mm long; upper twice as long as lower. Lemma 3–4 mm long, oblong, obtuse, with
7 distinct nerves and, along them, with obtuse tubercles. Stamens 3.
Anthers 1–1.7 mm long.

Silty river shoals. **East. Sib.**: ChI—Shi (Bol. Boty settlement on Shilka
river). —Far East, NE China (Manchuria). Described from Amur river
basin.

3. **G. lithuanica** (Gorski) Gorski 1849 Icon. Bot. Char. Cyp. Gram.
Lith.: tab, 20. —*G. remota* (Forsell.) Fries.

Rhizome short, with several slender fibrous roots. Stems 50–100 cm
or more tall, relatively slender, 2–4 mm diam. Leaf sheaths open in
upper part, finely scabrous. Leaf blades 4–8 mm broad, slender, green
on both surfaces, finely scabrous on upper surface, subglabrous beneath,
sometimes scabrous along veins. Ligules 2–3 mm long, slender, unevenly
flat truncate and lacerated at tip. Panicles 15–30 cm long, very lax, with
many filiform, usually flexuous scabrous branches, frequently nutant.
Spikelets 4–8 mm long, green or violet. Glumes unequal, 1.5–2.5 mm
long, subobtuse. Lemma 3–4 mm long, ovate or oblong, obtuse, finely
and obliquely dentate and narrowly scarious at tip, with 7 sharply scabrous nerves, frequently scabrous even between nerves. Stamens 2. Anthers 0.5–0.8 mm long.

Forest belt, marshes, forest valley meadows, banks of water reservoirs, wet coniferous and birch forests. **West. Sib.**: TYU—Khm, Tb, KU,
OM, TO, NO, KE, AL—Ba, Go (Ongudai village). **Cen. Sib.**: KR—Ve,
TU. **East. Sib.**: IR—An, Pr, BU—Se, Yuzh, YAK—Vi. —Europe, Caucasus, Far East, NE China (Manchuria), Korean peninsula, Japan. Map 257.

4. **G. maxima** (Hartman) Holmb. 1919 in Bot. Not. (Lund) 1919: 97.
—*G. aquatica* (L.) Wahlb.

Rhizome with many long shoots. Stems up to 1 m or more tall, 10–
12 mm diam, stout, coarse. Sheath scabrous; leaf blades 5–15 mm broad,
green on both surfaces, often with diffuse spinules on upper surface,
but without fine papillae, scabrous beneath. Panicles 10–40 cm long,
fairly dense, with 4–10 relatively thick, scabrous, suberect or declinate
branches in each whorl. Spikelets up to 8 mm long, green, frequently
brownish or violet. Glumes 2–4 mm long, subacute. Lemma 3–4 mm
long, obtuse, with 7 highly prominent nerves, covered at base with relatively short and thick spinules, membranous along margin and at tip.
Stamens 3. Anthers 0.9–1.6 mm long.

Banks of water reservoirs, marshy meadows. **West. Sib.**: TYU—Tb, KU, OM. —Europe, Mediterranean, North America (perhaps introduced). Described from Europe. Map 254.

5. **G. plicata** (Fries) 1842, Nov. Fl. Suec., Mantissa 3: 176.

Rhizome short, decumbent. Stems 30–70 cm tall, semidecumbent, rooting at nodes. Sheath laterally flattened, somewhat keeled, finely scabrous in upper part. Leaf blades 4–8 mm broad, relatively thick, finely scabrous on lower surface along margin and veins. Ligules 3–5 (7) mm long, deeply laciniate. Panicles 10–15 cm long, with branches weakly declinate in different directions; branches somewhat scabrous and rather thick, with few spikelets. Latter 8–15 mm long, narrowly cylindrical, pale green. Glumes 1.5–3.5 mm long, ovate. Lemma 3.5–5 mm long, with 7 nerves, oval-ovate, obtuse, scabrous, narrowly scarious at tip. Stamens 3. Anthers 0.8–1.4 mm long.

Wet sites on banks of water reservoirs. **West. Sib.**: NO (Chany Lake), KE. **East. Sib.**: BU—Yuzh (Tankhoi station). —Europe, Caucasus, Mid. Asia, Cen. and West. Asia, the Himalayas. Described from south. Sweden.

6. **G. spiculosa** (Fr. Schmidt) Roshev. 1929 in Fl. Zabaik. 1: 85.

Rhizome long, decumbent. Stems 50–100 cm or more tall, stout, 3–7 mm diam, rooting at lower nodes. Leaf sheaths open in upper part, glabrous or slightly scabrous. Leaf blades 3–5 (6) mm broad, grayish-green, convolute or flat, slender-acuminate, glabrous, smooth or more often with somewhat coarse, thick spinules along margin and thicker veins. Ligules 0.5–2.5 mm long, obtuse, orbicular, sometimes lacerated. panicles 10–25 cm long, with obliquely erect branches; branches smooth or with diffuse spinules. Spikelets 7–9 mm long, greenish-white. Glumes slightly unequal, 3.5–5 mm long, lanceolate, acuminate; upper almost as long as spikelet. Lemma 3.8–4.8 mm long, lanceolate, scarious at tip, with 7 slender and finely scabrous nerves, sometimes finely scabrous even between nerves. Stamens 3. Anthers 1.3–1.8 (2) mm long.

Banks of water reservoirs, swampy meadows, forms pure scrubs. **East. Sib.**: IR—An, Pr, BU—Se, Yuzh, ChI—Shi, YAK—Vi. —Far East, NE China (Manchuria). Described from Sakhalin island. Map 255.

7. **G. triflora** (Korsh.) Kom. 1934 in Fl. SSSR 2: 459, 758. —*G. aquatica* auct. p. p. —*G. aquatica* var. *debilior* auct. p. p. —*G. arundinacea* subsp. *triflora* (Korsh.) Tzvelev. —*G. lithuanica* auct. non Gorski.

Large plant with long thick rhizome and stout, thick stems up to 1.5 (2) m tall. Sheath smooth on slightly scabrous, frequently transversely sulcate. Leaf blades 4–12 mm broad, somewhat thick, usually slightly grayish or pale green on upper surface, with fine papillae sometimes interspersed with very short spinules, glabrous beneath or somewhat scabrous along veins. Ligules 2–4 mm long, flat, rounded at tip with short cusp in middle. Panicles 10–30 cm long, up to 20 cm broad, broadly spreading. Panicle branches weakly scabrous due to short diffuse spinules, often nearly smooth. Spikelets 5–8 mm long, frequently violet; glumes unequal, 1–3 mm long, obtuse or acute. Lemma 2–3 (3.5) mm long, ovate or oblong, with 7 nerves scabrous due to fine spinules, obtuse sometimes nearly truncate, frequently crenate, with or without scarious fringe. Stamens 3. Anthers (0.8) 0.9–1.2 (1.4) mm long. Krasnoyarsk region (Dubinino village), Tuva (Bel'bei village), and Buryat (Uoyan settlement), $2n = 20$.

Marshes, marshy and wet meadows, banks of water reservoirs, forest roadsides, clearances; enters forests. **West. Sib.**: TYU—Tb (environs of Tyumen'), OM (Ekaterininsk settlement), TO, NO, KE, AL—Go. **Cen. Sib.**: KR—Tn, Kha, Ve, TU. **East. Sib.**: IR—An, Pr, BU—Se, Yuzh, ChI—Ka, Shi, YAK—Vi, Al, Yan, Ko (Zyryanka settlement). —Urals, Far East, Mongolia, Nor. China. Described from Far East (Ivanovskoe village between Zeya and Bureya). Map 258.

51. Pleuropogon R. Br.

1. **P. sabinii** R. Br. 1824 in Parry, Jour. Voy. N.W. Pass. Suppl. to App. 11: 289.

Perennial, 10–40 cm tall, with long decumbent subsurface or submerged shoots. Stems erect, leaf blades 1–3 mm broad, linear, flat, glabrous. Ligules 2–4 mm long, membranous. Inflorescence secund cluster 3–10 cm long with 4–10 spikelets on glabrous nutant stalks. Spikelets 8–15 mm long, with 4–10 florets, rachilla glabrous, with joint under each floret. Glumes 2 mm long, membranous, violet, toothed at tip. Lemma and palea violet, white scarious along margin; lemma 3.5–5 mm long, coriaceous-membranous, with 7 scabrous nerves; palea nearly as long as lemma, with 2 nerves projecting like keels, bearing 2 each (1 shorter) awnlike 1–2.5 mm long appendages. Anthers 1.2–2 mm long. Caryopsis 2.5–3.5 mm long. Taimyr peninsula (Syndasko settlement), $2n = 42$.

Banks of rivers, lakes and brooks, swampy tundras. **West. Sib.**: AL—Go (between Dzhyumala and Ak-Kol rivers). **Cen. Sib.**: KR—Ta, Pu (Anabaro-Khatangsk interfluve region; Dyupkun and Sebyaki lakes). **East. Sib.**: YAK—Ar, Yan (Verkhoyansk mountain range, Altan river; Chuorka river). —Circumpolar distribution. Described from nor. Canada (Melville Island). Map 262.

52. Schizachne Hackel

1. **S. callosa** (Turcz. ex Griseb.) Ohwi 1933 in Acta Phytotax. et Geobot. (Kyoto) 2: 279. —*Avena callosa* Turcz.

Loosely cespitose plant with shortened slender rhizomes. Stems up to 70 cm tall, slender, delicate, scabrous under panicle. Leaves 1–3 mm broad, linear, flat or longitudinally convolute, glabrous beneath, scabrous with diffuse hairs on upper surface. Ligules 1–1.5 mm long, with short bristles. Panicles up to 10 cm long, narrow, racemose, almost secund, with scabrous branches. Spikelets 9–14 mm long, 3–4-flowered, rachilla with joint under each floret; glumes broadly lanceolate, membranous; lower 4–6 mm long, upper 6–8 mm long. Lemma 6–9 mm long, broadly lanceolate, without keels, scabrous along nerves, bidentate at tip, with straight 10–12 mm long awn arising from base of teeth. Palea 1/4 shorter than lemma. Callus with 1–1.75 mm long hairs. Anthers 1–2 mm long. Caryopsis 3.8–4.1 mm long.

Coniferous and mixed forests and scrubs along river valleys. **West. Sib.**: TYU—Tb (Kopotilovo village, Uspenka village), TO, AL—Go. **Cen. Sib.**: KR—Kha (Abaza town; Dzhoi river), Ve, TU (Bel'bei village). **East. Sib.**: IR—An, Pr, BU—Se, Yuzh, ChI—Shi (Sretensk town; Nerchinsk Zavod village, Sokhondo hill), YAK—Al (Uchur river lower course). — Europe, Far East, Japan, China, Mongolia. Described from Baikal. Map 263.

53. Melica L. —Melica

1. Lemma covered with somewhat long, distant hairs .... 2.
+ Lemma glabrous, smooth or scabrous, without hairs .... 4.
2. Lemma with sparse and not very long (1–1.5 mm) hairs along nerves in midback .... 3.
+ Lemma with dense and long (2–3.5 mm) cilia along marginal nerves .... 3. *M. transsilvanica.*
3. Panicles lax, spreading; branches long (up to 10 cm); spreading laterally at anthesis and thereafter. Leaves flat, green ....
.... 4. *M. turczaninowiana.*
+ Panicles narrow, spicate; branches short (1–2 cm), appressed to main rachis. Leaves convolute, grayish-green .... 5. *M. virgata.*
4. Panicles spicate, compressed, 0.5–3 cm broad, with many spikelets. Ligules of upper leaves 1–5 mm long .... 5.
+ Inflorescence secund, nutant clusters or racemose panicles with few (not more than 15) interrupted spikelets. Ligules of upper leaves not more than 0.5 mm long .... 2. *M. nutans.*
5. Spikelets large, 9–12 mm long, aggregated into long, secund dense panicles 1.5–3 cm broad. Leaves flat, broad (4–12 mm), green ....
.... 1. *M. altissin.a.*
+ Spikelets relatively small, 4–6 mm long; up to 1 cm broad in narrow compressed panicles. Leaves convolute, narrow (2–4 mm), grayish-green .... 5. *M. virgata.*

1. **M. altissima** L. 1753, Sp. Pl.: 66.

Rhizomatous plant, sometimes forming loose mats. Stems 40–200 cm or more tall; together with lower surface of leaf blades and sheaths coarsely scabrous due to somewhat thick sharp spinules. Leaves 3–10 (16) mm broad, flat. Ligules of upper leaves up to 5 mm long, oblong. Panicles 10–25 cm long, dense, with many spikelets, interrupted in lower part, with short (1–5 cm long), straight obliquely erect or appressed branches. Spikelets 8–12 mm long, nutant, green or violet, turning pale in fruit. Glumes 7–9 mm long, elliptical, nearly equal. Lemma in fruit oblong, apiculate, with broad scarious margin and several (7–13) scabrous nerves, of which 5–7 usually reaching tip. Anthers about 2 mm long. Altay (Katunsk Belki), $2n = 18$.

Rocks, rocky slopes, steppe scrubs, fringes of pine groves. West. Sib.: KU (Ozernaya village and Ukrainets, Zverinogolovsk region), KE, AL—Ba, Go. Cen. Sib.: KR—Ve. East. Sib.: BU—Yuzh (Koma village), YAK—Vi (Peledui settlement). —Europe (SE part), Caucasus, Mid. and West. Asia, West. China (Junggar). Described from Siberia. Map 264.

2. **M. nutans** L. 1753, Sp. Pl.: 66.

Plant with long rhizome, sometimes forming loose mats. Stems 20–70 cm tall, slender, sulcate, scabrous in upper part. Leaf blades 2–5 mm broad, flat, usually puberulent on upper surface, sometimes interspersed with sparse long hairs, rarely subglabrous, finely veined beneath, glabrous or with sparse slender spinules along veins. Sheath scabrous, lower ones purple-violet. Ligules of upper cauline leaves very short (0.1 mm long), sometimes almost wholly reduced. Clusters or racemose secund panicles with few (3–15) interrupted spikelets on short, usually nutant stalks. Spikelets 6–8 mm long, glumes elliptical or broadly ovate, obtuse, purple-violet. Lemma in fruit elliptical or elliptical-oblong, obtuse, thick-nerved at base due to numerous nerves, of which 5–7 reaching tip, finely scabrous or glabrous, opaque, slightly longer than glumes. Anthers 1–1.5 mm long.

Herbaceous coniferous and deciduous forests, forest glades and tall-grass meadows. West. Sib.: TYU—Tb, KU, OM, TO, NO, KE, AL—Ba, Go. Cen. Sib.: KR—Tn, Kha, Ve. East. Sib.: IR—An, Pr, BU—Se, Yuzh, YAK—Vi. —Europe, Caucasus, Mid. Asia, Mediterranean (north), the Himalayas, West. China, Korean peninsula, Japan. Described from Europe. Map 265.

3. **M. transsilvanica** Schur 1866, Enum. Pl. Transsilv.: 764. —*M. ciliata* auct non. L.

Stems 35–80 (100) cm tall, several, slender, glabrous, sometimes scabrous under panicle; together with radical leaves forming loose mat. Leaf blades 2–4 (5) mm broad, usually longitudinally folded or convolute, grayish-green, stiff, and densely pubescent on upper surface, sometimes with sparse long hairs; glabrous or scabrous beneath due to short spinules or tubercles, with thick midrib. Sheaths of lower leaves generally scabrous, rarely glabrous. Ligules of upper leaves 2–4 mm long, deeply lacerated. Panicles 3–7 (11) cm long, dense, cylindrical, subspicate, with several densely appressed branches. Spikelets 5–7 (9) mm long, young ones grayish-violet, later whitish or pale yellow. Glumes unequal; lower 3.5–5 mm long, acute, 1.5–2 times shorter than upper; latter lanceolate, acuminate. Lemma 4–6 mm long, as long or slightly shorter than upper glume, elliptical, scabrous on back due to short spinules or tubercles, with 7–9 nerves; of these, 1 or 2 marginal ones bearing slender and long (2–3.5 mm) cilia. Anthers 1–1.2 mm long. Altay (Chuisk road), $2n = 18$.

Rocks, rocky steppe slopes, steppe scrubs. **West. Sib.**: OM (environs of Omsk), TO (Uksunai river between Goldaevsk settlement and Tomsk zavod), KE, AL—Go. **Cen. Sib.**: KR—Kha (Barit settlement, Askizsk region), Ve. —Europe, Caucasus, Urals, Mid. and West. Asia, Mediterranean (Balkans), West. China (Junggar). Described from Rumania (Transylvania). Map 267.

4. **M. turczaninowiana** Ohwi 1932 in Acta Phytotax. Geobot. (Kyoto) 1 (2): 142.

Stems 30–100 cm or more tall and, together with abundant vegetative shoots, forming loose mats. Leaf blades 2–7 (12) mm broad, flat, glabrous, with distant sparse hairs either on whole upper surface or only along midrib; like sheath, scabrous or finely tuberculate beneath along veins. Ligules of upper leaves 2–3 (4) mm long, lacerated almost to base. Panicles large, up to 20 cm or more long, broadly spreading, lax. Spikelets 9–13 (15) mm long. Glumes 6.5–9 (10) mm long, scarious, brownish, nearly equal, oblong, obtuse, with distinct anastomosed nerves. Lemma of fertile florets 7–9 (11) mm long, mostly violet, with broad scarious margin at tip and numerous nerves, of which only 7 reach scarious margin; with long flexuous hairs in middle part along nerves, scabrous along back due to very short spinules or tubercles. Anthers 1.9–2.5 mm long.

Rocks, rock debris, rubble southern steppe slopes, steppe scrubs; enters forest. **Cen. Sib.**: TU (Systyg-Khem village, Todzhinsk region). **East. Sib.**: IR—An, Pr (Kulinga river in Kachugsk region), BU—Se (Vitim river near Polivtsevo village, Romanovka village, lower course of Garga river in Barguzinsk region), Yuzh (Kharatsai village on Dzhida river—class. hab.—and others), ChI—Shi.—Far East, Mongolia, NE China (Manchuria), Korean peninsula. Map 266.

5. **M. virgata** Turcz. ex Trin. 1831 in Mém. Sci. Pétersb., sér. 6 (1): 369.

Stems 50–80 (95) cm tall, several, slender, glabrous, densely foliated, aggregated into loose or fairly dense mats. Leaf blades 1–3 mm broad, grayish-green, usually convolute, slender-acuminate, glabrous on both surfaces, smooth on upper surface, sometimes with short spinules or tubercles beneath along veins. Ligules 1–1.5 mm long, obtuse, slightly lacerated. Panicles 10–25 cm long, compressed, narrow (up to 1 cm broad), spicate, interrupted, with short appressed branches. Spikelets

4–6 mm long, green or slightly violet, later pale, straw-brown. Glumes ovate, acuminate, unequal; lower about 2 mm long, 1.5–2 times shorter than upper. Lemma in fruit 3–4 mm long, oval or oblong, obtuse or acute, scabrous on back, with 7 nerves bearing isolated long hairs shedding over time. Anthers 1.2–1.5 mm long.

Rocky steppe slopes and their peaks. **East. Sib.**: BU—Yuzh (near Kharatsai village on Dzhida river—class. hab.—and others), ChI—Shi (Aksha town, Ust'-Ilya village on Onon river). —Mongolia, NE China (Manchuria). Map 268.

54. Molinia Schrank

1. **M. caerulea** (L.) Moench. 1794, Méth. Pl.: 183.

Perennial, 40–90 cm tall, with short decumbent subsurface shoots covered with scale leaves, forming mats. Stems erect, glabrous and smooth, with tuberous thickened lower internode. Leaves 3–9 mm broad, long, present only in lower part of stem, glabrous; diffusely pilose only in upper part. Ligules in form of hairy fimbrilla. Lax panicles 5–40 cm long, spicate. Spikelets 4–9 mm long, violet, occasionally greenish. Glumes coriaceous-membranous, lanceolate. Lemma 3–5 mm long, leptodermatous, acute or obtuse; palea 0.5 mm shorter than lemma. Callus glabrous or with few hairs. Anthers 1.5–3 mm long, violet.

Peat meadows, marshes, thin coniferous and mixed forests. **West. Sib.**: TYU—Tb, KU, OM (Motorovo village). —Europe, Caucasus, Mid. Asia, Asia Minor. Described from Europe. Map 269.

55. Nardus L.

1. **N. stricta** L. 1753, Sp. Pl.: 53.

Perennial, 10–40 cm tall, with extravaginal shoots forming compact mat and covered at base with whitish scales. Leaves setaceous, 0.3–0.5 mm diam, radical leaves grayish-green, somewhat arcuate. Ligules up to 2 mm long. Inflorescence a secund spike, 4–8 cm long, with sessile single-flowered, bluish, 5–7 mm long spikelets arranged singly in 1 or 2 close rows. Glumes reduced to narrow fringe. Lemma as long as spikelet, narrowly lanceolate, coriaceous-membranous, with 3 nerves projecting like keels, with straight 2–4 mm long terminal awn.

Hilly meadows, forest glades; ascends to high altitudes. **West. Sib.**. TYU—Khm (Sokur'ya river; upper course of Man'ya river, Sale-Uroika hill), KE (Kemerovo town). **East. Sib.**: IR—An (Khamar-Daban mountain range). —Europe, Caucasus, Asia Minor, Nor. America. Described from nor. Europe. Map 300.

56. Achnatherum Beauv.

1. Awns 1.2–2.5 cm long, twice geniculate, with contorted lower geniculation. Meadow-steppe plant .... 2.
+ Awns 0.5–1.5 cm long, nearly straight or somewhat curved. Desert-steppe plant .... 3. *A. splendens.*
2. Glumes acute; upper rather uniformly curved with spinules or bristles almost throughout surface. Blades of cauline leaves covered on upper surface with spinules and long hairs .... 1. *A. confusum.*
+ Glumes somewhat acuminate, with spinules only along keel, elsewhere glabrous. Blades of cauline leaves glabrous on upper surface or covered with spinules, very rarely with bristles, and without long hairs .... 2. *A. sibiricum.*

1. **A. confusum** (Litv.) Tzvelev 1977 in Probl. ekol., geobot., bot., geogr. i florist.: 140. —*Stipa confusa* Litv. —*S. sibirica* auct. p. p.

Stems up to 100 cm tall, forming small mats or single. Radical leaves reaching half stem length, tubularly convolute like cauline leaves, rarely flat, 2–6 mm broad, scabrous on outer surface, covered with spinules and soft long hairs within. Ligules up to 1 mm long. Panicles 15–20 cm long, fairly lax. Glumes broadly lanceolate, obtuse, 6–8 mm long, brownish-violet in most cases; upper, sometimes lower also, covered with spinules throughout surface. Lemma nearly as long or slightly shorter than glumes, pilose. Awns about 15 mm long, geniculate. Yakutia (environs of Yakutsk town), $2n = 24$.

Thin forests, steppified meadows, rocky slopes, among shrubs. **West. Sib.:** TO (Kuyuma river valley), NO (Kur'i river valley, Rassolkino settlement), KE, AL—Ba (Kolyvan' and Zmeinogorsk settlements), Go. **Cen. Sib.:** KR—Kha, Ve, TU. **East. Sib.:** IR—An, BU—Se, Yuzh, ChI—Ka, Shi, YAK—Vi. —Far East, Nor. China. Described from Altay. Map 271.

2. **A. sibiricum** (L.) Keng ex Tzvelev 1977 in Probl. ekol., geobot., bot. geogr. i florist.: 140. —*Stipa sibirica* (L.) Lam.

Stems 60–100 cm tall, usually few, forming small mats. Cauline leaves 2–6 (10) mm broad, stiff, convolute or flat, glabrous or scabrous on upper surface, without interspersed long hairs; glabrous beneath. Ligules 0.5–1 mm long. Panicles 10–25 cm long, narrow, compressed, Spikelets pale green, sometimes with violet tinge. Glumes 9–10 mm long, narrowly lanceolate, with spinules only along nerves, elsewhere glabrous. Lemma 5–7 mm long, pubescent throughout surface. Awns 15–20 mm long, geniculate twice. Baikal region (Lake Irkana), $2n = 44$.

Steppes, steppified meadows, forest glades, rocky slopes. **West. Sib.:** KE, AL—Ba (Charyshskoe and Berezovka villages between Allak and Kamen'), Go. **Cen. Sib.:** KR—Kha, Ve, TU. **East. Sib.:** IR—An, BU—Se, Yuzh, ChI—Shi. —Caucasus, Mid. Asia, Mongolia, China, Far East. Described from Siberia. Map 291.

3. **A. splendens** (Trin.) Nevski 1937 in Tr. Bot. in-ta AN SSSR, 1 (4): 224. —*Stipa splendens* Trin. —*Lasiagrostis splendens* (Trin.) Kunth.

Stems 0.5–2 m tall, very stout, smooth, forming compact tussocks, surrounded by lustrous remnants of leaf sheaths at base. Leaf blades grayish-green; stiff, usually tubularly convolute, rarely flat, 1–1.5 mm or 3–5 mm broad. Panicles with many spikelets, elongated, 15–45 cm long. Glumes scarious, whitish or, in lower part, violet, 5–6 mm long. Lemma nearly as long as glumes, membranous along margin, crinite. Awns 6–10 mm long, slightly curved, covered with spinules. Altay (Chibit village), $2n = 24$.

Solonetzes, solonetz steppes, rocky and rubble slopes, semideserts. **West. Sib.**: NO (Vishnevka and Vdovino villages), AL—Ba, Go. **Cen. Sib.**: KR—Kha, Ve, TU. **East. Sib.**: IR—An, BU—Yuzh, ChI—Shi. —European USSR, Mid. and Cent. Asia, China. Described from Transbaikal. Map 273.

57. Ptilagrostis Griseb.

1. Panicles broadly spreading, with long (more than 1.5 cm) flexuous branches. Leaf blades scabrous on outer surface along veins due to fairly dense spinules .... 2.
+ Panicles weakly spreading, with short (0.1–1 (1.5) cm long) branches. Leaf blades smooth on outer surface (rarely with diffuse, very short spinules) .... 2. *P. junatovii.*
2. Panicle base without bract or cilia. Anthers glabrous at tip .... 3.
+ Panicle base with small lanceolate bract or cilia. Anthers with hairy
· tufts at tip .... 1. *P. alpina.*
3. Spikelets 5–7 (9.5) mm long, 8–20 in panicle. Awns 20–25 (35) mm long. Lemma 4–5.5 mm long .... 3. *P. mongholica* s. str.
+ Spikelets 4.5–5 mm long, 20–30 in panicle. Awns 15–20 mm long. Lemma 3.3–4 mm long .... 4. *P. mongholica* subsp. *minutiflora.*

1. **P. alpina** (Fr. Schmidt) Sipl. 1970 in Spisok rast. Gerb. fl. SSSR 18: 60. —*Stipa alpina* (Fr. Schmidt) V. Petrov.

Plants 15–35 cm tall, forming small compact mats. Radical leaves numerous, setaceous, 0.3–0.4 mm diam; like sheaths, weakly scabrous due to diffuse short spinules, ovate in cross section, with 7–9 vascular bundles. Ligules of upper cauline leaves 1.5–3 mm long. Panicles broadly spreading, with few florets and 6–14 spikelets. Panicle base with lanceolate bract or only cilia. Spikelets 3.5–5 mm long, violet. Glumes obtuse. Awn 15–20 mm long, flexuous or indistinctly geniculate, plumose. Stanovoi upland (Yuzhno-Muisk mountain range), $2n = 22$.

Alpine meadows, tundras, dwarf arctic birch groves in high-altitude belt. **East. Sib.**: BU—Se, ChI—Ka. —Far East, China (NE). Described from Bureya upper courses. Map 275.

2. **P. junatovii** Grubov 1955 in Bot. mat. (Leningrad) 17: 3. —*Stipa mongholica* Turcz. ex Trin. f. *minor* Krylov.

Plants 10–20 cm tall, forming small compact mats. Radical leaves numerous, slender, convolute, setaceous, glabrous and smooth; leaf cross section pentagonal with 5 vascular bundles. Leaf ligules of vegetative shoots very short, of cauline leaves up to 2 mm long. Inflorescence 2.5–6 cm long, rather compressed, with 10–15 dark violet spikelets, with short appressed glabrous (0.1) 0.5–1.5 cm long branches. Panicle base often with lanceolate bract with fimbriated fringe. Lower half of lemma crinite; upper half covered with short appressed spinules. Awn 14–20 mm long, geniculate. Anthers pilose at tip. Stanovoi upland (Yuzhno-Muisk mountain range), $2n = 22$.

Rocky slopes, arid tundras in high-altitude belt. **West. Sib.:** AL—Go. **Cen. Sib.:** TU. **East. Sib.:** IR—An, BU—Se, Yuzh. —Mongolia. Described from Khangai. Map 274.

3. **P. mongholica** (Turcz. ex Trin.) Griseb. s. str. 1852 in Ledeb. Fl. Ross. 4: 447. —*Stipa mongholica* Turcz. ex Trin. —*S. czekanowskii* V. Petrov.

Compactly cespitose plant 20–60 cm tall. Radical leaves numerous, setaceous, pentagonal in cross section, with 3 vascular bundles and 5 sclerenchyma strands, sharply scabrous along veins on outer surface. Ligules of cauline leaves with spinules outside; leaf ligules of vegetative shoots elongated, glabrous. Inflorescence 18–20-flowered. Panicle branches subglabrous. Spikelets single-flowered, 5–7 (9.5) mm long. Lemma with long appressed hairs almost up to top. Awn 20–25 (35) mm long, plumose, geniculate. Altay (Aktash settlement), $2n = 22$.

Tundras, meadows, rocky slopes, rocks predominantly in high-mountain belt. **West. Sib.:** AL—Go. **Cen. Sib.:** KR—Pu (Khaya-Kyuel' and Darima lakes, Medvezhei river midcourse), Ve (Bol. Tepsel' river upper course), TU. **East. Sib.:** IR—An (Dzhiginai river at confluence with Oka river—class. hab.—and others), BU—Se, Yuzh, ChI—Ka (Sr. Sakukan river), Shi (Kukun river), YAK—Ol (Arga-Sala river), Vi (Daldyn and Mogdy rivers). —Mid. Asia, West. China, Mongolia. Described from East. Sayan. Map 272.

4. **P. mongholica** subsp. **minutiflora** (Titov ex Roshev.) Tzvelev 1974 in Novosti sist. vyssh. rast. 11: 7. —*P. mongholica* var. *minutiflora* Titov ex Roshev.

Cespitose plant 40–60 cm tall. Radical leaves numerous. Panicles broadly spreading, with 20–30 spikelets. Latter grayish or with violet tinge, 4.5–5 mm long. Awns 15–20 mm long, plumose.

Marshes in steppe belt. **Cen. Sib.:** KR—Kha (Ulen' and Karo valleys—class. hab.—and others). —Endemic.

58. Stipa L.

1. Awns scabrous or puberulent (hairs not longer than 0.8 mm) .... 2.
+   Awns distinctly plumose, crinite (hairs longer than 1 mm) ...... 8.
2. Awns covered with 0.3–0.8 mm long hairs .... 3.
+   Awns covered with up to 0.1 (0.2) m long spinules .... 4.
3. Hairs on lemma in rows. Glumes (16) 20–28 mm long .... 5. *S. consanguinea.*
+   Hairs uniformly covering lemma surface. Glumes 12–15 mm long .... 12. *S. korshinskyi.*
4. Lemma with distinct corona of hairs at awn base .... 5.
+   Lemma without corona of hairs at awn base, sometimes few spinules only in marginal joints. Leaf blades scabrous on outer surface due to acute tubercles or subglabrous; puberulent on inner surface, interspersed with much longer hairs along margin .... 4. *S. capillata.*
5. Leaf blades 0.3–1 mm diam., outer surface glabrous and smooth .... 6.
+   Leaf blades 0.3–0.6 mm diam., outer surface with short spinules interspersed with small stiff bristles .... 18. *S. praecapillata.*
6. Awn 8–18 (20) cm long, lower contorted section 2–4.5 cm long .... 7.
+   Awn 20–28 cm long, lower contorted section (up to first geniculation) 5.0–8.0 cm long .... 9. *S. grandis.*
7. Leaf blades puberulent on inner surface, interspersed with very long hairs. Lemma 12–18 mm long. Stem nodes almost invariably concealed under sheaths .... 1. *S. baicalensis.*
+   Leaf blades puberulent on inner surface, not interspersed with long hairs. Lemma (7) 9–12 mm long. Stem nodes dark-colored, not covered by sheaths .... 13. *S. krylovii.*
8 (1). Awn geniculate once .... 9.
+   Awn geniculate twice .... 11.
9. Lower contorted section of awn pilose .... 10.
+   Lower contorted section of awn glabrous, smooth. Awn 7–13 cm long .... 11. *S. klemenzii.*
10. Plant 5–20 (24) cm tall. Leaf blades curved acinaciform or contorted, shorter than stems. Awn 4–6 cm long .... 8. *S. glareosa.*
+   Plant 25–35 cm tall. Leaf blades straight, as long or longer than stems. Awn 6–10 cm long. Plant of sand dunes .... 2. *S. barchanica.*

11. Awn longer than 7 cm; glabrous in lower part, usually smooth, rarely rather scabrous due to sharp tubercles or short spinules .... 12.

+ Awn 3–7 cm long, densely pilose in lower contorted section .... 15. *S. orientalis.*

12. Lemma uniformly covered with hairs throughout periphery and almost up to top with short hairs at awn base .... 13.

+ Lemma with hairs in oblong rows. Hairy corona at awn base lacking .... 14.

13. Sheath of cauline leaves glabrous .... 14. *S. lessingiana.*

+ Sheath of cauline leaves very short but densely pilose .... 3. *S. brauneri.*

14. Leaves glabrous or covered with tubercles, stiff bristles or spinules .... 15.

+ Leaves fairly densely covered with slender and soft distant, 0.5–1 mm long, hairs. Marginal band of hairs 1–1.5 mm short of reaching awn base .... 6. *S. dasyphylla.*

15. Marginal band of hairs on lemma reaching, or not more than 1 mm short of reaching awn base .... 16.

+ Marginal band of hairs on lemma 2–6 mm short of reaching awn base .... 18.

16. Lemma 12–16 (20) mm long. Leaves scabrous on outer surface. Sheaths of lower leaves of vegetative shoots with short pubescence .... 17.

+ Lemma 20–25 mm long. Leaves and sheath glabrous .... 19. *S. pulcherrima.*

17. Leaves of vegetative shoots 0.7–1 mm diam, only with acute tubercles on outer surface, without bristles or spinules .... 7. *S. glabrata.*

+ Leaves of vegetative shoots 0.3–0.7 mm diam, with acute tubercles and scattered stiff bristles on outer surface .... 21. *S. zalesskii.*

18. Ligules of leaves of vegetative shoots 0.3–3 (4) mm long, often variable in length within the same mat .... 19.

+ Ligules of all leaves of vegetative shoots poorly visible, not more than 0.2 mm long. Leaves narrow, trichoid at ends, as long or longer than inflorescence .... 20. *S. tirsa.*

19. Leaves 0.6–1 mm diam, smooth or weakly scabrous on outer surface .... 20.

+ Leaves 0.4–0.7 mm diam, highly scabrous on outer surface due to dense sharp tubercles, spinules or stiff bristles .... 10. *S. kirghisorum.*

20. Sheaths of upper cauline leaves somewhat scabrous due to

extremely fine sharp tubercles ....
.... 17. *S. pennata* subsp. *sabulosa.*
+ Sheaths of upper cauline leaves smooth ....
.... 16. *S. pennata* s. str.
1. **S. baicalensis** Roshev. 1929 in Izv. Glavn. bot. sada SSSR 28: 380.
—*S. attenuata* P. Smirnov.

Stems 50–100 (110) cm high, forming compact mats. Leaves long, reaching inflorescences, convolute setaceous, glabrous on outer surface, rarely very weakly scabrous, covered with dense short and diffuse long hairs on inner surface. Panicles 10–40 cm long. Lemma 12–18 mm long, with short hairs under awn. Awns 8–18 (20) cm long, geniculate twice, scabrous.

Steppes, steppified meadows, rocky slopes and rocks. **West. Sib.**: AL—Go (Elekmonar settlement). **Cen. Sib.**: KR—Kha, Ve, TU. **East. Sib.**: IR—An, BU—Se, Yuzh (Tataurovo station—class. hab.—and others), ChI—Shi. —Far East, Mongolia, Manchuria. Map 277.

2. **S. barchanica** Lomonosova sp. nova.

Planta perennis 25–35 cm alta caespites plus minusve compactos formans. Caules rectivel subgeniculati, breviter pilosi. Folia radicalia vaginis flavidogriseis longis nitidis margine dense pilosis caulinum superius vagina dilatata ubique scabra, inflorescentiam totam vel partem ei us inferiorem tantum amplectente, ligulis brevissimis margine dense pilosis, laminis setiformibus 0.5–0.8 mm in diam. levibus vel scabriusculis 25–35 cm longis rectis vel subincurvatis, cauli aequilongis vel eo longioribus. Inflorescentiae breves. Lemmata ca 10 mm longa. Aristae per totam longitudinem plumosae semel geniculatae, 6–10 mm longae.

In arenis mobilibus crescit.

T y p u s: regio Tuvaensis, lacus Teri–Chol (Erzin), in arenis mobilibus. 29 VI 1947, K. Sobolevskaja.

A f f i n i t a s. A *Stipa glareosa* P. Smirnov oecologia, foliis rectis longis ac magnitudine omnium partium plantae distinguitur.

A r e a g e o g r a p h i c a: RSSA Tuva.

Perennial, 25–35 cm tall, forming somewhat compact mats. Stems straight or barely geniculate, with short pubescence. Sheath of radical leaves yellowish-gray, long, lustrous. Sheath of upper cauline leaf enlarged, scabrous throughout surface, covering entire inflorescence or only its lower part. Leaf ligules very short, densely pubescent along margin. Leaf blades setaceous, 0.5–0.8 mm diam, smooth or weakly scabrous, 25–35 cm long, straight or poorly curved, as long or longer than stem. Inflorescence short. Lemma about 10 mm long. Awns plumose throughout length, geniculate once, 6–10 cm long.

Sand dunes. **Cen. Sib.**: TU (Tere-Khol' Lake in Erzinsk region—class. hab.—and others). —Endemic.

3. **S. brauneri** (Pacz.) Klokov 1976 in Novosti sist. vyssh. i nizsh. rast.: 21. —*S. lessingiana* subsp. *brauneri* Pacz.

Grayish-green plant forming small loose mats. Stems 20–50 cm tall. Leaf sheaths shortly villous. Leaves setaceous, 0.2–0.6 mm diam, long, with few spinules or glabrous on outer surface. Leaf ligules of vegetative shoots very short, rather densely pilose along margin. Lemma 8–11 mm long, uniformly pubescent almost throughout surface, with hairy corona under awn. Awns 15–25 cm long, plumose.

Steppes. **West. Sib.**: KE (environs of Novokuznetsk, Pinegino village). —Crimea. Described from Crimea.

4. **S. capillata** L. 1762, Sp. Pl. 2: 116.

Compactly cespitose plant 40–80 cm tall. Stem nodes concealed by sheaths. Leaves 0.5–1.3 mm diam, setaceous (sometimes flat and then up to 2.5 mm broad), long, scabrous on outer surface due to sharp tubercles or almost smooth, covered with short and long hairs on inner surface. Leaf ligules of vegetative shoots 0.6–2 (2.5) mm long. Panicles 10–25 cm long. Lemma 10–13 mm long, without hairy corona at awn base; sometimes only few spinules seen at joint of lemma to awn. Awns 10–20 cm long, geniculate twice, scabrous. Altay (Katunsk Belki), $2n = 44$.

Steppes, steppified meadows. **West. Sib.**: TYU—Tb (between Ikovsk and Karavashnin, and Klepikov and Simakov villages; Zamiralovo village), KU (Paderinsk, Lesnikovo, Zamanilki villages), OM, TO (Ilovka and Karpysak villages, between Bezgolosova and Plotova villages), NO, KE, AL—Ba, Go. **Cen. Sib.**: KR—Kha, Ve, TU. **East. Sib.**: IR—An, BU—Se, Yuzh, ChI—Shi (Petrovsk-Transbaikal, El'tsovo village), YAK—Vi. —Europe, Caucasus, Mid. Asia, West. Asia, the Himalayas, Junggar, Mongolia (west). Described from central Europe. Map 280.

5. **S. consanguinea** Trin. et Rupr. 1842, Sp. Gram. Stip.: 78.

Stems 15–50 cm tall, straight or geniculate. Leaves 0.3–0.8 mm diam, grayish-green, setaceous, smooth on outer surface. Leaf ligules of vegetative shoots 0.1–0.4 mm long, of cauline leaves 1–1.5 mm long. Panicles compressed, 6–8 cm long. Lemma 8.5–10 mm long, with 7 rows of hairs and corona at awn base. Awns 8–11 cm long, geniculate once or indistinctly twice, covered throughout length with short, 0.3–0.7 mm long hairs.

Steppes, rocky slopes. **West. Sib.**: AL—Go (Toboshak hill, Tarkhata and Sebistei rivers). —Mongolia (NW). Junggar. Described from Altay.

6. **S. dasyphylla** (Lindem.) Trautv. 1884 in Tr. Peterb. bot. sada 9: 350.

Stems 35–80 cm tall, forming rather compact mats. Leaf blades 0.6–1.2 mm diam, covered on both surfaces with long (0.5–1.5 mm), soft, distant or semidistant hairs. Leaf ligules of vegetative shoots 1–3 mm long, of cauline leaves 2.5–5 mm long. Leaf sheaths longer than internodes. Lemma 18–22 (24) mm long; marginal band of hairs reaching or 1–1.5 mm short of reaching awn base. Awns geniculate twice, plumose, up to 45 cm long.

Steppes, thin forests, forest fringes. **West. Sib.**: TYU—Tb (Tyumen' town), KU (Vargashi village, between Mostovsk and Belaya villages), OM (Sargatsk and Achair villages), KE (Novorossiika and Karakany settlements), AL—Ba. **Cen. Sib.**: KR—Kha (Moskva village), Ve (Eliseevka settlement). —Europe. Described from Khar'kov region. Map 282.

7. **S. glabrata** P. Smirnov 1928 in Roshevits, Fl. Yugo-Vostoka 2: 115.

Compactly cespitose plant 90–90 cm tall. Leaf blades filiform, 0.8–1 mm diam, densely covered on outer surface with sharp tubercles, sometimes interspersed with sparse bristles; covered on inner surface with short and long hairs, often forming hairy fringe outside along margin. Lemma 18–20 mm long, with marginal hairy band, reaching awn base or not more than 1 mm shorter. Awns geniculate twice, 25–40 cm long, plumose.

Steppes. **West. Sib.**: KU (between Barnaul'sk and Ryamova, and Pimernovka and Chesnokovsk villages; Mal. Dubravnaya village), AL—Ba (Topol'naya village). —Europe. Described from Europe.

8. **S. glareosa** P. Smirnov 1929 in Feddes Repert. 26: 264. —*S. caucasica* subsp. *glareosa* (P. Smirnov) Tzvelev. —*S. caucasica* subsp. *desertorum* (Roshev.) Tzvelev. —*S. orientalis* var. *humilior* Krylov.

Grayish-green plant forming small mats. Stems 5–20 (25) cm high, often geniculate. Leaves glaucescent, usually curved, shorter than stem, covered somewhat with spinules throughout length or only in lower part. Ligules about 1 mm long, densely pilose. Panicles short, with closely aggregated spikelets. Lemma 7–9 mm long. Awns 4–6 cm long, pilose throughout length, geniculate once. Altay (Kosh-Agach village), $2n = 44$.

Steppes, rocky slopes. **West. Sib.**: AL—Go. **Cen. Sib.**: TU. **East. Sib.**: IR—An. —Cen. Asia. Described from Mongolia. Map 278.

9. **S. grandis** P. Smirnov 1929 in Feddes Repert. 26: 267.

Stems 60–100 cm high, aggregated into loose mats. Leaves of vegetative shoots nearly as long as stems, sometimes half as long. Leaf blades 0.6–0.7 mm diam, glabrous outside, densely covered within with short spinules and few long hairs. Leaf ligules of vegetative shoots 0.3–0.5 mm long. Lemma 15–17 (18) mm long. Awns (20) 22–28.5 mm long, lower contorted part (up to first geniculation) 5–8 cm long.

Steppes, on sand. **Cen. Sib.**: KR—Kha, Ve (Krivoe village), TU (Kyzyl town environs). **East. Sib.**: BU—Se (Kharkhushun village), Yuzh, ChI—Shi. —Mongolia, China (NE). Described from Mongolia. Map 279.

10. **S. kirghisorum** P. Smirnov 1925 in Feddes Repert. 21: 233.

Stems 30–60 cm tall, forming compact mats. Leaf sheaths longer than internodes, glabrous, smooth or somewhat scabrous. Leaves long, nearly as long as stems, convolute, setaceous, 0.4–0.7 mm diam, densely covered with sharp tubercles on outer surface; with short hairs or spinules on inner surface. Panicles 10–15 cm long. Lemma 14–18 mm long, without hairy corona under awn. Awns 20–26 cm long, geniculate twice, glabrous in lower part, plumose in upper.

Steppes, rocky slopes and rocks. **West. Sib.**: OM (Isil'-Kul' and Vorontsovka villages), AL—Go. **Cen. Sib.**: TU (Khadyn Lake). —Mid. Asia, Junggar, Mongolia (west). Described from East. Kazakhstan. Map 283.

11. **S. klemenzii** Roshev. 1924 in Bot. mat. (Leningrad) 5: 12. —*S. gobica* auct. non Roshev.

Cespitose plant 10–30 cm tall. Stem base covered with lustrous sheaths. Leaves setaceous, 0.4–0.7 mm in diam, as long as or longer than stem, glabrous on outer surface, covered with very short spinules on inner surface. Leaf ligules up to 1 mm long, densely covered with hairs. Leaf sheaths longer than internodes, upper 10–16 cm long, inflated and concealing panicles, drawn into awnlike terminal cusp. Lemma 8–11 mm long, without hairy corona at awn base. Awns 7–13 cm long, geniculate once, lower contorted part 1.5–2 cm long, glabrous; upper part plumose with 5–6 mm long hairs.

Rocky and sandy steppes. **Cen. Sib.**: TU (Erzin village). **East Sib.**: BU—Yuzh (Borgoi village, Dyresui area), ChI—Shi (Krasnyi Velikan settlement). —Mongolia. Described from Mongolia.

12. **S. korshinskyi** Roshev. 1916 in Fedchenko. Fl. Aziatsk. Ross. 12: 163.

Stems 35–70 cm tall, aggregated into somewhat compact mats. Leaves of vegetative shoots setaceous, 0.3–0.6 mm diam, glabrous or slightly scabrous on outer surface, several, reaching 1/3 or 1/2 of stem length. Panicles 10–20 cm long, narrow, compressed. Glumes 12–16 mm long, narrowly lanceolate. Lemma 8–9 mm long, uniformly covered with hairs almost up to top, not forming distinct oblong rows, with corona at awn base. Awns 9–12 cm long, geniculate twice, covered with 0.3–0.8 mm long hairs.

Rocky steppes. **West. Sib.**: KU (between Shemetov and Yasnaya Polyana, Stepnaya village), OM (environs of Omsk, Lezhanka village), AL—Ba. —Urals (south), Transvolga, Mid. Asia. Described from Kazakhstan.

13. **S. krylovii** Roshev. 1929 in Izv. Glavn. bot. sada SSSR 28: 379 p. p. —*S. decipiens* P. Smirnov. —*S. capillata* var. *coronata* Roshev. f. *glabrifolia* Krylov. —*S. densiflora* P. Smirnov. —*S. densa* P. Smirnov.

Compactly cespitose plant (20) 30–60 (90) cm tall. Stems often geniculate. Stem nodes usually dark-colored, not covered by sheaths. Leaves bristly, 0.3–0.5 mm diam, 2 or 3 times shorter than stems, rarely nearly as long, usually glabrous, smooth on outer surface, puberulent on inner,

234

without interspersed long hairs. Lemma 9–12 mm long, with hairy corona at awn base. Awns 8–15 (20) cm long, geniculate twice, scabrous. Yakutia (environs of Yakutsk), $2n = 44$.

Steppes, rocky slopes, steppified meadows. **West. Sib.**: AL—Go. **Cen. Sib.**: KR—Kha, Ve, TU. **East. Sib.**: IR—An, BU—Se, Yuzh (hills between Temnik and Dzhida—class. hab.—and others), ChI—Shi, YAK—Vi, Yan (Verkhoyansk settlement, midcourse of Sartang river). —Mid. and Cen. Asia, Nor. China. Map 281.

Plants with leaves covered by few spinules are found in Transbaikal.

14. **S. lessingiana** Trin. et Rupr. 1842, Sp. Gram. Stip.: 79.

Grayish-green plant 30–70 cm tall, forming compact mats. Leaf sheath glabrous. Leaves setaceous, 0.5–0.6 (0.8) mm diam, shorter than stem, sharply scabrous, ligules very short, ciliate along margin. Lemma 8–10 mm long, with uniform, somewhat dense pubescence throughout surface. Awns 15–25 cm long, plumose, with hairs of plume up to 3 mm long.

Steppes. **West. Sib.**: TYU—Tb (Larinskoe village), KU (Ukrainka village), OM (Poltavskoe village), NO (Erestnaya village), AL—Ba. — Europe, Caucasus, Mid. Asia, West. Asia, Junggar. Described from Orenburg region. Map 284.

15. **S. orientalis** Trin. 1829 in Ledeb., Fl. Alt. 1: 83.

Stems 25–35 cm tall, forming compact mats. Leaf blades straight or slightly curved, 0.3–0.5 mm diam, rather scabrous on outer surface, densely puberulent on inner. Sheath of lower leaves with fine pubescence; of upper usually glabrous. Leaf ligules of vegetative shoots elongated, narrowed upward, sparsely pilose along margin. Panicles 5–10 cm long, compressed. Lemma 5–8 mm long, densely pubescent at base, with rows of hairs above reaching tip, or almost so, with hairy corona under awn. Awns 5–7 cm long, geniculate twice, plumose throughout length.

Rocky steppes, rocks, rocky slopes. **West. Sib.**: AL—Go. **Cen. Sib.**: KR—Kha, Ve, TU. —Mid. and West. Asia, West. China, Mongolia. Described from East. Kazakhstan. Map 285.

16. **S. pennata** L. s. str. 1753, Sp. Pl.: 78, p. p. —S. joannis Čelak.

Stems 30–80 cm tall, aggregated into large compact mats. Sheath of cauline leaves nearly as long as internodes, glabrous, smooth. Leaf blades often longitudinally folded, rarely flat, 0.6–1 mm diam, glabrous or weakly scabrous. Leaf ligules of vegetative shoots 1–3 mm long. Lemma 15–20 mm long. Awns 20–40 cm long, plumose.

Meadow steppes, forest fringes. **West. Sib.**: TYU—Tb, KU, OM, TO, NO, KE, AL—Ba, Go. **Cen. Sib.**: KR—Kha, Ve, TU. **East. Sib.**: IR—An, BU—Yuzh. —Europe, Mediterranean, Mid. Asia. Described from Cent. Europe. Map 286.

17. **S. pennata** subsp. **sabulosa** (Pacz.) Tzvelev 1973 in Novosti sist. vyssh. rast. 10: 80. —*S. borysthenica* Klokov ex Prokudin. —*S. pennata* auct. non L.

Cespitose plant 50–70 cm tall. Leaf sheaths of vegetative shoots glabrous or pubescent; sheath of upper leaves of flowering shoots covered with fine, sharp, somewhat curved tubercles or tuberculate spinules. Leaf ligules of vegetative shoots 1.5–2 (3) mm long. Lemma 15–20 mm long, marginal hairy band (4) 5–6 mm short of reaching awn base. Awns 20–35 cm long.

Sand, sandy steppes, pine groves. **West. Sib.**: AL—Ba. **Cen. Sib.**: KR—Kha, Ve, TU. —South. European USSR, Mid. Asia. Described from Khersonsk region. Map 287.

18. **S. praecapillata** Alechin 1926 in Alechin and Smirnov. Kratk. predv. otchet o rabotakh Nizhnegorod. geobot. eksped. 1925: 171. —*S. capillata* var. *coronata* Roshev. *f. praecapillata* (Alechin) Krylov. —*S. sareptana* subsp. *praecapillata* (Alechin) Tzvelev. —*S. sareptana* var. *kasakorum* Roshev. —*S. krylovii* var. *scabrida* Roshev.

Compactly cespitose grayish-green plant 30–70 cm tall. Leaves setaceous, slender, 0.3–0.6 mm diam, with short spinules on inner surface most often not interspersed with long hairs, covered with fine spinules and sometimes even sharp tubercles on outer surface. Leaf ligules of vegetative shoots 0.2–1.5 mm long. Lemma 9–11 mm long, with short hairs under awn. Awns 10–15 (20) cm long, scabrous.

Steppes, rocky slopes. **West. Sib.**: KU, OM (Poltavskoe and Russkaya Polyana villages; Vorontsovka village), AL—Ba. **Cen. Sib.**: TU (Kholu river). —East European USSR. Described from Gor'kovsk region. Map 288.

Differs from related species *S. sareptana* A. Becker distributed in Europe and Mid. Asia in absence of distinctly developed hairy corona under awn and pubescence of leaf blades which, in *S. sareptana*, are densely covered with bristles.

19. **S. pulcherrima** C. Koch 1848 in Linnaea 21: 440.

Stems 60–100 cm tall, forming large mats. Leaf sheaths longer than internodes, glabrous, smooth or slightly scabrous. Leaves filiform, 1–1.5 mm diam, glabrous, smooth, rarely scabrous on outer surface; densely pilose within. Leaf ligules of vegetative shoots 1–2 mm long. Lemma 20–25 mm long, with marginal bands of hairs reaching up to awn base. Awns 40–50 cm long, geniculate twice, plumose.

Steppes. **West. Sib.**: TYU—Tb (Afonino village), KU (Lesnikovo and Shemetovo villages), OM (Syropyatskoe village). —Europe, Caucasus, Mid. Asia, Mediterranean, West. Asia. Described from Transcaucasus. Map 289.

20. **S. tirsa** Steven 1857 in Bull. Soc. Nat. Moscou 30 (2): 115 p. p. — *S. stenophylla* (Lindem.) Trautv.

Compactly cespitose plant (40) 50–70 cm high. Leaves numerous, equal to or longer than stems, 0.2–0.5 mm diam, highly scabrous on outer surface, acuminate, capilliform. Leaf ligules of vegetative shoots very short, barely visible, 0.1–0.2 mm long. Lemma 18–20 mm long, uniformly pubescent beneath, with 7 rows of hairs above, 1–3 mm short of reaching awn base. Awns 35–45 cm long, geniculate twice; glabrous in lower contorted part, plumose above.

Steppes, rocky and melkozem (fine earth) slopes. **West. Sib.**: TYU—Tb (between Zaroslavsk and Kutluk, and Lebedev and Dubynsk villages; Yur'evo village), KU, OM. —Europe, Mediterranean, Mid. Asia, West. Asia. Described from Ukraine. Map 290.

21. **S. zalesskii** Wilensky 1921 in Dnevn. 1 Vseross. s"ezda russk. botanikov: 41. —*S. rubens* P. Smirnov.

Stems 40–75 cm tall, forming loose mats. Sheath of lower leaves with dense fine pubescence, of upper subglabrous. Leaves longitudinally convolute, 0.3–0.7 mm diam, covered with spiny tubercles and bristles on outer surface, with short spinules interspersed with long hairs on inner. Leaf ligules of vegetative shoots (1.5) 2–3 mm long. Lemma 17–19 mm long, with discoid pubescence at base and 7 rows of hairs above; 2 extreme ones of latter reaching or not more than 1 mm short of reaching awn base. Awns 20–35 cm long, geniculate twice, plumose.

Steppes, melkozem-covered rocky slopes. **West. Sib.**: KU, OM, NO, KE (Karakany village, between Kamyshina and Pish'l villages; Ust'-Serta village), AL—Ba, Go. **Cen. Sib.**: KR—Kha, Ve, TU. —Europe (south), Caucasus, Mid. Asia, Junggar, Mongolia (west). Described from Saratov region. Map 292.

59. Phragmites Adanson

1. **P. australis** (Gav.) Trin. ex Steudel 1841 in Nomencl. Bot., ed. 2 (2): 324. —*P. communis* Trin.

Stems 0.8–3.5 m tall, straight, glabrous, foliated up to inflorescence, with hollow decumbent rhizome. Leaves 0.5–2.5 cm broad, stiff, glaucescent or pale green, glabrous, sometimes covered on lower surface with sparse long hairs, sharply scabrous along margin. Ligules in form of fringes of short dense hairs. Panicles spreading, sometimes rather compact, light brown, 10–30 cm long. Panicle branches scabrous, bases of primary branches with dense long accumbent hairs. Spikelets 3–7-flowered, dark- or brown-violet, 9–12 mm long. Glumes brownish, lower broadly lanceolate, half of upper one, narrowly lanceolate. Lemma and palea strongly unequal, lemma drawn into long slender subulate cusp, palea obtuse, scabrous along keels. Anthers 1.5–2.5 mm long.

Banks of rivers, lakes, marshes, marshy meadows and solonchaks. **West. Sib.**: TYU—Khm, Tb, KU, OM, TO, NO, KE, AL—Ba, Go. **Cen. Sib.**: KR—Tn, Kha, Ve, TU. **East. Sib.**: IR—An, Pr, BU—Se, Yuzh, ChI—Ka, Shi, YAK—Vi, Al. —Eurasia, North and South America, Africa, Australia. Described from Australia.

60. Aeluropus Trin.

1. **A. intermedius** Regel 1869 in Bull. Soc. Nat. Moscou, 41 (4): 292. —*A. litoralis* auct. p. p. —*A. litoralis* var. *dasyphyllus* (Trautv.) Roshev. — *A. litoralis* subsp. *intermedius* (Regel) Tzvelev.

Stems 20–60 cm tall, golden-yellow, decumbent, branched, rooting at nodes, with ascending flowering shoots. Vegetative stems foliated to very tip, with many remnants of dead leaves at base. Leaves glaucescent green, alternate, stiff, longitudinally folded, lyrate, glabrous or with sparse hairs. Leaf sheaths short, covered with long slender hairs. Ligules short, densely covered with erect hairs. Panicles 2–5 cm long, spicate, with close branches with spikelets in 2 proximate rows. Spikelets 4–10-flowered, 3.5–4.5 mm long, light green, yellow-brown on maturation, laterally compressed, subsessile (on short stalks). Glumes obtuse, keeled, scarious along margin, with 3–5 indistinct nerves. Lemma glabrous, with 7–9 indistinct nerves, keeled, obtuse, with corona of short hairs on callus. Anthers 1–1.3 mm long.

Solonchak and saline meadows. **West. Sib.**: TYU—Tb (Yalutorovsk region), NO (Barlakol' village), AL—Ba. —Europe, Mid. Asia. Described from East. Kazakhstan.

61. Enneapogon Desv. ex Beauv.

1. **E. borealis** (Griseb.) Honda 1936 in Rep. First Sci. Exped. Manch., sect. IV (4): 101. —*Pappophorum boreale* Griseb.

Annual, finely cespitose, 4–30 cm tall. Stems geniculate at nodes; covered with soft clavate hairs like narrowly linear, longitudinally convolute leaves. Ligules in form of transverse scar. Panicles compressed spikelike, 1–4 cm long, with very short branches. Spikelets 3.5–11 mm long, 2–flowered; lower of them bisexual, upper staminate. Cleistogamous flowers growing sometimes in sheaths of reniform shoots at base of mats. Glumes coriaceous-membranous, lanceolate, puberulent; upper as long as spikelet, lower slightly shorter. Lemma 1.3–3.2 mm long, broadly ovate with 9 nerves, transforming into straight awn 1.5–2 times longer than lemma; plumose in lower part, scabrous in upper. Anthers 0.3–1 mm long. Caryopsis 1.2–2 mm long.

Rocks and rocky steppe slopes. **Cen. Sib.**: KR—Ve (Chingis cliff near Minusinsk), TU. **East. Sib.**: BU—Yuzh (Selenga river—class hab. —and others), ChI—Shi (northern bank of Zun-Torei Lake). —Mid. Asia, Mongolia. China. Map 294.

Genetically related to *E. desvauxii* Beauv. distributed in steppes of North and South America, differing in more profuse pubescence of rachilla segments.

62. Cleistogenes Keng

1. Lemma 4.5–7.5 mm long, with small lobes at tip; cusp or straight awn arising between lobes .... 2.
+ Lemma 3–4.5 mm long, without lobes at tip, acute or with up to 0.5 mm long cusp. Stems suberect .... 2. *C. songorica.*
2. Stem segment between upper node and base of panicle long, nearly as long as rest of stem or more than 1/2. Awns of lemma in chasmogamic spikelets 2.5–6 mm long. Dry stems serpentine ....
.... 3. *C. squarrosa.*
+ Length of stem segment between upper node and panicle base nearly indistinguishable from that of internodes, invariably less than 1/3 rest of stem. Awns of lemma in chasmogamic spikelets up to 2 (3) mm long. Stems straight or slightly curved ....
.... 1. *C. kitagawae.*

1. **C. kitagawae** Honda 1936 in Rep. First Sci. Exped. Manch., Sect. 4 (4): 99. —*Diplachne sinensis* auct. non Hance. —*C. serotina* auct. non Keng —*C. chinensis* auct. non Maxim, nec Keng —*Diplachne serotina* auct. non L., nec Link.

Plant 30–65 cm tall, somewhat violet, with numerous straight, densely foliate stems aggregated into loose mats. Shortened vegetative shoots absent or few. Stems slightly curved in upper half. Leaf blades dark green, 1–5 mm broad, readily perishing on desiccation. Ligules short, comprise profuse fine cilia. Panicles 4–10 cm long. Lemma crinite along edges, linear-lanceolate, with straight 0.5–2 (3) mm long awn arising from recess between teeth.

Rocky slopes and rocks, in steppes. **West. Sib.:** AL—Go. **Cen. Sib.:** KR—Kha, Ve, TU (Bel'bei settlement, Il'inka village). **East. Sib.:** IR—An, BU—Se (lower course of Garga river), Yuzh, ChI—Shi. —Mongolia, Manchuria, Far East. Described from NE China. Map 296.

2. **C. songorica** (Roshev.) Ohwi 1942 in Journ. Jap. Bot. 18: 540. — *Diplachne songorica* Roshev.

Plant 20–25 cm tall, forming compact mats with many short vegetative shoots, not longer than half of stem. Leaf blades linear, acuminate, up to 2 mm broad, stiff, highly declinate from stem. Leaf ligules short, ciliate, with sparse long hairs laterally. Panicles 2–4.5 cm long, spreading, with few spikelets and highly scabrous branches. Spikelets 3–5 (10) mm long, 2–5-(10)-flowered, somewhat violet, awnless, appressed-pilose along nerves and in lower part.

Steppes, rocky slopes. **Cen. Sib.:** TU. —Mid. Asia, Mongolia, Manchuria. Described from East. Kazakhstan.

3. **C. squarrosa** (Trin.) Keng 1934 in Sinensia 5: 149. —*Diplachne squarrosa* (Trin.) Maxim.

Plant (10) 15–35 cm tall, forming somewhat lax mats. Stems spreading, several, highly flexuous in lower part on desiccation, highly flattened, convex on one side and slightly grooved on other. Sheaths light green, recurved away from stem in middle cauline leaves. Panicles with few flowers, lax. Lemma lanceolate, with awn nearly as long or half of lemma.

Steppes, rocky slopes, sand. **West. Sib.**: OM (between Cherlak and Bol. Atmas, Krasnoyarsk and Cherlak villages), NO (Nov. Sharap, Barabka and Marksist villages), KE (Pesterevo village), AL—Ba, Go. **Cen. Sib.**: KR—Kha, Ve, TU. **East. Sib.**: IR—An, BU—Se, Yuzh, ChI—Shi, YAK—Vi (environs of Yakutsk and Olekminsk: Kil'demtsy village). — Mid. Asia, Mongolia, West. China, Manchuria, Far East. Described from Kazakhstan. Map 295.

63. Tripogon Roemer et Schultes

1. **T. chinensis** (Franchet) Hackel 1903 in Bull. Herb. Boiss.: 2 (3): 503.

Plant 10–15 (20) cm tall, forming small dense mats. Leaf blades setaceous, 0.5–0.7 mm broad, pilose along margin. Ligules short, ciliate. Spikes secund, thin. Spikelets linear-lanceolate, 3–4- flowered, in 2 close rows. Glumes unequal, shorter than spikelet. Lemma lanceolate, with short hairs at base; 0.9–1.7 mm long terminal awn arising between 2 short awnlike acuminate lobes. Anthers 0.8–1.3 mm long.

Rocky slopes and rocks. **East. Sib.**: ChI—Shi. —Mongolia (east), Far East, East. China, Korea. Described from Nor. China. Map 293.

64. Eragrostis Wolf

1. Spikelet stalks without cyathiform (in form of tubercles with broad but shallow pit at center) glands .... 2.
+ Spikelet stalks with 1–3 cyathiform glands. Panicles glabrous at lower nodes. Lemma ovate or broadly ovate, obtuse .... 3. *E. minor.*
2. All parts of plant without cyathiform glands .... 3.
+ Stems, sheaths, leaf blades somewhat covered with cyathiform glands ....1. *E. amurensis.*
3. Panicles broadly spreading, glabrous at nodes. Stalks of lateral spikelets up to 8 mm long .... 2. *E. imberbis.*
+ Panicles at lower nodes crinite. Stalks of lateral spikelets 1.5–6 mm long .... 4. *E. pilosa.*

1. **E. amurensis** Probat. 1981 in Bot. zhurn. 66 (11): 1591. —*E. pilosa* auct. non Beauv. p. p.

Plant 10–30 cm tall, with many stems, straight or geniculate at lower nodes; like sheath, somewhat covered with cyathiform glands. Leaf blades up to 3 mm broad, acuminate, glandular along keel, sometimes also along main veins. Panicles (5) 10–20 cm long, somewhat spreading, pilose or glabrous at lower nodes. Stalks of lateral spikelets 2–12 mm long. Spikelets 3–5 mm long, 1–1.3 mm broad, grayish-green, with 5–10 florets.

Sand, fields, and roadsides. **West. Sib.**: TO (Fokinsk settlement, Chilino village), KE (Kondoma and Itkary villages), AL—Ba (Severnyi settlement between Klyucha and Vasil'chakova). **Cen. Sib.**: KR—Kha, Ve, TU (Targalyk, Chaa-Khol' and Aryg-Bazhi settlements). **East. Sib.**: IR—Pr (Nakanno and Nepa settlements), BU—Se (Barguzin settlement), Yuzh, ChI—Shi, YAK—Vi (Peledui settlement), Al (Amga river estuary). —Far East, Manchuria. Described from Amur region. Map 299.

2. **E. imberbis** (Franchet) Probat. 1985, Sosud. rast. sov. Dal'n. Vostoka 1: 353. *E. pilosa* subsp. *imberbis* (Franchet) Tzvelev.

Plant 20–40 cm tall, with stems geniculate at lower nodes. Leaf blades 3–4 mm broad, acuminate, scabrous on upper surface. Panicles long (10–20 cm), broadly spreading, with very slender branches, glabrous at nodes. Spikelets linear, grayish-lilac, 3–8 mm long on stalks. Glumes acute, keeled.

Sand and pebble riverbeds. **East. Sib.**: ChI—Shi (Abagaitui and Kailastui settlements). —Far East, Mongolia (east), Manchuria. Described from NE China.

3. **E. minor** Host 1809, Gram. Austr. 4: 15.

Annual, with spreading, ascending stems (5) 10–40 cm tall. Leaf blades 2–5 mm broad; lower surface covered like sheaths with divergent distant long hairs. Ligules short, ciliate. Panicles 5–15 cm long, spreading, with horizontally declinate, rather thick branches, without long hairs at nodes. Spikelets grayish, 5–15-flowered, 4–10 mm long, up to 2 mm broad. Stalks of spikelets, with cyathiform glands like other plant parts. Glumes 1.3–1.5 mm long, acute. Lemma with 3 sharp nerves, broadly ovate, obtuse.

Roadsides, fields, rubble slopes. **West. Sib.**: AL—Go. **Cen. Sib.**: KR—Ve (Minusinsk town), TU. **East. Sib.**: BU—Yuzh, ChI—Shi (Borzya station, Chita town, Dureny settlement). —Europe, Caucasus, Mediterranean, Mid. and West. Asia, Mongolia, Japan, China, North America. Described from Italy. Map 297.

4. **E. pilosa** (L.) Beauv. 1812, Ess. Agrost.: 162.

Annual, 15–50 cm tall. Stems smooth, geniculate at lower nodes. Sheath glabrous, sheath blade joints crinite. Panicles 10–25 cm long, spreading, with slender branches, crinite at lower nodes. Spikelets dark gray, 4–10-flowered, 3–5 mm long, 1–1.5 mm broad. Glumes up to 1.5 mm long, lanceolate. Lemma broadly lanceolate, with 3 nerves, keeled.

Roadsides, fields, rubble slopes, river sand. **West. Sib.**: KU (environs of Kurgan, Vargashi village), TO, NO (environs of Novosibirsk, Chany Lake), KE (between Kaltan and Ashmarina, Mrassu river, Azhendarovo village). **Cen. Sib.**: KR—Kha, Ve, TU (Senek and Baryk interfluve region). **East. Sib.**: IR—An (Taishet settlement), BU—Se

(Kurumkan settlement), Yuzh, ChI—Shi. —Europe, Mediterranean, Mid. Asia, Central and South. Asia, Far East. Introduced in many other countries. Described from Italy. Map 298.

65. Crypsis Aiton

1. Inflorescence more or less terminal, rarely axillary, compact oblong or oval, spicate, 2.5–4 times longer than broad. Main rachis distinctly visible .... 2.
+ Inflorescence axillary, capitate or semiglobose, shorter than broad. Main rachis not visible .... 1. *C. aculeata.*
2. Sheath of terminal leaves longer than blades, narrow or barely enlarged, away from inflorescence base. Panicles cylindrical or elongate-ovate .... 2. *C. alopecuroides.*
+ Sheath of terminal leaves shorter than blade, greatly enlarged and located at base of oblong-ovate inflorescence, marking its lower part .... 3. *C. schoenoides.*

1. **C. aculeata** (L.) Aiton 1789, Hort. Kew. 1: 48.

Stems 1–35 cm tall, procumbent, several, somewhat branched, straight or geniculate, strongly unequal in length. Leaf blades 0.5–4 cm long, 1–3 mm broad, stiff, grayish-green, strongly (almost at right angle) declinate from stem, glabrous or with diffuse long hairs on both surfaces, acute. Sheath broad, somewhat flattened and declinate from stem, glabrous, with long cilia in sheath blade joints, considerably shorter (by 2 or 3 times) than blades. Inflorescence 5–15 mm broad, almost as high, capitate or semiglobose, main rachis of inflorescence not visible; sheath of 2 or 3 upper leaves proximate at inflorescence base and covering it considerably from all sides. Spikelets 3–4 mm long, subsessile. Glumes scarious, unequal, lanceolate, keeled, with very short cilia along keel. Lemma slightly longer than glumes. Stamens 2. Anthers 0.7–1.2 mm long.

Sandy banks of salt lakes, solonchaks, saline meadows. **West. Sib.:** AL—Ba. —Europe, Caucasus, Mid. Asia, Mediterranean, West. Asia, West. China, Mongolia. Described from Mediterranean. Map 276.

2. **C. alopecuroides** (Piller et Mitt.) Schrader 1806, Fl. Germ. 1: 167. —*Heleochloa alopecuroides* (Piller et Mitt.) Host ex Roemer.

Stems 10–30 cm long, procumbent, sometimes erect, many, geniculate. Leaf blades 2–6 cm long, 1.5–3 mm broad, flat or longitudinally folded, glaucous green, rather stiff, scabrous on upper surface due to very short spinules or hairs. Sheath narrow, cylindrical, usually longer than blades, slightly flattened. Inflorescence 2–5 cm long, 3–5 mm broad, cylindrical, slightly narrow in lower part; sheaths of upper leaves not reaching inflorescences. Spikelets about 2 mm long, on short stalk. Glumes narrowly lanceolate, keeled on back and with short cilia along keel. Lemma slightly longer than spikelets. Stamens 3. Anthers up to 1 mm long.

Wet saline sandy sites. **West. Sib.**: AL—Ba (Lokot' village on Alei river). —Europe, Caucasus, Mid. and West. Asia, Mediterranean. Described from Czechoslovakia.

3. **C. schoenoides** (L.) Lam. 1791, Tabl. Encycl. Méth. Bot. 1: 166. — *Heleochloa schoenoides* (L.) Host ex Roemer.

Stems 5–25 cm long, flattened along ground, several, straight or geniculate at nodes. Leaf blades 2–6 cm long, 1.5–3 mm broad, flat or longitudinally folded, grayish-green, rather stiff, scabrous on upper surface due to short acute hairs and with scattered long hairs interrupted below. Sheath enlarged, inflated, usuₐlly shorter than blade, flattened. Inflorescence 8–20 mm long, 5–7 mm broad, ovate or oblong-oval; sheaths of upper leaves located at base of inflorescence, masking its lower part. Spikelets about 2 mm long, glumes slightly unequal, acuminate, with short cilia along keel. Lemma longer than glumes. Stamens 3. Anthers 0.6–1.2 mm long.

Wet sandy saline sites. **West. Sib.**: OM (Glyadinsk village), AL—Ba (Veselyi Yar village on Alei river). —Europe, Caucasus, Mid. and West. Asia, Mediterranean, the Himalayas, West. China, Mongolia. Described from the Mediterranean.

66. Arundinella Raddi

1. **A. anomala** Steudel 1854 in Syn. Pl. Glum. 1: 116. —*A. hirta* var. *ciliata* (Thunb.) Koidz.

Perennial, with horizontal rhizomes bearing shoots, covered with white scales. Stems 30–80 cm tall, straight. Leaves 10–17 cm long, 3–8 mm broad, linear, acuminate, covered like sheath with cilia along keels, sometimes with diffuse hairs throughout surface. Ligules short, 2–4 mm long, pilose. Panicles up to 11 cm long, oblong, somewhat interrupted in lower part. Rachis and branches of panicles covered with short spinules. Spikelets 3–4.5 mm long, 2–flowered, greenish-reddish. Lower florets staminate, upper bisexual. Glumes with 3–5 nerves, often extended into terminal cusp, with few hairs. Lemma of bisexual floret with hairs at base and 0.5 mm long terminal awn. Anthers 1–1.5 mm long. Caryopsis up to 1.5 mm long.

Southern steppe rocky slopes. **East. Sib.**: ChI—Shi. —Far East, Japan, China, South Asia. Described from Japan. Map 270.

67. Panicum L.

1. Panicles somewhat compressed, nutant. Glumes yellow or orange, 1.7–2.2 mm broad .... 1. *P. miliaceum* s. str.

+ Panicles broadly spreading, not nutant. Glumes brownish cinnamon, 1.3–1.6 mm long .... 2. *P. miliaceum* subsp. *ruderale*.
1. **P. miliaceum** L. s. str. 1753, Sp. Pl.: 58.
Annual, 50–100 cm tall. Leaves broadly linear, sheaths and leaf blades covered with slender squarrose hairs. Ligules short, covered with long hairs along margin. Panicles spreading or somewhat compressed, nutant, with slender branches terminating in spikelets. Spikelets 4–5 mm long, 1.5–2.3 mm broad, oblong-ovate, with 2 florets one underdeveloped and represented by lower glume which is broadly ovate, acuminate, 5-nerved, almost half length of spikelet. Glumes of developed floret similar (one upper glume and one lemma), with extended tip and 9–12 distinct nerves. Lemma and palea of fertile floret chondroid, ovate, glabrous, lustrous, yellow or orange. Anthers 1–1.7 mm long.

Cultivated plant but sometimes found as weed on roadsides and in other plantations. **West. Sib.:** TYU—Tb, KU, OM, NO, AL—Ba, Go (Chemal village). **Cen. Sib.:** TU. **East. Sib.:** IR—An, BU—Yuzh (Ust'-Kyakhta), ChI—Shi (environs of Chita, Ulety settlement). —Europe, Caucasus, Mid. Asia. Far East. Described from India.

2. **P. miliaceum** subsp. **ruderale** (Kitag.) Tzvelev 1968 in Novosti sist. vyssh. rast. 1968: 18.
Annual, 50–100 cm tall. Leaves linear, acuminate. Leaf blades glabrous, sheath covered with long squarrose hairs. Panicles broadly spreading, not nutant, with slender branches terminating in spikelets. Latter oblong-ovate, 4–5 mm long, 1.5–2 mm broad. Lemma and palea of fertile floret chondroid, glossy, brownish-cinnamon.

Cultivated plant, sometimes found as weed in other plantations. **West. Sib.:** AL—Ba (Berezovka settlement). —Mid. and West. Asia, West. China, Mongolia, East. Asia. Described from Nor. China.

68. Echinochloa Beauv.

1. Palea of sterile floret at least half as long as awnless or awned lemma (or nearly as long). Only some spikelets with joint at base and shedding readily in fruit .... 2.
+ Palea of sterile floret absent or present as barely visible scale; lemma usually with long (3–5 cm) violet awn. Nearly all spikelets with joint at base and shedding readily in fruit .... 1. *E. caudata*.
2. Primary branches of panicle usually with shortened branches of second order only at base; elsewhere, spikelets forming nearly regular rows .... 2. *E. crusgalli*.
+ Primary branches of panicles with numerous, highly shortened branches of second order; thus spikelets not forming regular rows .... 3. *E. occidentalis*.

1. **E. caudata** Roshev. 1934 in Fl. SSSR 2: 35. —*E. crusgalli* subsp. *caudata* (Roshev.) Tzvelev.

244

Annual, green or slightly violet. Stems up to 100 cm or more tall, 3–8 cm diam, together with sheaths glabrous and smooth. Leaf blades (3) 5–12 (18) mm broad, flat, glabrous, sharply scabrous along margin and sometimes even along veins. Ligules absent. Panicles 10–20 cm long, with hispid main rachis and branches covered additionally with profuse, uniformly distributed, long (longer than spikelets) cilia. Spikelets 2.5–3 mm long, with long violet awn (3–5 cm) arising from tip of lemma of sterile floret, palea almost wholly reduced. Glumes unequal; upper thrice longer than lower; like lemma of sterile floret, covered with fine spinules and diffuse bristles of different size. Lemma and palea of fertile floret 2–2.5 (3) mm long, glabrous, smooth, ovate, acuminate, glossy, with dingy violet tinge. Anthers 0.4–0.5 mm long.

Banks of water reservoirs. **East. Sib.**: ChI—Shi (Argun' river—class. hab., Abagaitui and Kailastui villages). —Far East, eastern part of Mongolia and China, Japan, Korean peninsula.

2. **E. crusgalli** (L.) Beauv. 1812, Ess. Agrost.: 161. —*Panicum crusgalli* L.

Annual, dark green plant. Stems 10–60 (80) cm tall; glabrous and smooth like sheath. Leaf blades 4–10 (15) mm broad, flat, glabrous, sharply scabrous along margin. Ligules absent. Panicles 5–15 (20) cm long, narrow, with appressed or slightly declinate, alternate and often interrupted hispid or sharply scabrous branches, with clusters of very long cilia at point of emergence of secondary branches. Spikelets 2.5–3.5 mm long, awnless or with awn of different length (up to 2–4 cm). Glumes unequal; lower 2 or 3 times shorter than upper. Lemma of sterile florets and upper glume nearly equal, with 5–7 nerves, finely scabrous, covered with stiff cilia along nerves. Palea of sterile florets scarious, up to 1.5 times shorter than lemma. Lemma and palea of fertile florets glabrous, smooth, glossy, ovate, light in color. Anthers 0.4–0.8 mm long.

As weed along field borders and roadsides, sometimes on riverbed sand, banks of water reservoirs, wet meadows. **West. Sib.**: TYU—Tb, KU, OM, TO, NO, KE, AL—Ba. **Cen. Sib.**: KR—Kha, Ve, TU. **East. Sib.**: IR—An, BU—Yuzh, ChI—Shi. —Widely distributed in moderately warm, subtropical and tropical regions of both hemispheres. Described from Europe and the USA.

3. **E. occidentalis** (Wiegand) Rydb. 1931, Brittonia 1: 82. —*E. spiralis* Vasinger. —*E. crusgalli* subsp. *spiralis* (Vasinger) Tzvelev.

Annual, light or grayish, green, with many stems 30–100 cm tall. Sheath, like stem, glabrous, smooth. Leaf blades 5–15 mm broad, flat, glabrous, with thickened whitish margin, smooth or barely scabrous due to diffuse short spinules or tubercles. Ligules lacking. Panicles 13–20 cm long, 3–5 cm broad, dense, usually with branches obliquely declinate upward, dense clusters of 2 or 3 emerging from main rachis; branches sharply scabrous, with longer cilia at point of branching. Primary branches further branched, hence spikelets not in regular rows but somewhat in spiral form. Spikelets 2.5–3 mm long, awnless or with 1.5–2 cm long awn. Lemma 2 or 3 times shorter than palea. Lemma of sterile florets and upper glume nearly equal, with 5–7 nerves, finely scabrous and, along nerves hispid. Palea of sterile florets nearly as long as lemma. Lemma and palea of fertile floret glabrous, smooth, glossy, white. Anthers 0.4–0.7 mm long.

Wet marshy meadows, along field borders. **West. Sib.**: AL—Ba. **East. Sib.**: BU—Yuzh. —Tropical, subtropical and temperate warm regions of both hemispheres. Described from North America (Illinois state).

69. Eriochloa Kunth

1. **E. villosa** (Thunb.) Kunth 1829, Révis. Gram. 1: 30.

Annual with branched stems 30–80 cm tall, geniculate at base. Leaves 4–8 mm broad, flat, puberulent. Spikelets 4–5 mm long, ovate, single-flowered, aggregate in secund clusters arising from main rachis of inflorescence, pilose. Inflorescence branches also with dense long hairs. Glumes ovate, whitish green or violet, slightly pilose, with 5 nerves. Lemma and palea green, coriaceous, transversely rugose, chondroid during defloration.

As weed in potato fields. **West. Sib.**: OM (environs of Omsk). —Caucasus, West. Asia, Far East, East. and South. Asia. Described from Japan.

70. Digitaria Hall.

1. Spikelets 1.8–2 mm long. Upper glume and lemma of reduced floret rather profusely pubescent with short and long simple hairs. Plants forming procumbent shoots rooting at nodes .... 1. *D. asiatica.*

+ Spikelets 2.2.–2.5 mm long. Upper glume profusely pubescent. Lemma of reduced floret rather pubescent with clavate hairs near margin and between nerves. Plants without procumbent shoots rooting at nodes .... 2. *D. ischaemum.*

1. **D. asiatica** Tzvelev 1963 in Bot. mat. (Leningrad) 22: 64.

Annual, 15–30 cm tall, forming compact mats, often with elongated, decumbent shoots rooting at nodes. Leaves linear-lanceolate, shortened, glabrous, with hairs at base. Ligules 1–1.6 mm long. Panicles with 2–5 linear, long digitate (sometimes, lower branches in axils of root leaves) branches; spikelets 1.8–2 mm long, lanceolate-ovate, arranged in groups of 2 or 3 in 2 close rows, with upper fertile bisexual floret and lower sterile floret with developed lemma. Lower glume reduced, upper pilose, with 3–5 nerves. Lemma of reduced floret with 5–7 nerves, rather pilose between nerves and close to margin, smooth along nerves; pubescence of glumes and lemma consisting of simple hairs.

Banks of rivers in forest belt. **East. Sib.**: IR—An (Tal'niki settlement, Mal. Belaya river). —Europe, Caucasus, Asia. Described from Japan.

2. **D. ischaemum** (Schreber) Muehl. 1817, Deser. Gram. 131. —*D. linearis* (Krocker) Crepin p. p. —*Panicum lineare* Krocker.

Annual, 10–50 cm tall, forming compact mats. Leaves linear-lanceolate, glabrous. Ligules 1–2 mm long. Panicles with 2–6 spicate branches; spikelets 2.2–2.5 mm long, lanceolate-ovate, in groups of 2 or 3; upper glume profusely pilose, with 3–5 nerves; lemma of reduced floret with 5–7 nerves, rather pubescent with short and long clavate hairs.

Meadows, dry exposed slopes. **West. Sib.**: AL—Go (Uznezya river, Chemal village). **Cen. Sib.**: Kha (Saragash village), KR—Ve (Krasnoturansk town). **East. Sib.**: ChI—Shi. —Europe, Caucasus, Asia, North America. Introduced in many countries. Described from FRG. Map 306.

71. Setaria Beauv.

1. Upper glume nearly as long as spikelet. Lemma of fertile floret almost smooth, indistinctly punctate-rugose. Bristles green, violet or dark purple .... 2.
+ Upper glume shorter than spikelet by 1/3 or 1/2. Lemma of fertile floret sharply transversely rugose. Bristles golden-yellow ....
.... 1. *S. pumila.*
2. Stems often several, usually branched from base, with 3–5 nodes. Leaf blades 2–15 mm broad. Panicles 2–12 cm long .... 3.
+ Stems single or few, usually not branched, with 5–10 nodes. Leaf blades 6–20 mm broad. Panicles 5–20 cm long ....
.... 5. *S. viridis* subsp. *pycnocoma.*
3. Panicles cylindrical, nearly uniformly broad or slightly enlarged in middle part. Bristles turned laterally, often flexuous, green or violet .... 4.
+ Panicles oval or obovate, rarely oblong, narrowing wedgelike toward base. Bristles straight, obliquely erect, yellowish, green or dark purple .... 4. *S. viridis* subsp. *purpurascens.*
4. Stems 10–70 cm tall, erect. Panicles 3–12 cm long, 7–15 (20) mm broad, dense, compact, cylindrical. Leaves and sheaths green ....
.... 2. *S. viridis* s. str.
+ Stems 5–25 (40) cm tall, ascending, sometimes geniculate in lower part. Panicles 1–5 (7) cm long, 5–7 mm broad, narrowly cylindrical, interrupted in lower part. Leaves and sheaths, frequently violet .....
.... 3. *S. viridis* subsp. *glareosa.*

1. **S. pumila** (Poiret) Schultes 1824 in Schultes et Schultes fil., Mantissa 2: 274. —*S. glauca* auct. p. p.

Annual, (10) 15–50 (70) cm tall. Stems straight, single, rarely poorly branched from base, glabrous; scabrous and densely puberulent only under inflorescence. Sheaths glabrous, margin also glabrous. Leaf blades 2–7 mm broad, flat, grayish-green, glabrous, slightly scabrous, with diffuse long hairs on upper surface near sheath. Ligules consist of rows of dense short, rather stiff hairs. Panicles 1–6 cm long, spicate, cylindrical, golden-yellow, rachis puberulent (villous). Panicle branches usually with solitary fully developed spikelet, rarely additionally with 1 or 2 underdeveloped and 4–12 bristles 3–8 mm long. Spikelets 2.8–3.4 mm long, easily shedding. Upper glume 1/3 or 1/2 shorter than spikelet. Lemma of fertile floret highly transverse-rugose, exserted from glume. Anthers 0.8–1.2 mm long, dark brown or purple-violet.

As weed in farms, borders, fallow lands, roadsides, less often on sand and pebble riverbeds, pine forests. **West. Sib.:** AL—Ba, Go, NO (Barabka village, Iskitimsk region), KE (Kondoma village, Tashtagol'sk region). **Cen. Sib.:** KR—Kha, Ve. **East. Sib.:** ChI—Shi (near Chita and Sretensk Experimental Farm, introduced). —Europe, Caucasus, Mediterranean, Mid. and West. Asia, Far East, the Himalayas, Mongolia, China, Japan, Korean peninsula, North and South America. Described from Europe. Map 301.

2. **S. viridis** (L.) Beauv. s. str. 1812, Ess. Agrost.: 51.

Annual, 10–70 cm high, highly bushy from base, less often with few or stray stems, scabrous only under inflorescence. Sheaths glabrous, lower ones often with diffuse hairs, ciliate along margin. Leaf blades 2–12 mm broad, flat, acute, reaching or exceeding inflorescence base, scabrous along veins due to extremely fine spinules. Ligules in form of bands of dense cilia about 1 mm long. Panicles 0.7–1.2 cm broad, cylindrical or oval, dense, green or violet (var. *weinmannii* (Roemer et Schultes) Borb.), rachis with distant hairs. Panicle branches with aggregate spikelets 2–2.5 mm long, surrounded by 3–8 (10) mm long scabrous bristles, 2 or 3 on each spikelet. Upper glume nearly as long as spikelet. Lemma of fertile floret weakly punctate-rugose. Anthers 0.4–0.7 mm long.

Weed in crops, along field borders, roadsides, inhabited sites; wild only in southern regions on sand and pebble riverbeds, floodplain and dry valley meadows, sandy steppes, rocky-rubble slopes and sand dunes. **West. Sib.:** TYU—Tb, KU, OM, NO, KE, AL—Ba, Go. **Cen. Sib.:** KR—Kha, Ve, TU. **East. Sib.:** IR—An, Pr, BU—Se, Yuzh, ChI—Shi. Warm and temperate belts of both hemispheres (Holarctic). Described from Cen. Europe. Map 302.

3. **S. viridis** subsp. **glareosa** (V. Petrov) Peschkova, comb. nova. — *S. glareosa* V. Petrov 1930, Fl. Yakut. 1: 18. —*S. viridis* auct. quoad pl.

Plant 5–25 (40) cm high, cespitose. Stems procumbent or ascending, sometimes geniculate, strongly unequal in length within a given mat. Leaf blades 2–6 mm broad, flat, green or violet, acute, usually short, very rarely reaching inflorescence base. Panicles 0.5–0.7 mm broad, narrowly cylindrical, often interrupted, green or with violet tinge in lower part. Bristles in each branch few and rather flexuous.

Sand and pebble beds along river valleys. **Cen. Sib.**: KR—Tn, Ve. **East. Sib.**: IR—An, Pr, BU—Se, Yuzh, ChI—Shi, YAK—Vi, Al (Amga river near ferry on Ust'-Maisk road—class. hab.—and others). —Endemic. Map 305.

S. viridis subsp. *viridis* intermediate to subsp. *glareosa* is found in southern hilly regions of West. Siberia and in Tuva hills.

4. **S. viridis** subsp. **purpurascens** (Maxim.) Peschkova comb. nova. —*S. viridis* var. *purpurascens* Maxim. 1859, Prim. Fl. Amur.: 330. —*S. weinmannii* auct. non Roemer et Schultes.

Plants 10–60 cm tall, with many stems. Stems straight or geniculate, purple at base and sometimes throughout. Leaf blades 2–10 mm broad, flat, green or purple, relatively short, far short of reaching inflorescence base. Panicles oblong or obovate, 1.5–4 (8) cm long, 1–2 cm broad, dense, compact, narrowed wedgelike at panicle base due to flabellate arrangement of bristles without spikelets. Bristles yellowish-green or dark purple, several, slender, divaricate.

Dry steppes, on stony and rocky slopes. ChI—Shi (SE). —Far East, NE China (Manchuria), Japan, Korean peninsula. Described from Bureya river estuary. Map 303.

5. **S. viridis** subsp. **pycnocoma** (Steudel) Tzvelev 1968 in Novosti sist. vyssh. rast. 5: 19.

Plants 50–100 cm tall, with 5–10 nodes. Stems single or few, straight. Leaf blades 10–20 mm broad, up to 40 cm long, supporting and exceeding inflorescences. Panicles up to 20 cm long, green, very dense, straight or somewhat curved, cylindrical, slightly lobed, lower branches up to 3 cm long.

As weed along field borders (probably introduced). East. Sib.: ChI—Shi (Byankino village, Nerchinsk region). —Eurasia. Described from Japan.

S. italica is cultivated for fodder and human consumption. Usually not found outside fields.

72. Spodiopogon Trin.

1. **S. sibiricus** Trin. 1820, Fund. Agrost.: 192.

Perennial 50–100 cm tall, with decumbent subsurface shoots; surface shoots covered with coriaceous scale leaves. Leaves 5–15 mm broad, broadly linear, flat, stiff, with spare long hairs on upper surface. Ligules 0.4–1 mm long, membranous. Panicles 10–20 cm long, fairly dense, with glabrous smooth branches. Spikelets 4–6.5 mm long, 2-flowered, disposed in groups of 2 or 3. Upper floret bisexual, lower sterile or staminate. Glumes 3; lower leptodermatous, densely covered with long soft hairs; inner scarious. Lemma membranous, bifid, with awn in incision. Awn 10–12 mm long, contorted and geniculate. Anthers 2–3 mm long. Caryopsis 2–2.5 mm long.

Steppe rocky slopes and dry valley meadows. **East. Sib.:** IR—An, BU—Yuzh, ChI—Shi. —Mongolia, East. Asia. Described from Irkutsk region. Map 304.

Plant Distribution
Maps

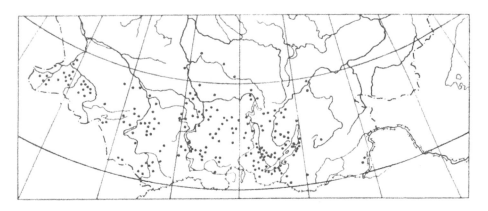

1. *Brachypodium pinnatum.*

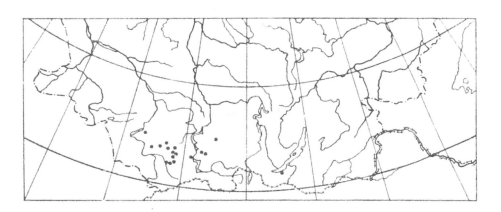

2. *Brachypodium sylvaticum.*

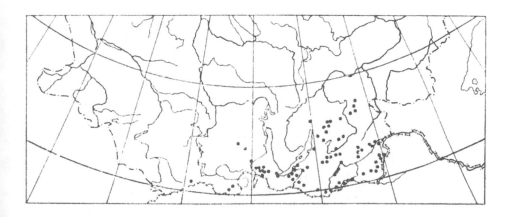

3. *Elymus confusus.*

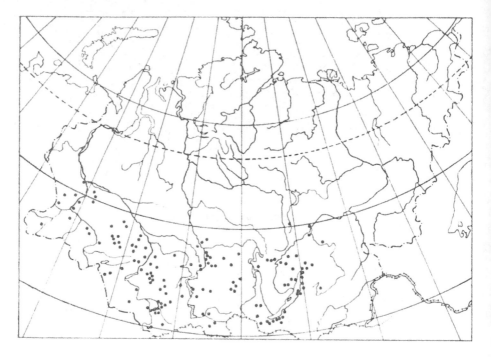

4. *Elymus caninus.*

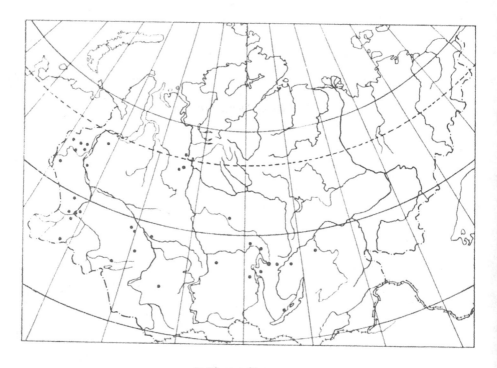

5. *Elymus fibrosus.*

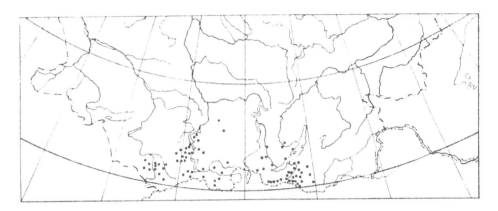

6. *Elymus exselsus.*

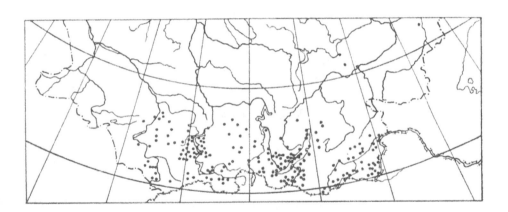

7. *Elymus gmelinii.*

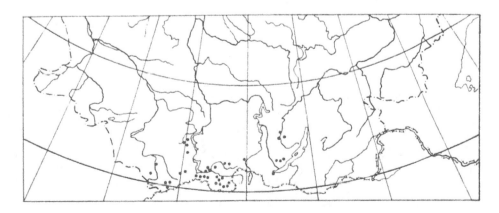

8. *Elymus komarovii.*

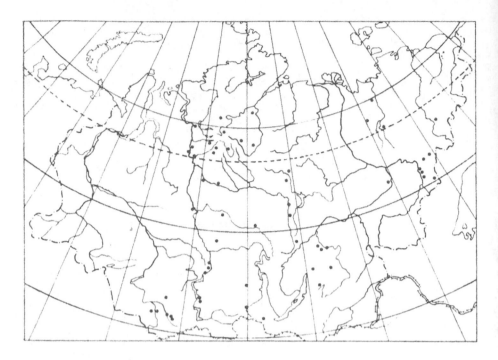

9. Elymus jacutensis.

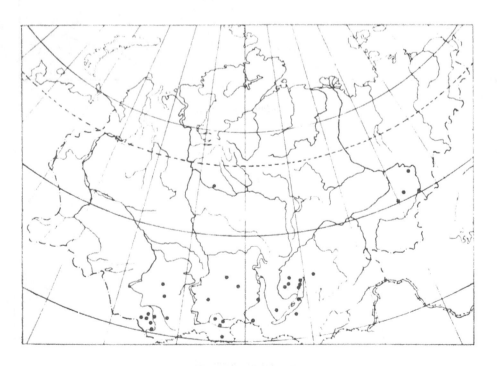

10. Elymus ircutensis.

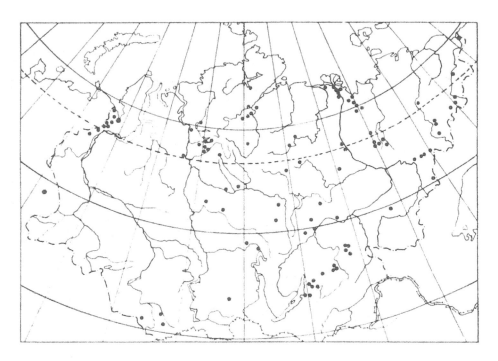

11. *Elymus kronokensis.*

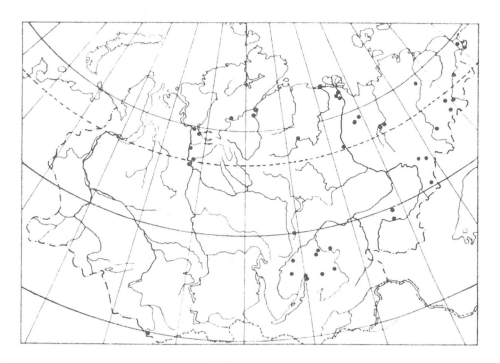

12. *Elymus macrourus.*

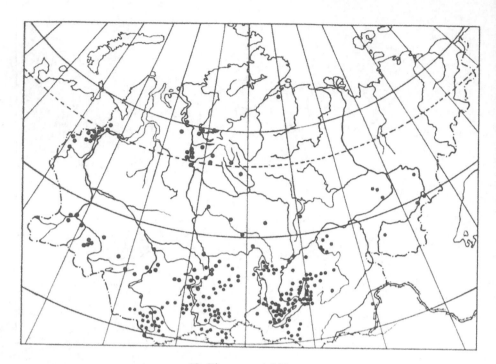

13. *Elymus mutabilis.*

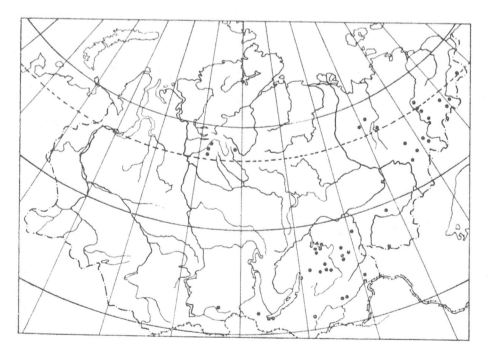

14. *Elymus pubiflorus.*

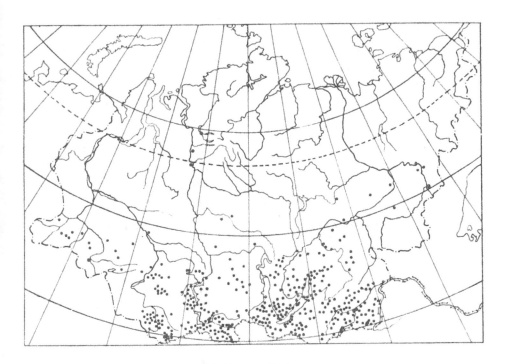

15. *Elymus sibiricus.*

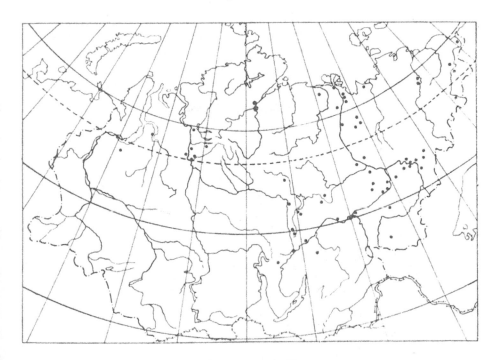

16. *Elymus subfibrosus.*

260

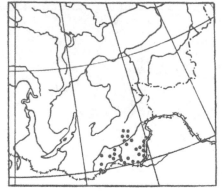

17. *Elymus dahuricus.*

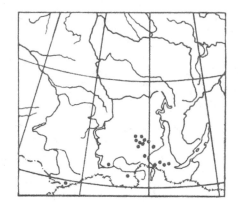

18. *Elymus sajanensis.*

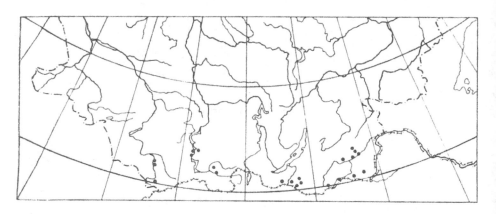

19. *Elymus pendulinus.*

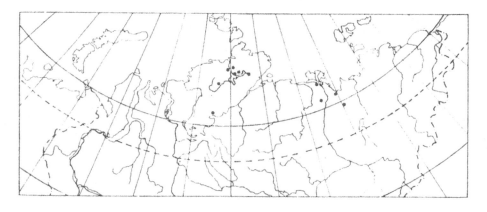

20. *Elymus vassiljevii.*

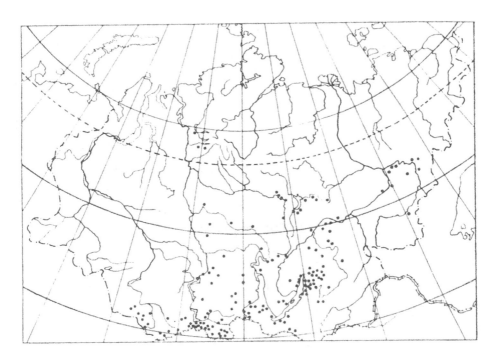

21. *Elymus transbaicalensis.*

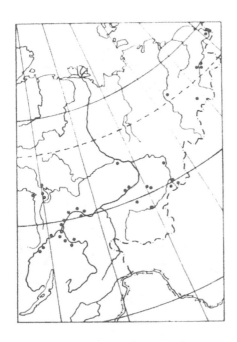

22. *Elytrigia jacutorum.*

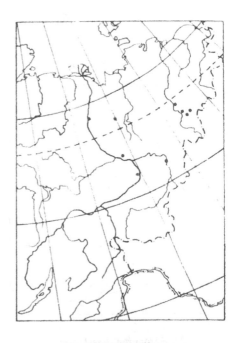

23. *Elytrigia villosa.*

262

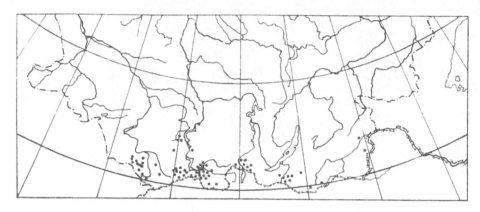

24. Elytrigia gmelinii.

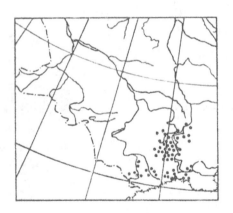

25. Elytrigia geniculata.

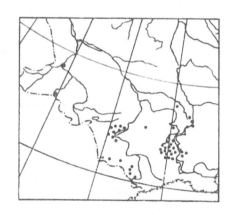

26. Elytrigia lolioides.

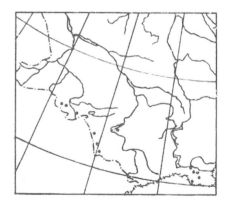

27. Agropyron desertorum.

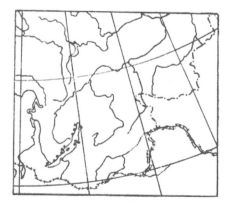

28. Agropyron distichum.

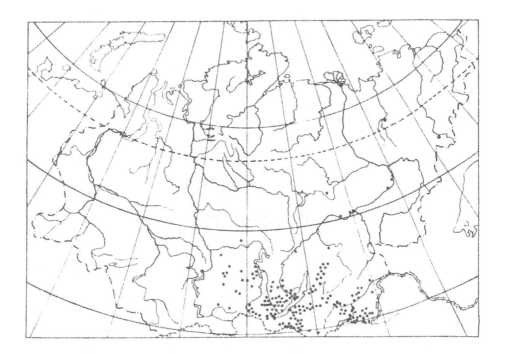

29. Agropyron cristatum.

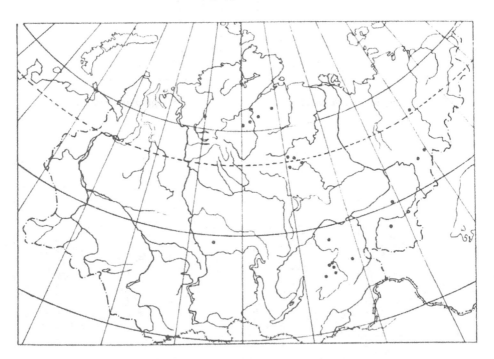

30. Hystrix sibirica.

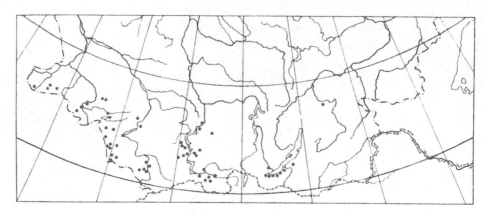

31. Agropyron pectinatum.

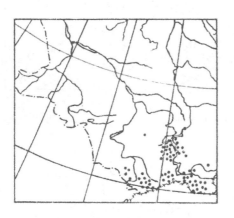

32. Agropyron kazachstanicum.

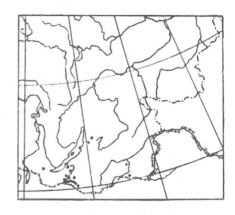

33. Agropyron michnoi.

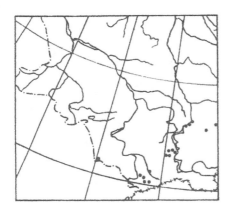

34. Agropyron pumilum.

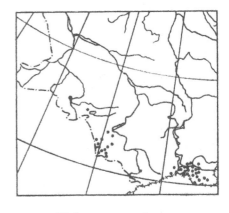

35. Leymus angustus.

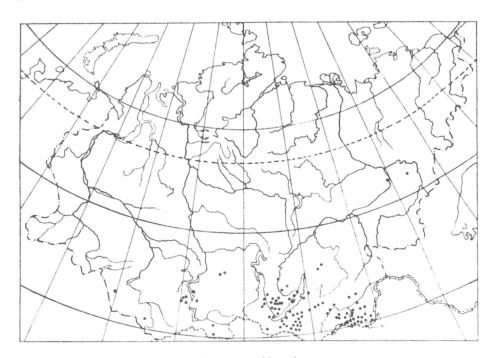

36. Leymus chinensis.

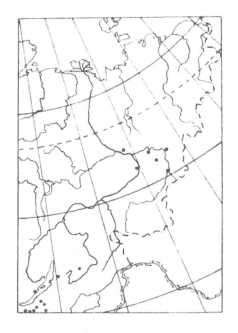

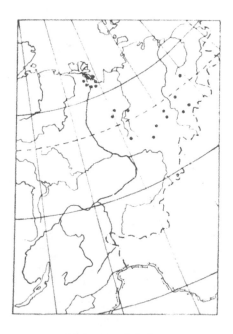

37. Leymus buriaticus.

38. Leymus interior.

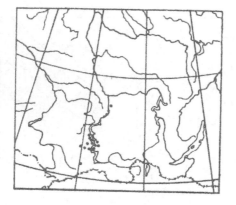

39. *Leymus chakassicus.*

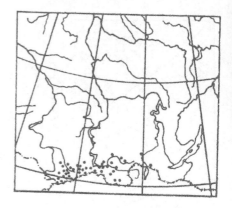

40. *Leymus dasystachys.*

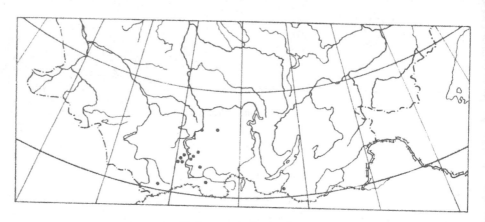

41. *Leymus jenisseiensis.*

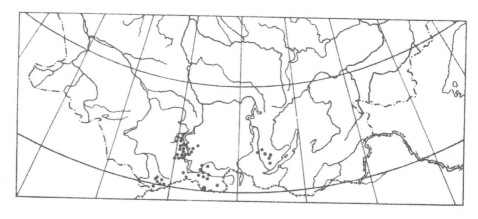

42. *Leymus ordensis.*

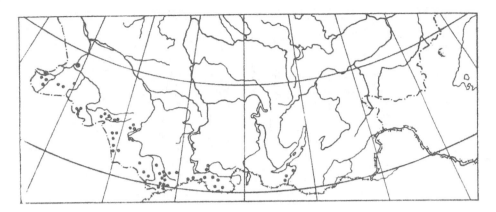

43. Leymus paboanus.

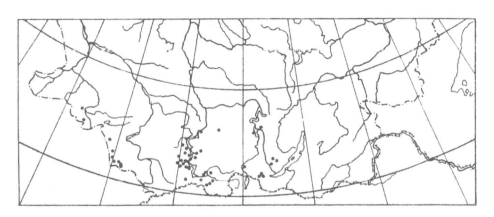

44. Leymus ramosus.

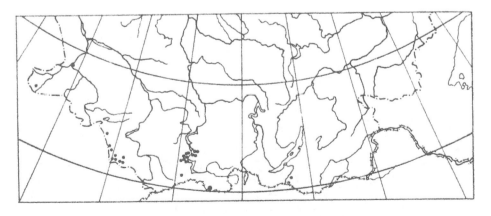

45. Leymus racemosus subsp. *crassinervius.*

268

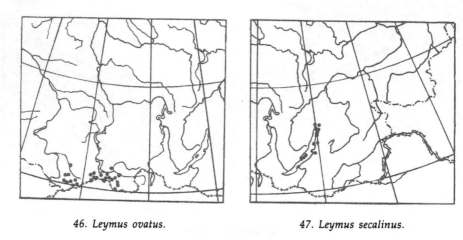

46. *Leymus ovatus.*   47. *Leymus secalinus.*

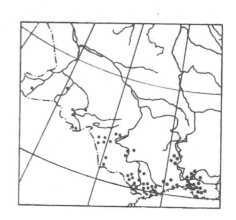

48. *Leymus sphacelatus.*   49. *Psathyrostachys Juncea.*

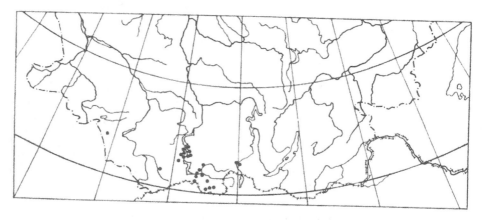

50. *Leymus tuvinicus.*

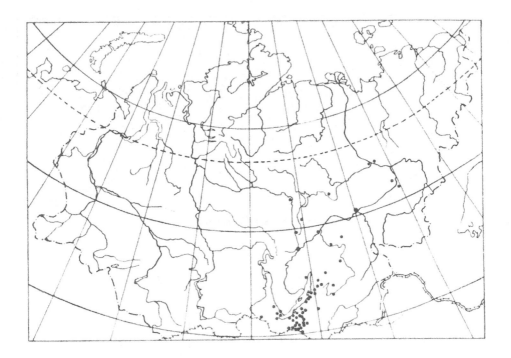

51. Leymus littoralis.

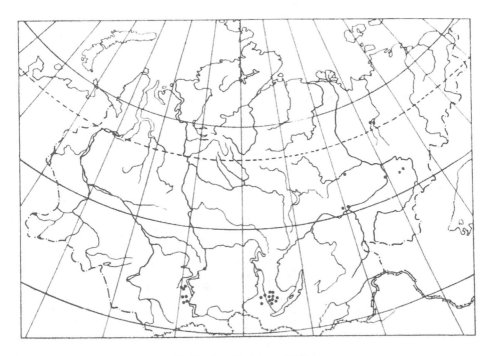

52. Psathyrostachys caespitosa.

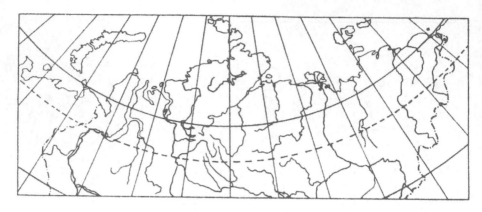

53. Leymus villosissimus.

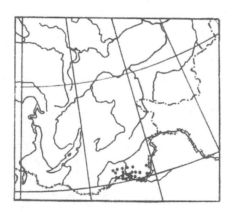

54. Hordeum macilentum.

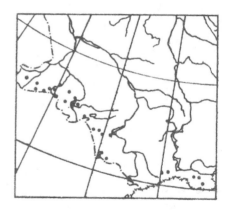

55. Hordeum nevskianum.

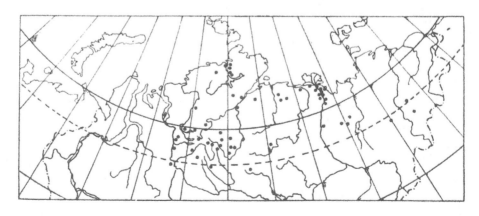

56. Bromopsis taimyrensis.

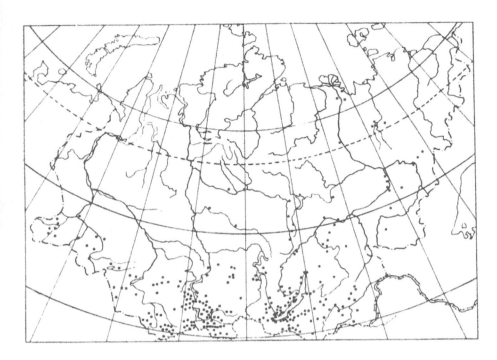

57. *Hordeum brevisubulatum.*

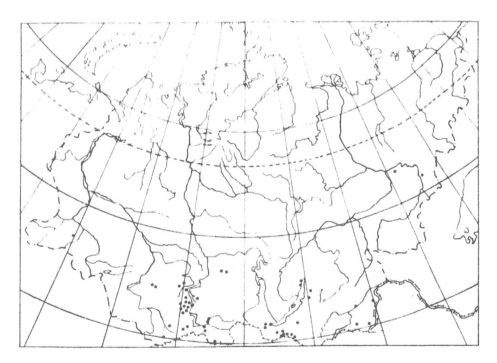

58. *Hordeum roshevitzii.*

272

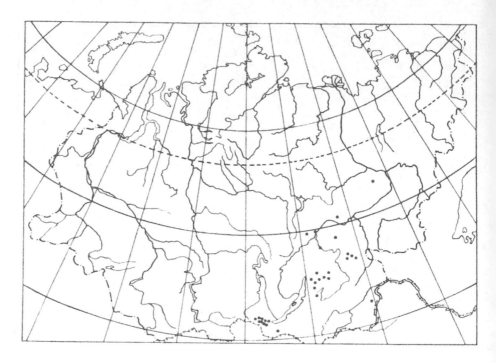

59. *Bromopsis alpina.*

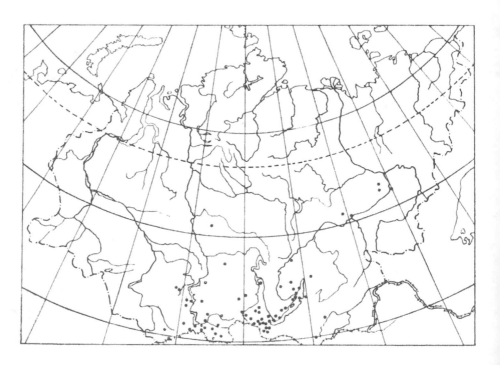

60. *Bromopsis austrosibirica.*

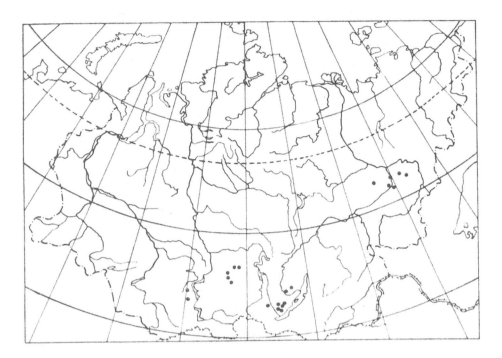

61. *Bromopsis karavajevii.*

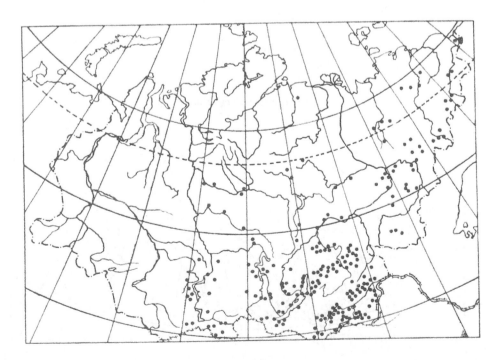

62. *Bromopsis sibirica.*

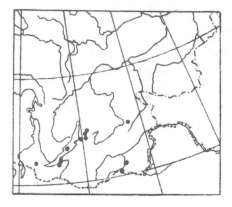

63. *Bromopsis korotkiji.*

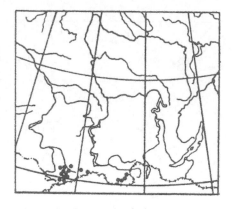

64. *Bromopsis altaica.*

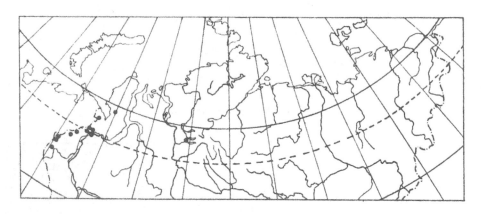

65. *Bromopsis vogulica.*

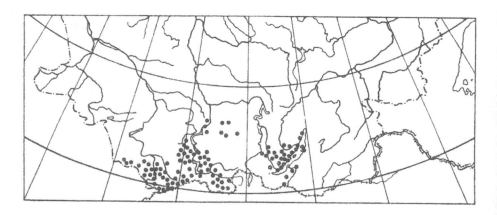

66. *Helictotrichon altaicum.*

275

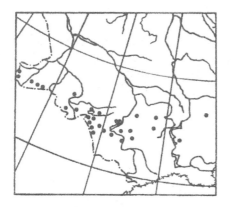

67. *Helictotrichon desertorum.*

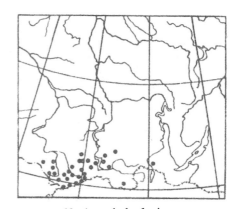

68. *Avenula hookeri.*

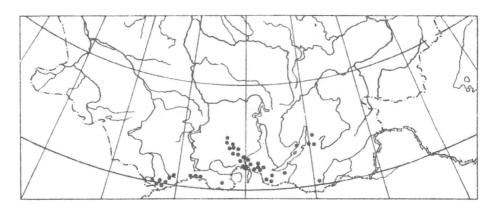

69. *Helictotrichon mongolicum.*

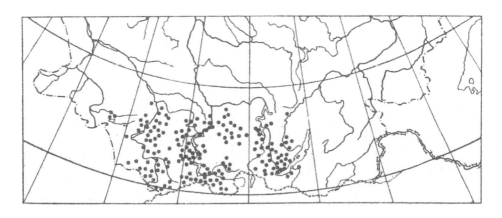

70. *Avenula pubescens.*

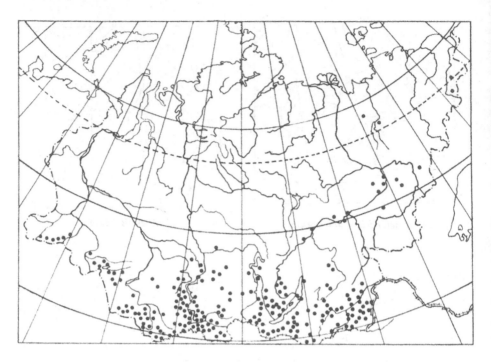

71. *Avenula hookeri* subsp. *schelliana.*

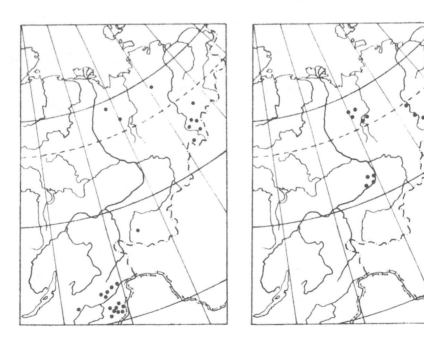

72. *Avenula dahurica.*

73. *Helictotrichon krylovii.*

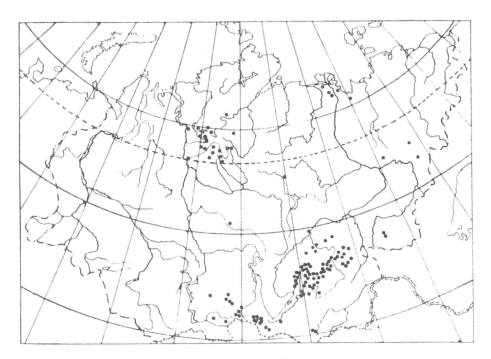

74. Trisetum agrostideum.

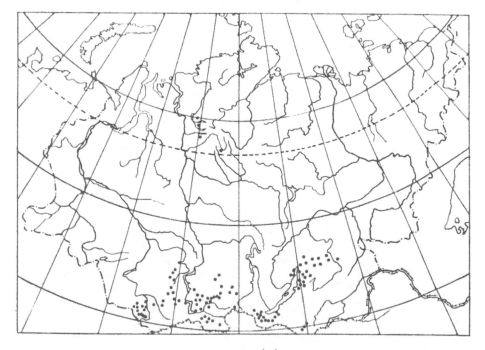

75. Trisetum altaicum.

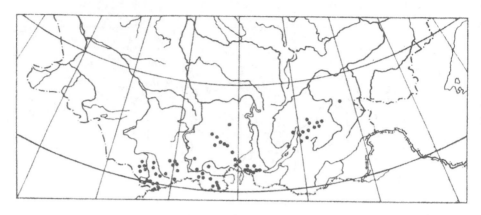

76. Trisetum mongolicum.

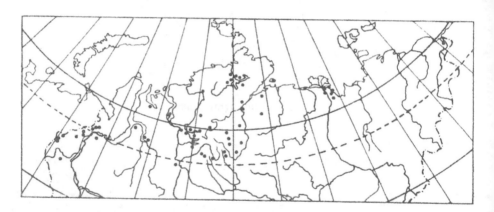

77. Trisetum spicatum.

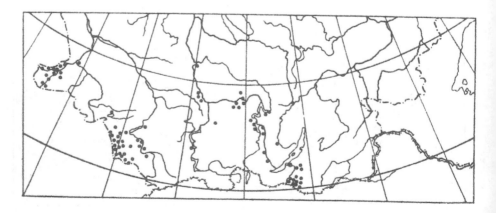

78. Koeleria glauca.

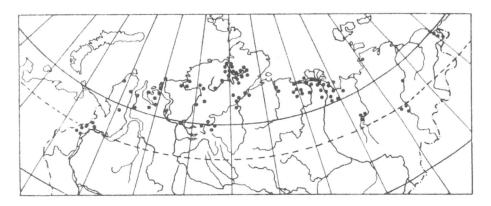

79. Koeleria asiatica.

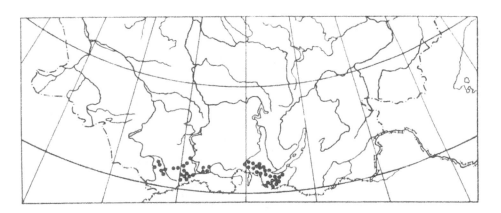

80. Koeleria atroviolacea.

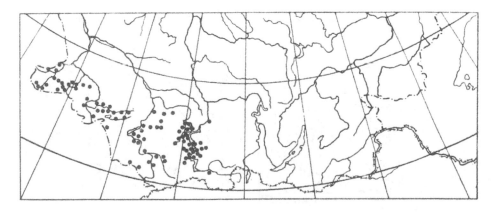

81. Koeleria delavignei.

280

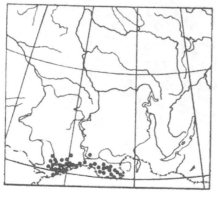

82. *Koeleria altaica.*

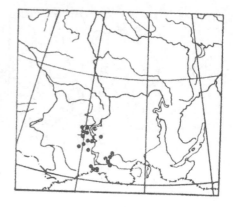

83. *Koeleria chakassica.*

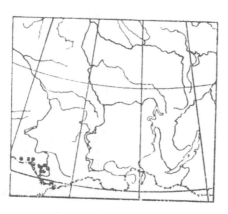

84. *Koeleria ledebourii.*

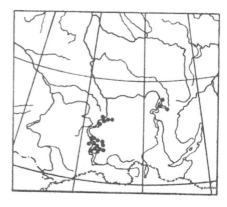

85. *Koeleria thonii.*

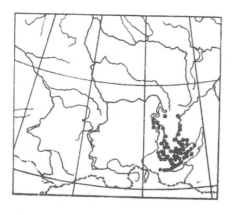

86. *Koeleria cristata* subsp. *hirsutiflora.*

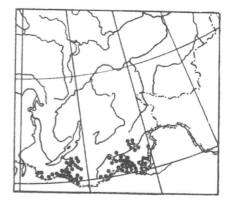

87. *Koeleria cristata* subsp. *mongolica.*

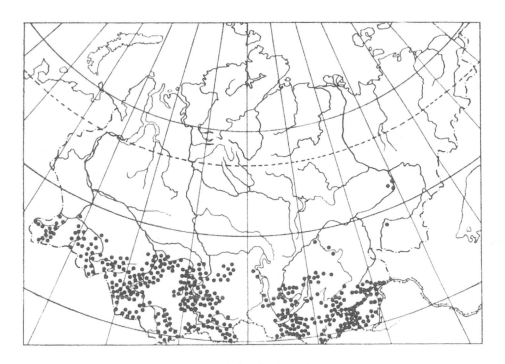

88. Koeleria cristata s. str.

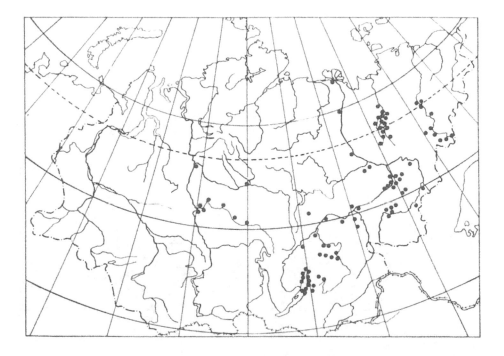

89. Koeleria cristata subsp. *seminuda.*

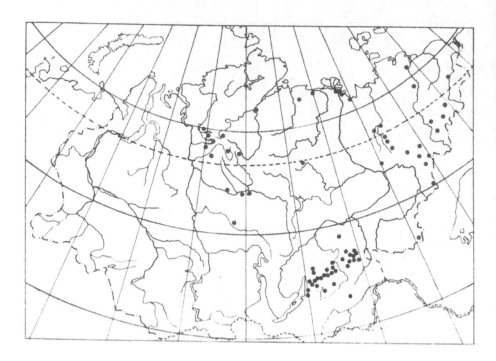

90. *Trisetum molle.*

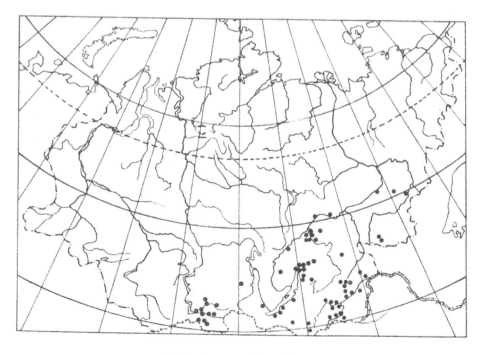

91. *Deschampsia sukatschewii.*

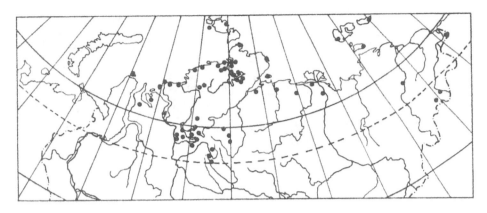

92. Deschampsia borealis.

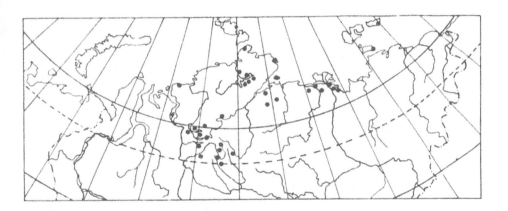

93. Deschampsia brevifolia.

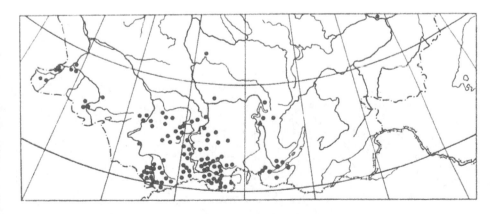

94. Deschampsia cespitosa.

284

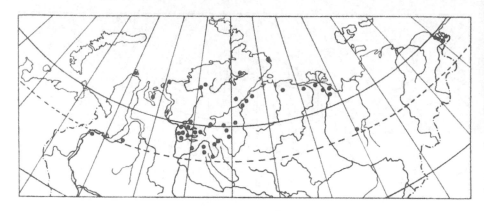

95. Deschampsia glauca.

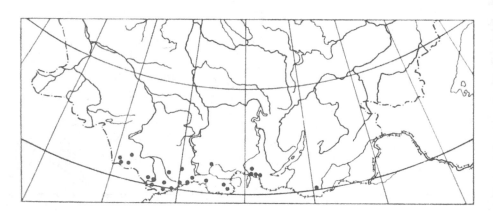

96. Deschampsia koelerioides.

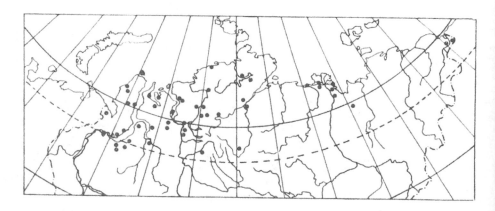

97. Deschampsia obensis.

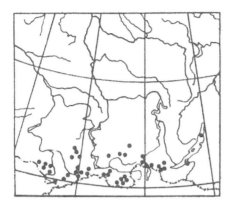

98. *Deschampsia altaica.*

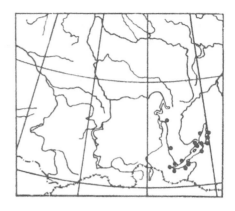

99. *Deschampsia turczaninowii.*

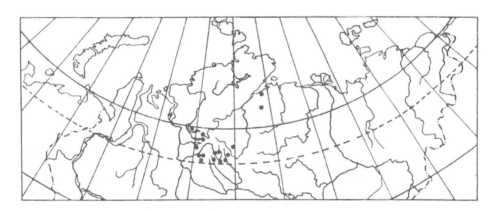

100. *Deschampsia vodopjanoviae.*

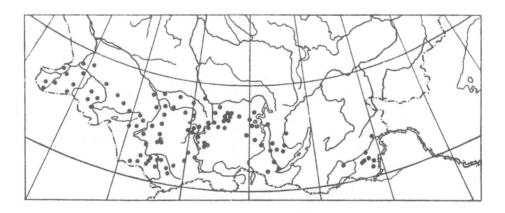

101. *Calamagrostis arundinacea.*

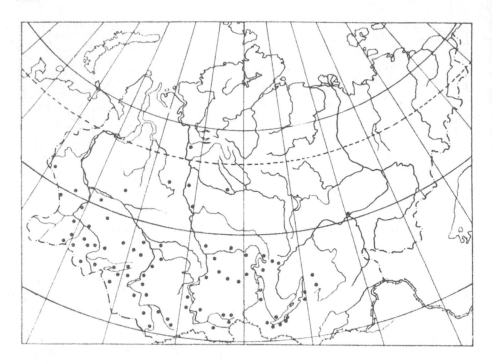

102. Milium effusum.

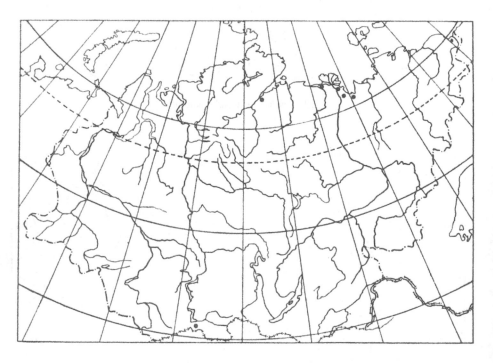

103. Calamagrostis deschampsioides.

287

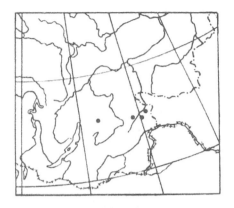

104. *Calamagrostis arctica.*

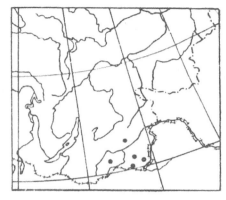

105. *Calamagrostis barbata.*

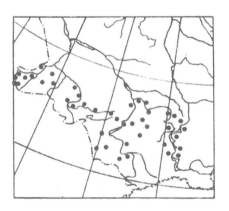

106. *Calamagrostis canescens.*

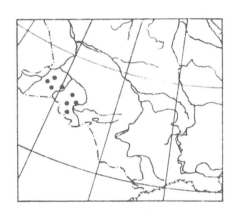

107. *Calamagrostis chalybaea.*

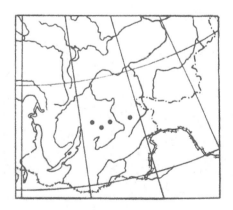

108. *Calamagrostis kalarica.*

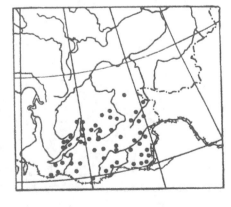

109. *Calamagrostis korotkyi.*

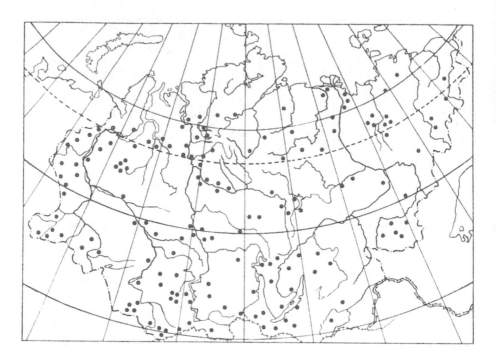

110. *Calamagrostis langsdorfii.*

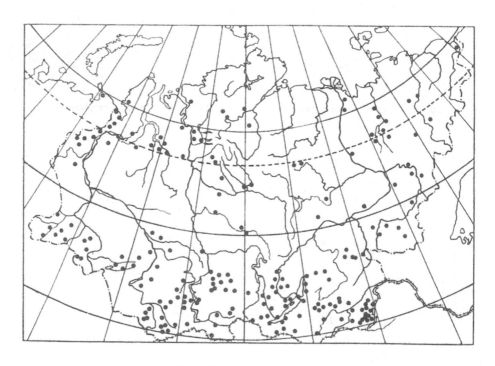

111. *Calamagrostis neglecta.*

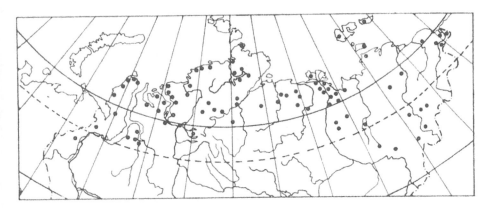

112. *Calamagrostis holmii.*

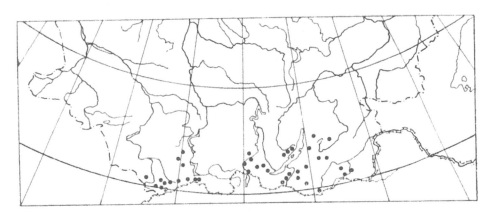

113. *Calamagrostis macilenta.*

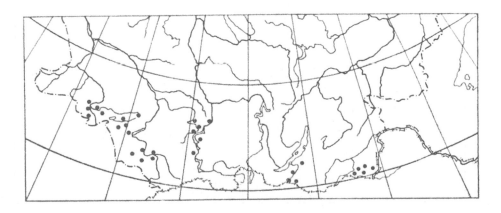

114. *Calamagrostis macrolepis.*

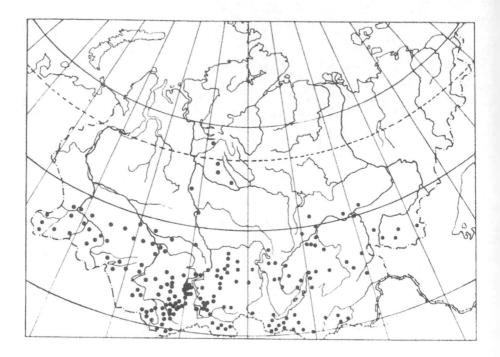

115. Calamagrostis obtusata.

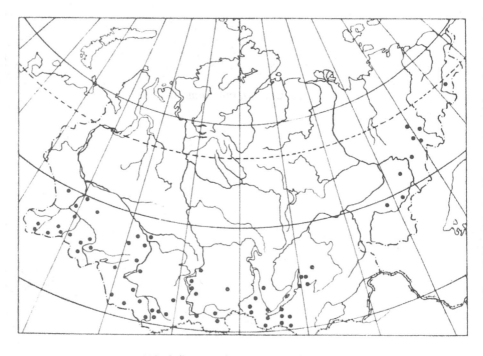

116. Calamagrostis pseudophragmites.

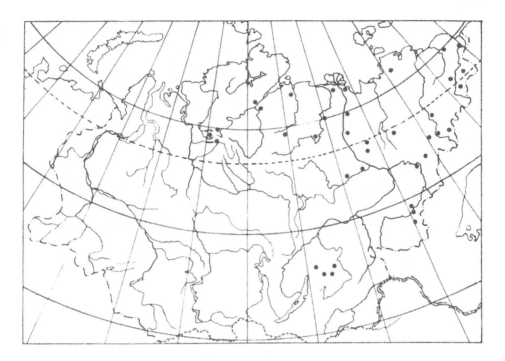

117. *Calamagrostis purpurascens.*

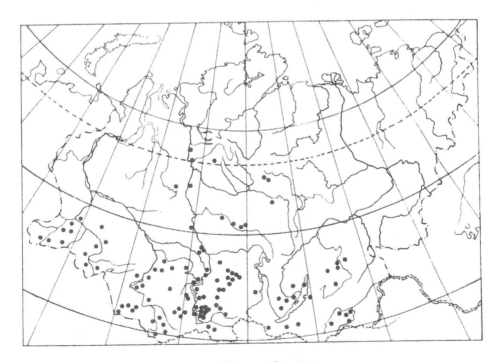

118. *Calamagrostis purpurea.*

292

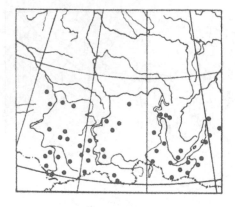

119. *Calamagrostis pavlovii.*

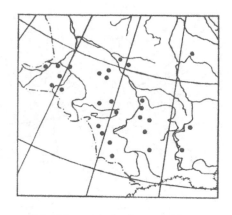

120. *Calamagrostis phragmitoides.*

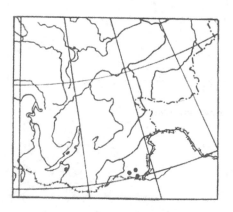

121. *Calamagrostis salina.*

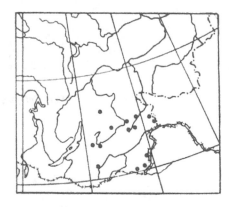

122. *Calamagrostis tenuis.*

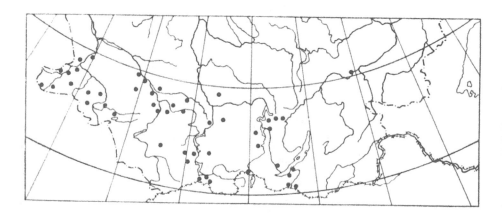

123. *Apera spica-venti.*

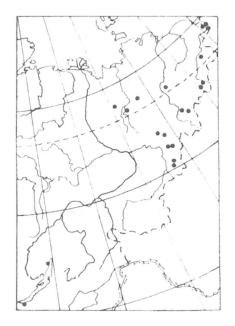

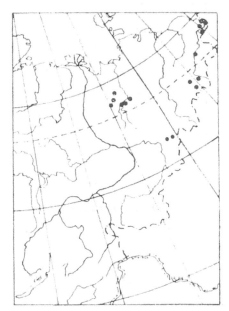

124. *Agrostis anadyrensis.*

125. *Hierochloë annulata.*

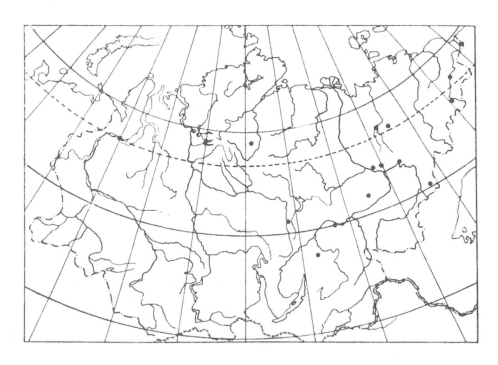

126. *Agrostis jacutica.*

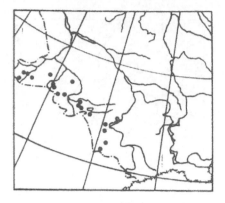
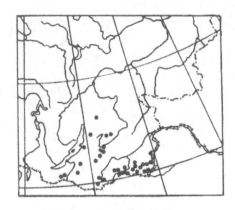

127. *Agrostis albida.*

128. *Agrostis divaricatissima.*

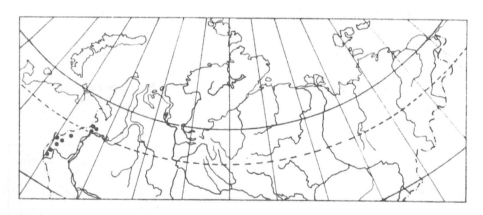

129. *Agrostis borealis.*

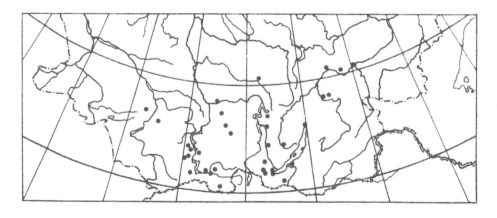

130. *Agrostis sibirica.*

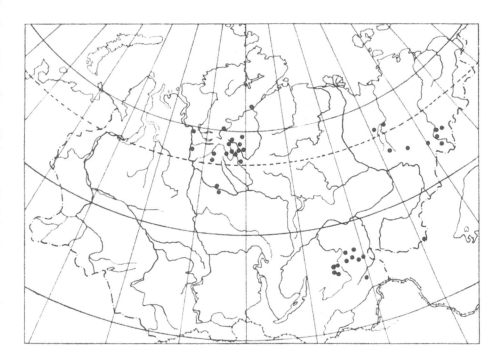

131. Agrostis kudoi.

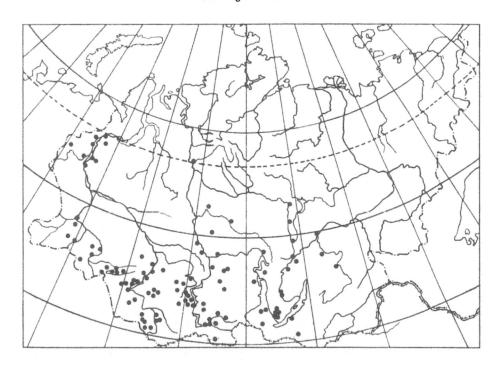

132. Agrostis stolonifera.

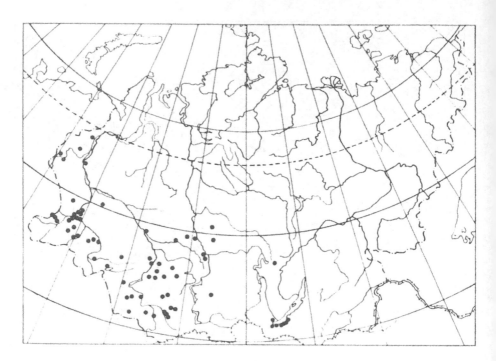

133. Agrostis tenuis.

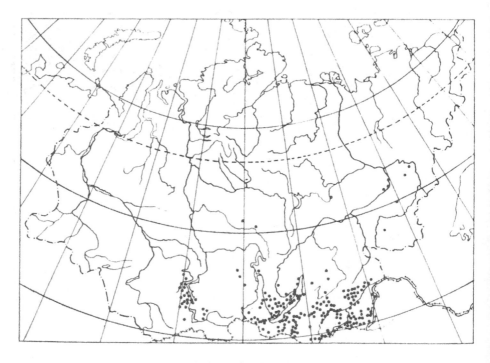

134. Agrostis trinii.

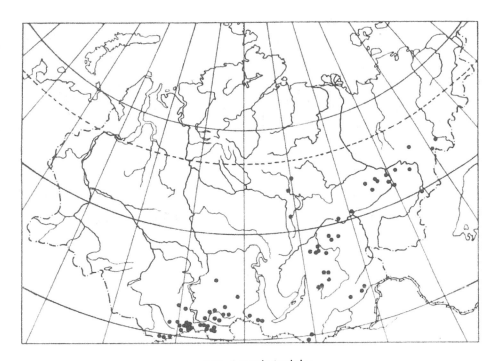

135. *Agrostis tuvinica.*

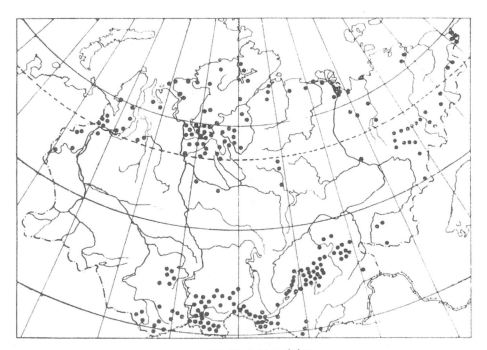

136. *Hierochloë alpina.*

298

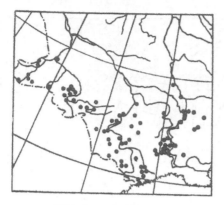

137. *Agrostis vinealis.*

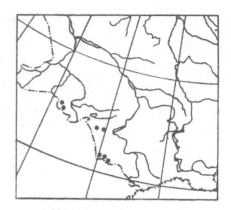

138. *Hierochloë repens.*

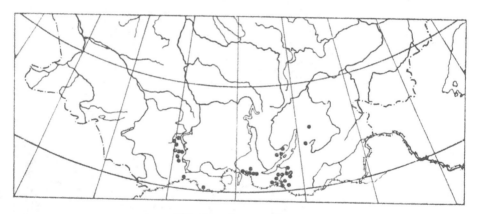

139. *Agrostis mongolica.*

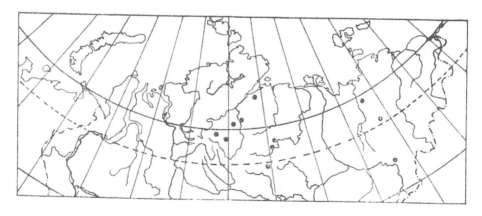

140. *Limnas malyschevii.*

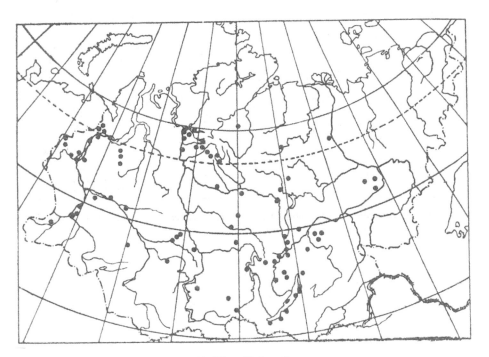

141. *Hierochloë arctica.*

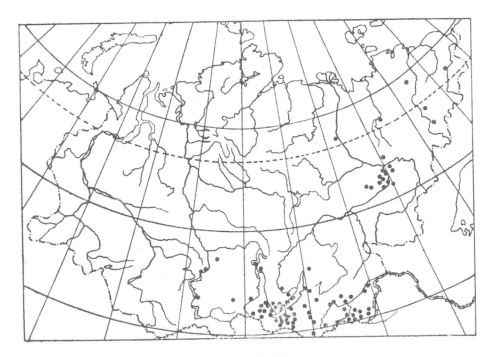

142. *Hierochloë glabra.*

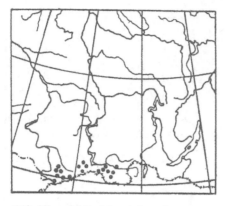 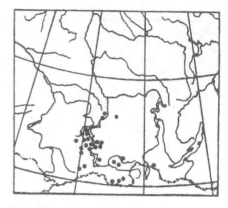

143. *Hierochloë glabra* subsp. *bungeana.*   144. *Hierochloë glabra* subsp. *chakassica.*

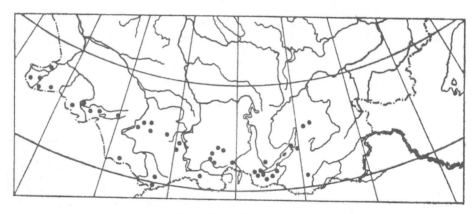

145. *Hierochloë odorata.*

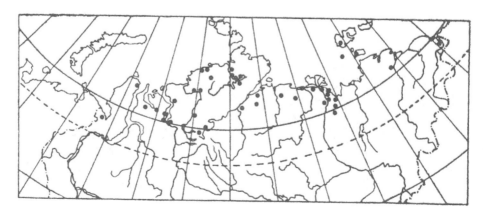

146. *Hierochloë pauciflora.*

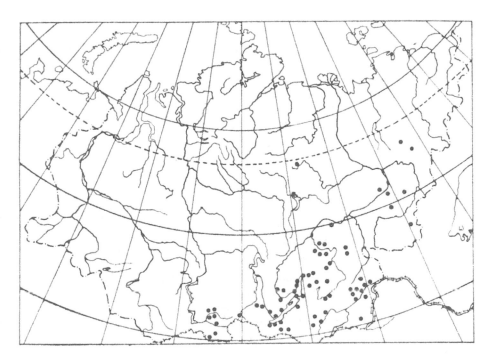

147. *Hierochloë ochotensis.*

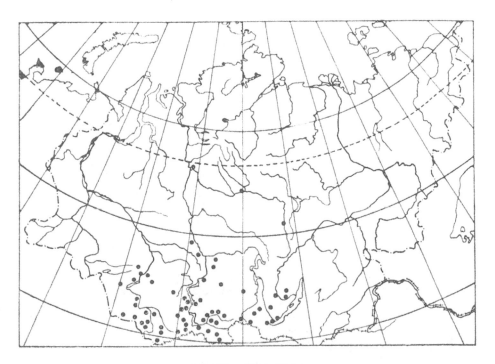

148. *Hierochloë sibirica.*

302

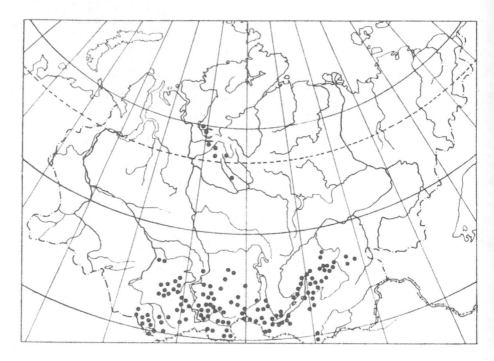

149. Anthoxanthum alpinum.

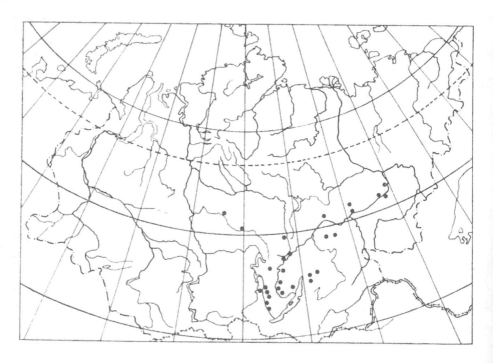

150. Limnas stelleri.

304

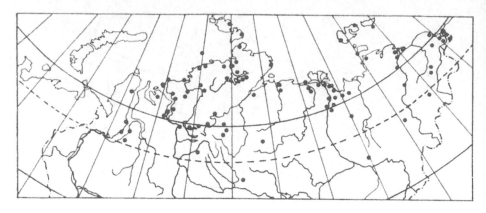

154. *Alopecurus alpinus.*

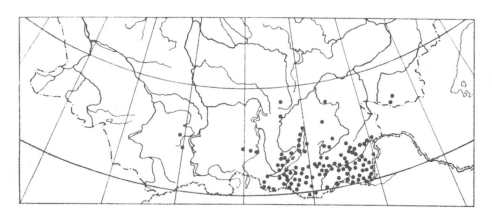

155. *Alopecurus brachystachyus.*

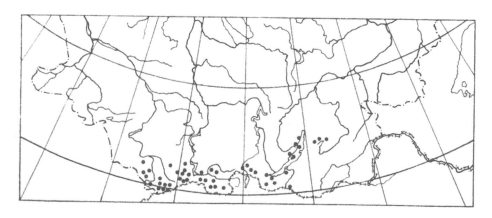

156. *Alopecurus turczaninovii.*

305

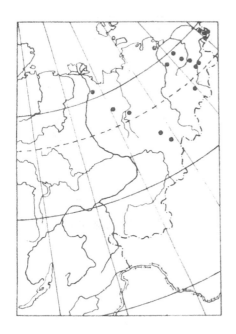

157. *Alopecurus roshevitzianus.*

158. *Hyalopoa lanatiflora.*

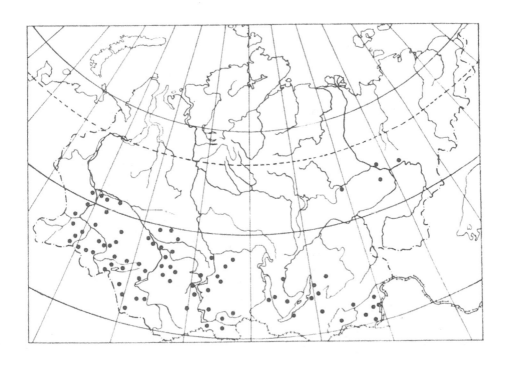

159. *Scolochloa festucacea.*

306

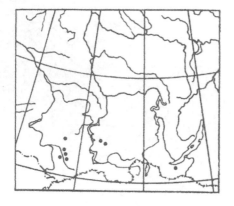

160. *Festuca altissima.*

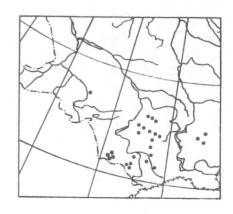

161. *Festuca gigantea.*

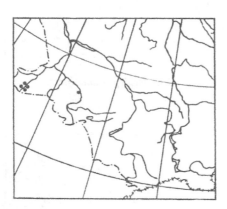

162. *Festuca arundinacea.*

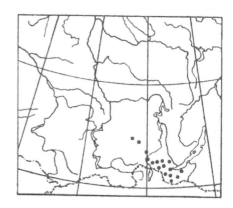

163. *Festuca komarovii.*

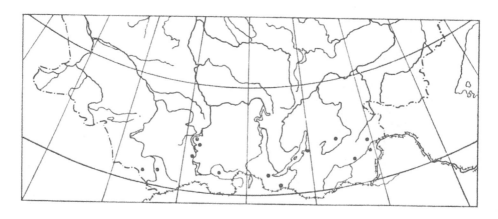

164. *Festuca extremiorientalis.*

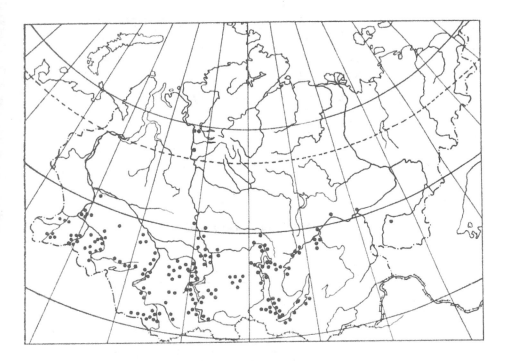

165. *Festuca pratensis.*

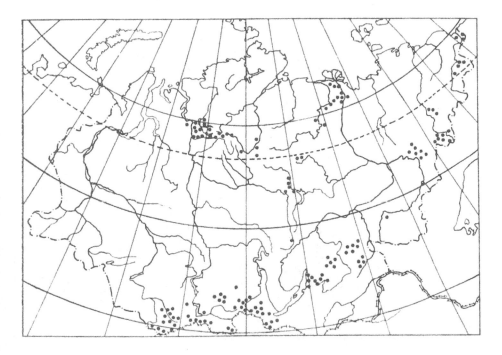

166. *Festuca altaica.*

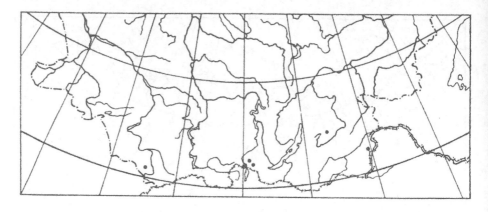

167. Festuca hubsugulica.

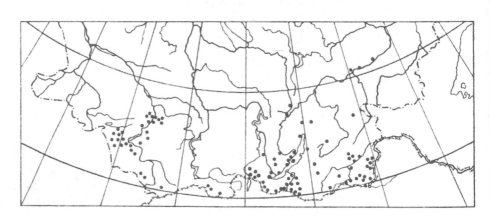

168. Festuca sibirica.

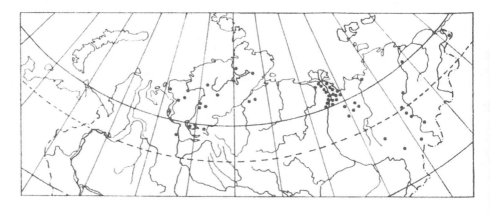

169. Festuca auriculata.

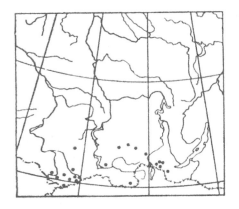

170. Festuca tristis.

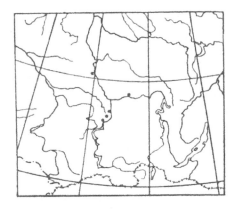

171. Festuca beckeri.

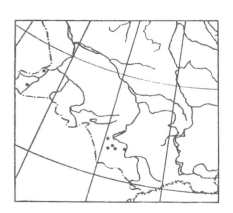

172. Festuca beckeri subsp. *polesica.*

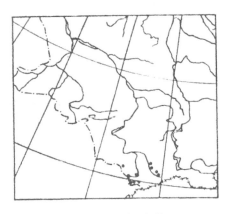

173. Festuca borissii.

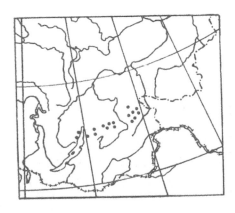

174. Festuca chionobia.

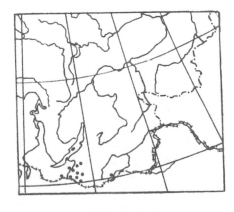

175. Festuca dahurica.

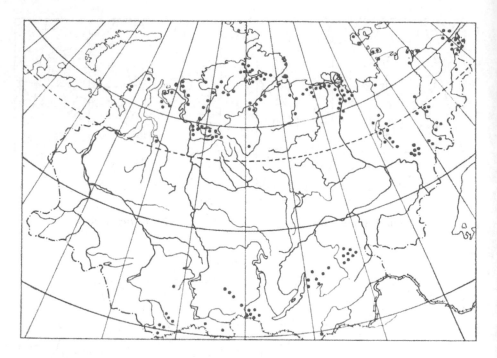

176. Festuca brachyphylla.

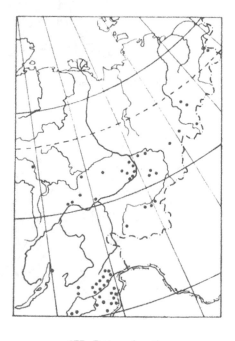

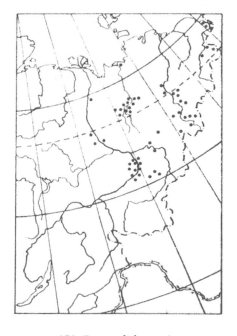

177. Festuca jacutica.

178. Festuca kolymensis.

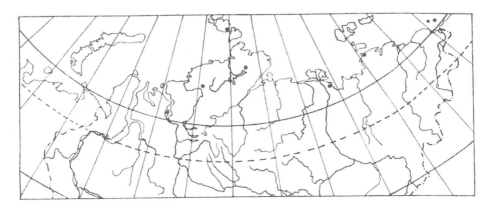

179. Festuca hyperborea.

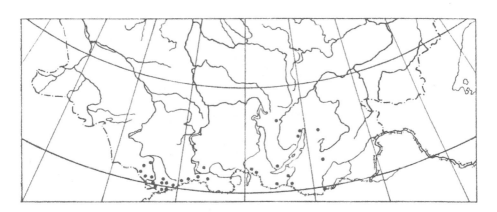

180. Festuca kryloviana.

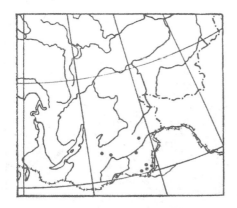

181. Festuca litvinovii.

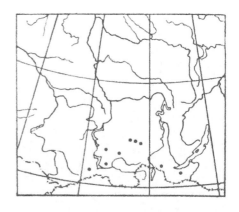

182. Festuca malyschevii.

312

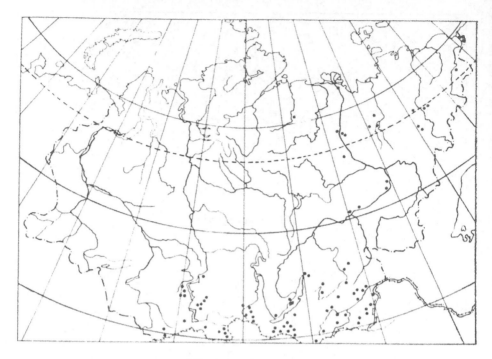

183. Festuca lenensis.

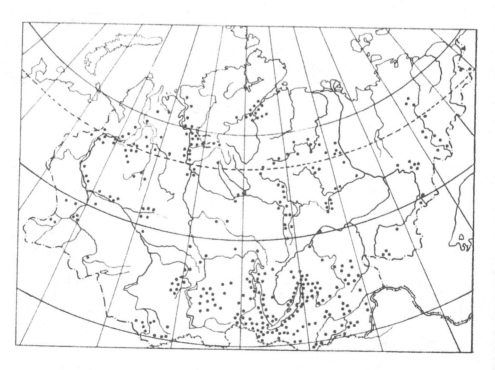

184. Festuca ovina.

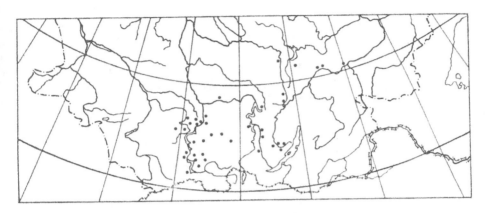

185. *Festuca pseudosulcata.*

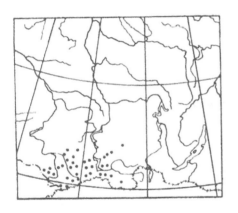

186. Festuca ovina subsp. *sphagnicola.*

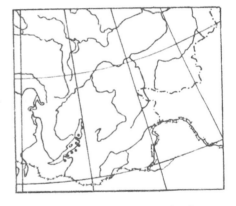

187. Festuca ovina subsp. *vylzaniae.*

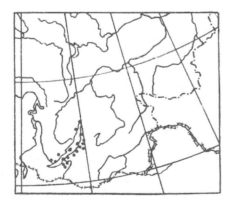

188. Festuca rubra subsp. *baicalensis.*

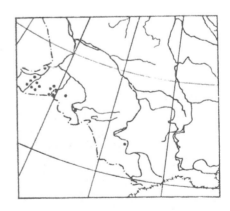

189. *Festuca rupicola.*

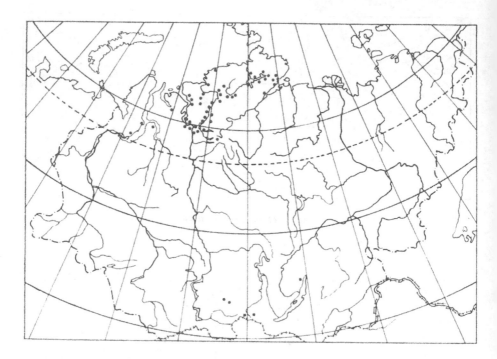

190. Festuca viviparoidea.

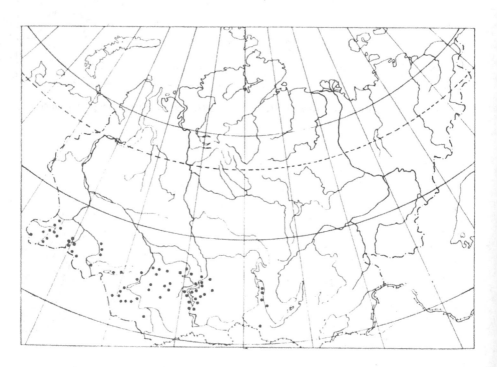

191. Festuca pseudovina.

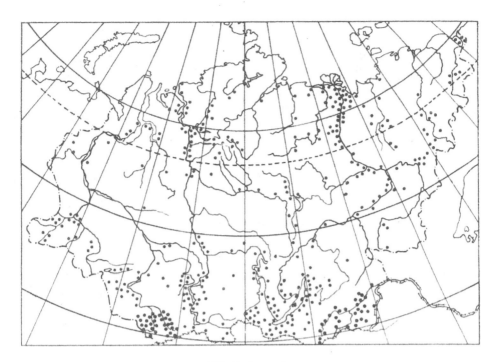

192. *Festuca rubra.*

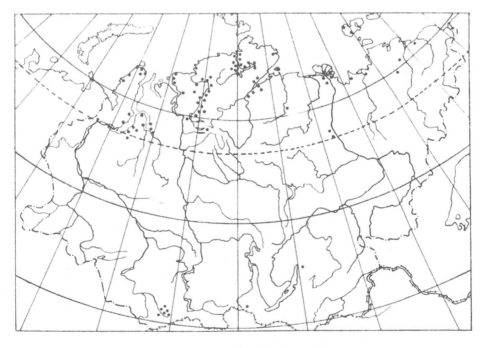

193. *Festuca rubra* subsp. *arctica.*

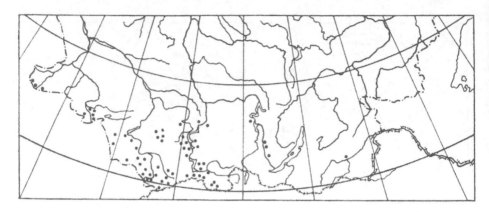

194. Festuca valesiaca.

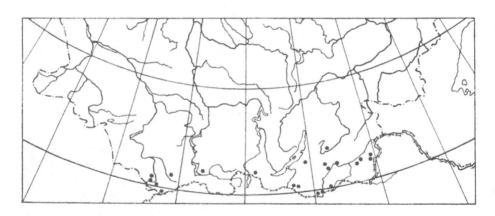

195. Festuca valesiaca subsp. *hypsophila.*

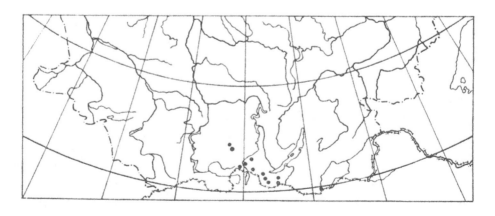

196. Festuca venusta.

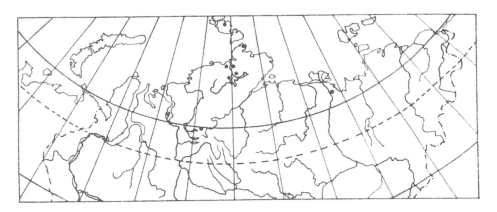

197. *Poa abbreviata.*

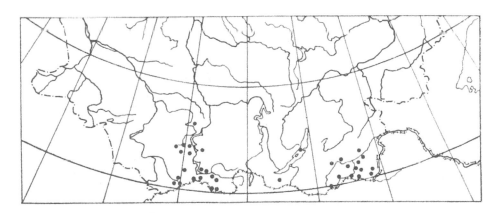

198. *Poa argunensis.*

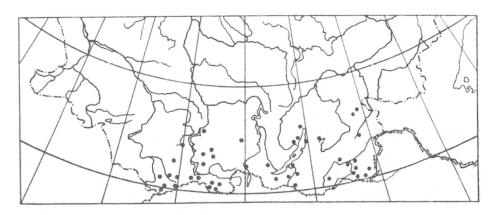

199. *Poa attenuata.*

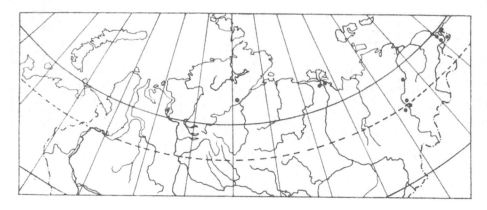

200. Poa filiculmis.

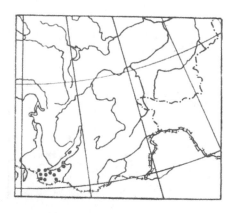

201. Poa ircutica.

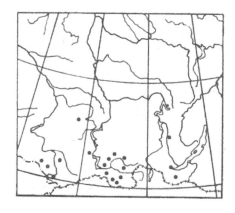

202. Poa krylovii.

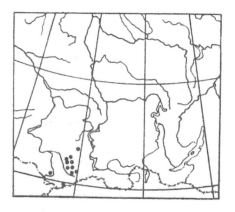

203. Poa mariae.

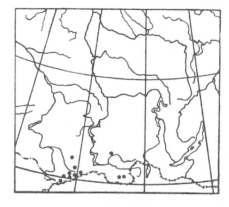

204. Poa litvinoviana.

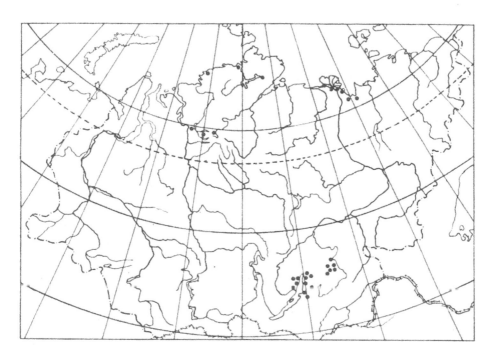

205. *Poa paucispicula.*

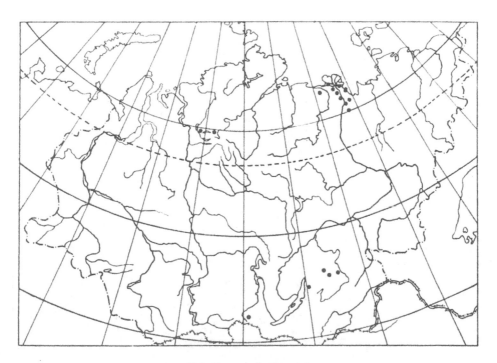

206. *Poa pseudoabbreviata.*

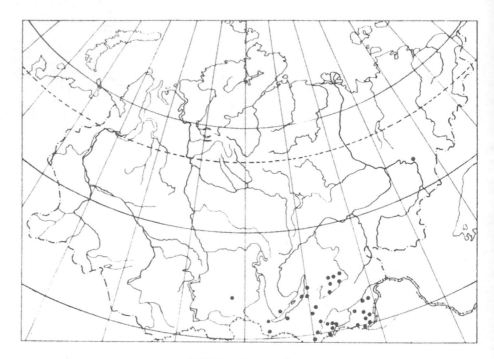

207. *Poa skvortzovii.*

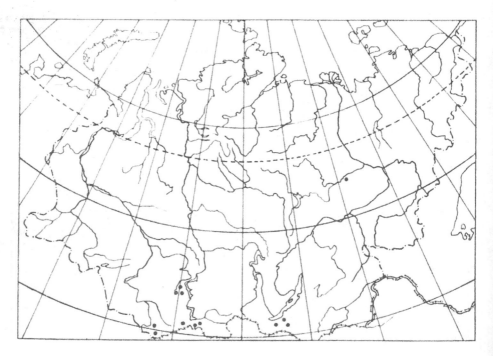

208. *Poa sabulosa.*

321

209. *Poa pruinosa.*

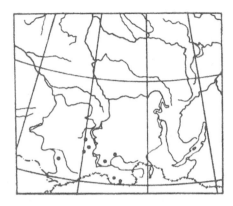

210. *Poa reverdattoi.*

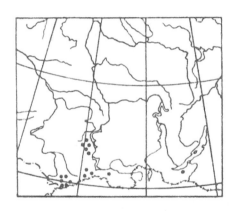

211. *Poa schischkinii.*

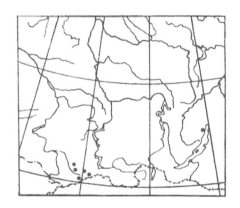

212. *Poa sobolevskiana.*

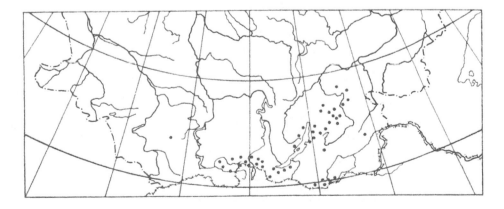

213. *Poa smirnovii.*

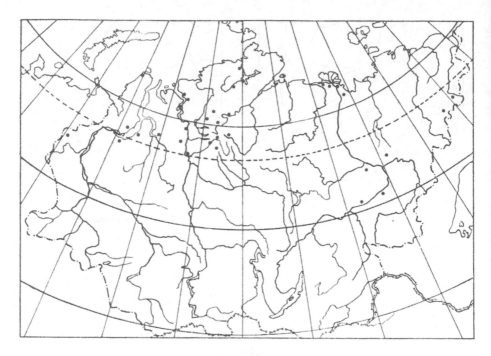

214. *Poa sublanata.*

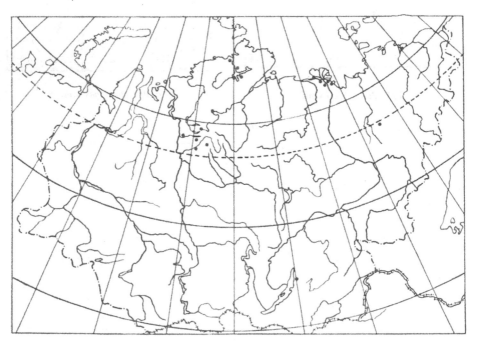

215. *Poa tolmatchewii.* .

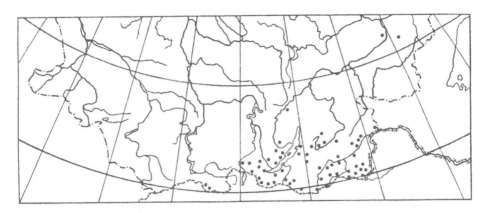

216. *Poa subfastigiata.*

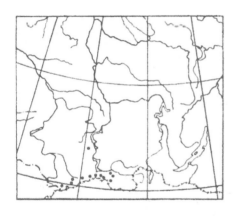

217. *Poa tianschanica.*

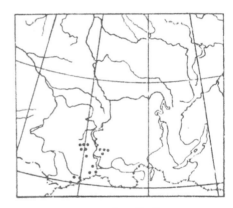

218. *Poa tibetica.*

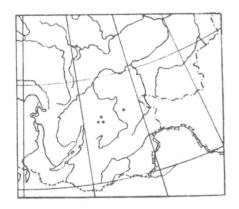

219. *Hyalopoa lanatiflora* subsp. *ivanoviae.*

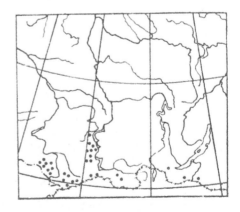

220. *Catabrosa aquatica.*

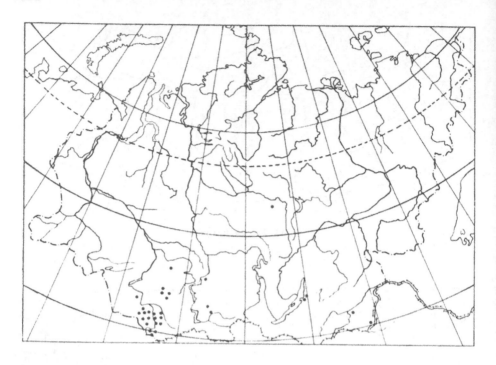

221. *Poa urssulensis.*

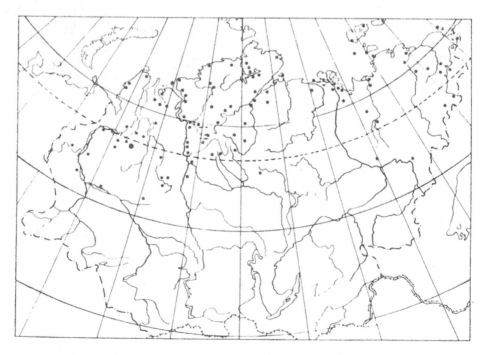

222. *Arctophila fulva.*

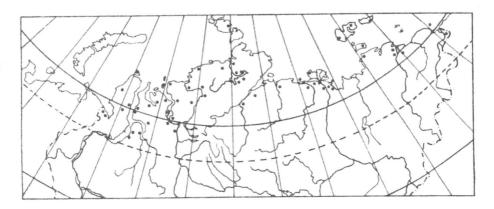

223. *Dupontia fisheri.*

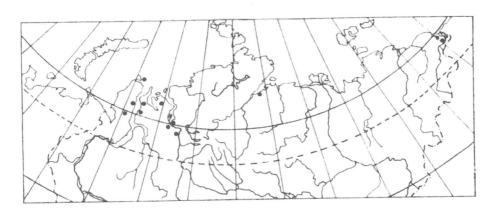

224. *Dupontia fisheri* subsp. *pelligera.*

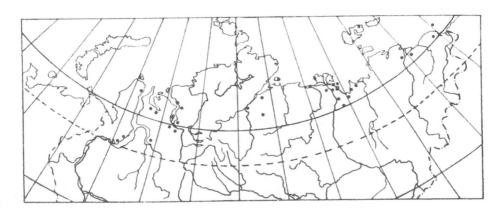

225. *Dupontia psilosantha.*

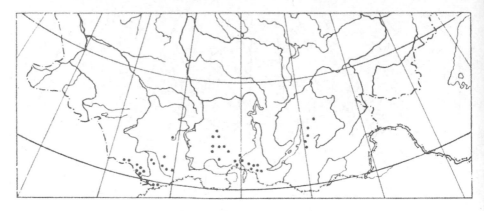

226. *Paracolpodium altaicum.*

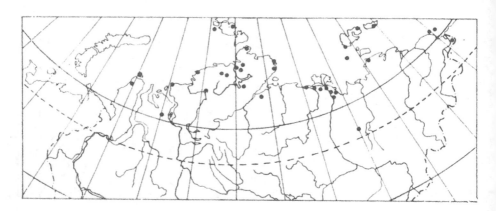

227. *Phippsia algida.*

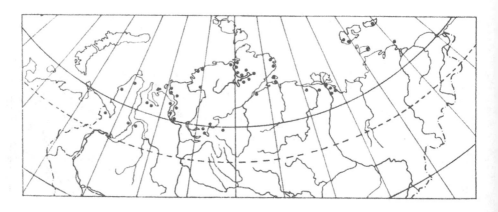

228. *Phippsia concinna.*

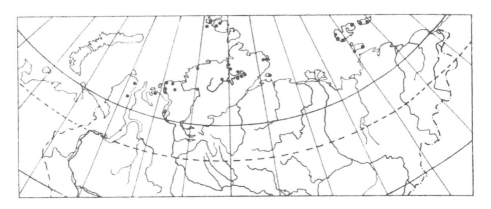

229. *Puccinellia angustata.*

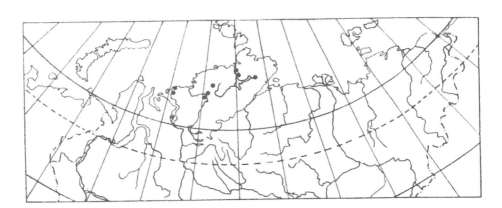

230. *Puccinellia byrrangensis.*

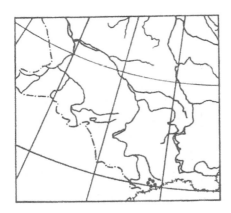

231. *Puccinellia altaica.*

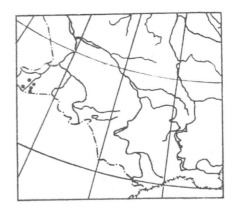

232. *Puccinellia dolicholepis.*

328

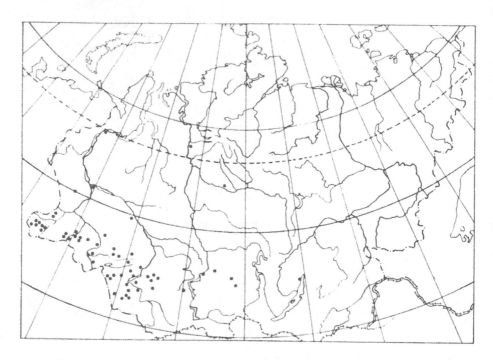

233. Puccinellia distans.

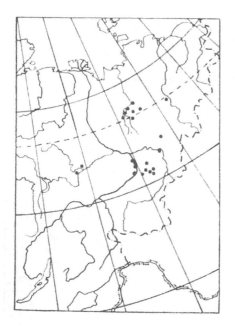

234. Puccinellia jacutica.

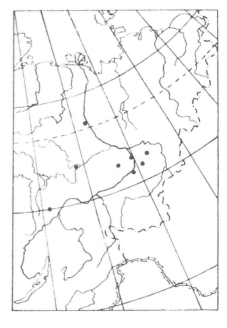

235. Puccinellia interior.

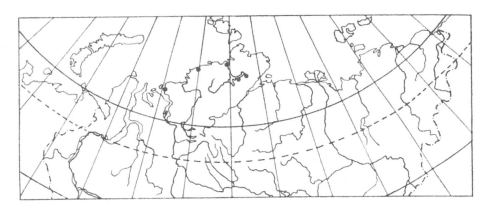

236. *Puccinellia gorodkovii.*

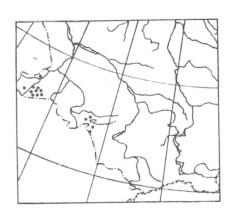

237. *Puccinellia gigantea.*

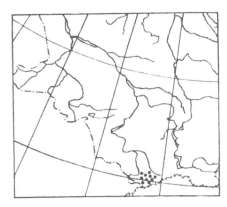

238. *Puccinellia kalininiae.*

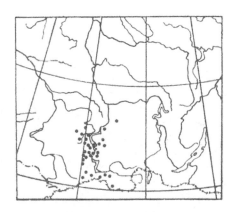

239. *Puccinellia kreczetoviczii.*

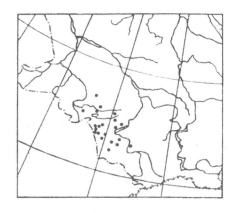

240. *Puccinellia kulundensis.*

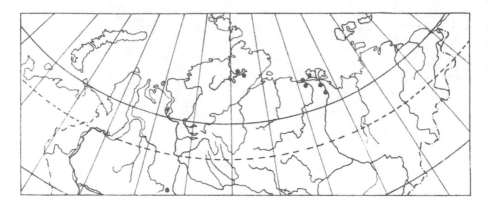

241. *Puccinellia lenensis.*

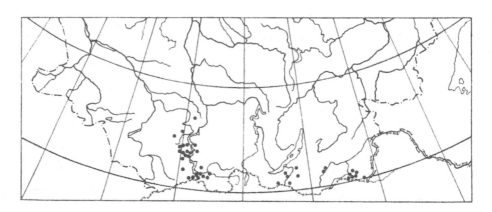

242. *Puccinellia macranthera.*

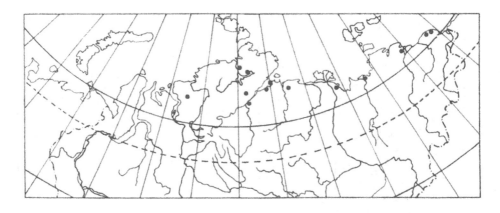

243. *Puccinellia neglecta.*

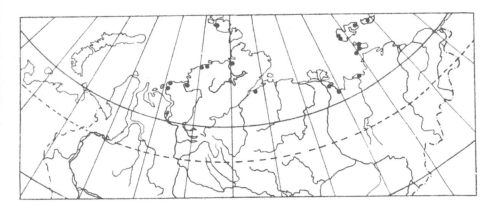

244. *Puccinellia phryganodes.*

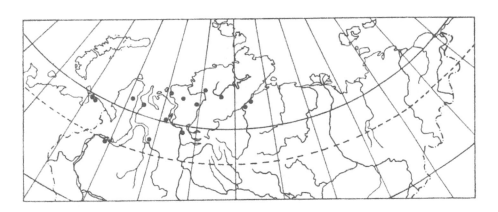

245. *Puccinellia sibirica.*

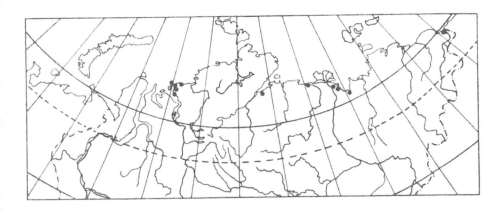

246. *Puccinellia tenella.*

332

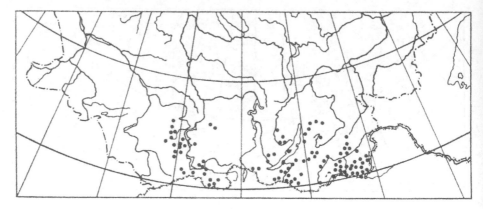

247. Puccinellia tenuiflora.

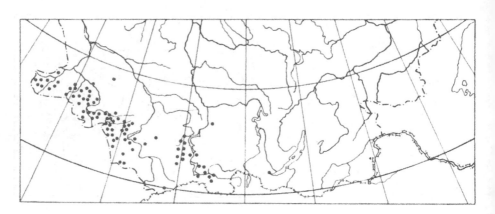

248. Puccinellia tenuissima.

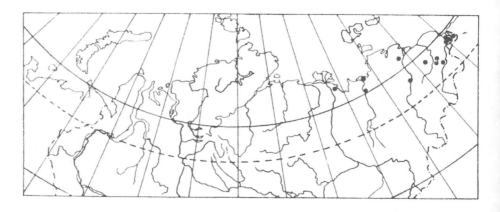

249. Puccinellia borealis.

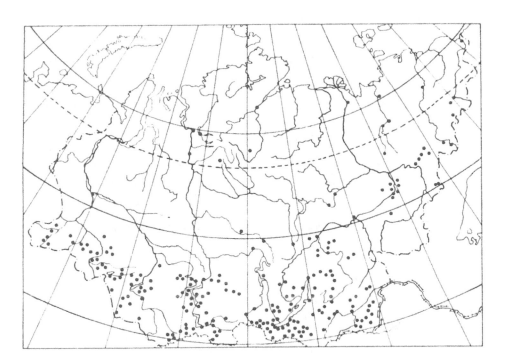

250. *Puccinellia hauptiana.*

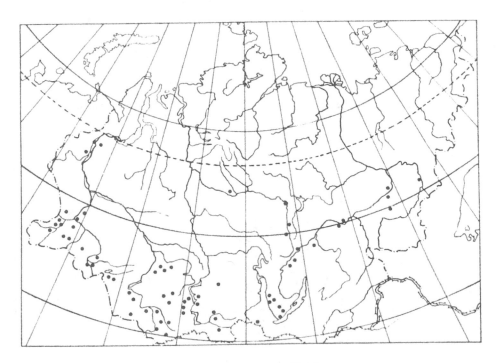

251. *Beckmannia eruciformis.*

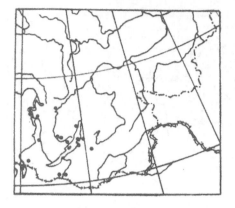

252. Puccinellia mongolica.

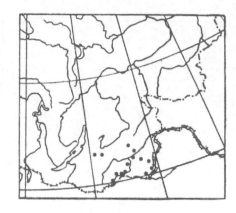

253. Phalaroides japonica.

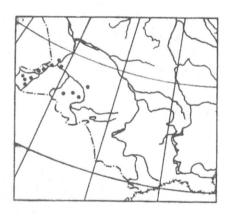

254. Glyceria maxima.

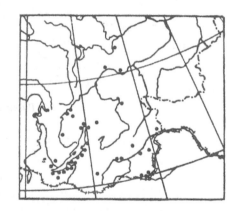

255. Glyceria spiculosa.

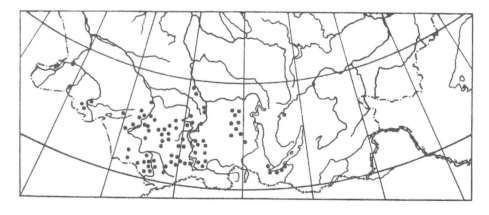

256. Dactylis glomerata.

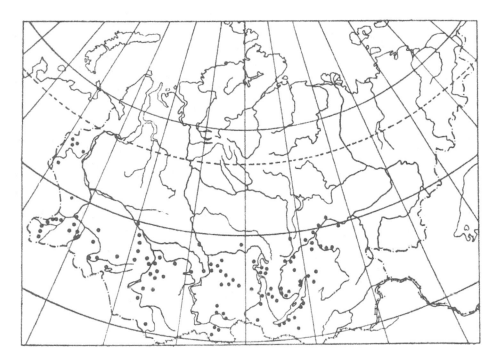

257. *Glyceria lithuanica.*

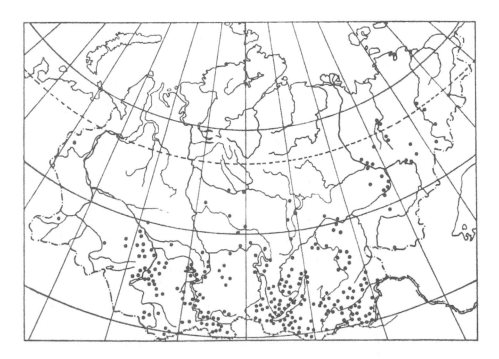

258. *Glyceria triflora.*

336

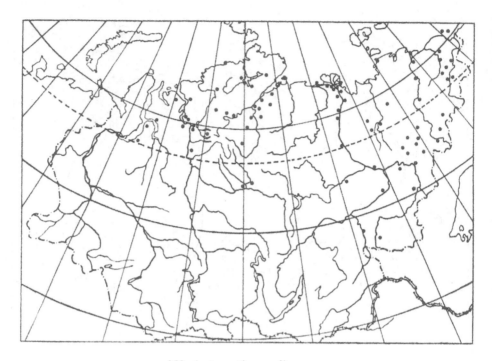

259. *Arctagrostis arundinacea.*

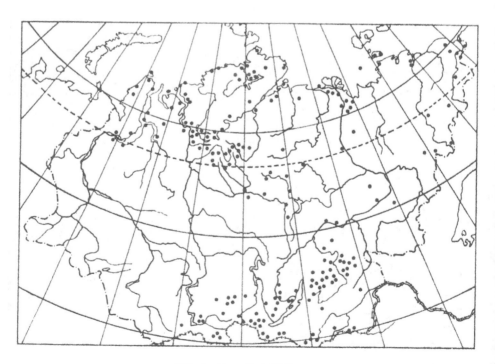

260. *Arctagrostis latifolia.*

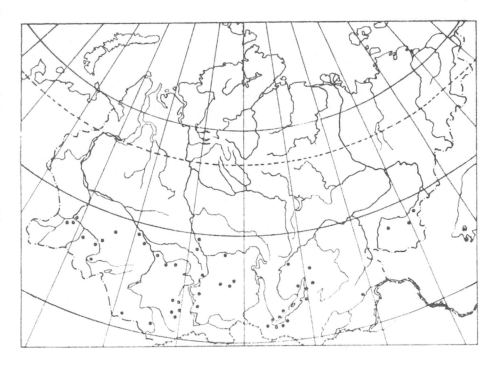

261. *Cinna latifolia.*

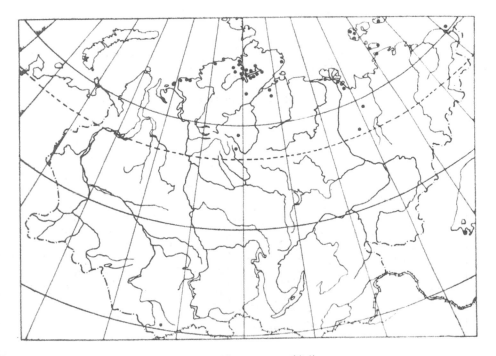

262. *Pleuropogon sabinii.*

338

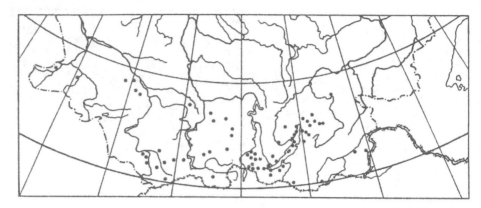

263. *Schizachne callosa.*

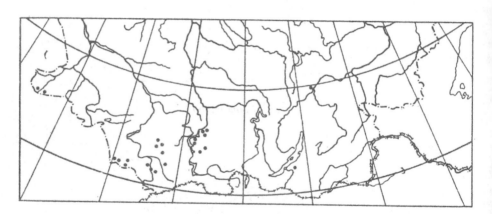

264. *Melica altissima.*

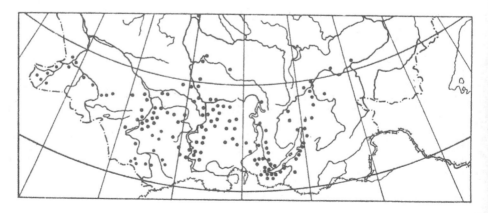

265. *Melica nutans.*

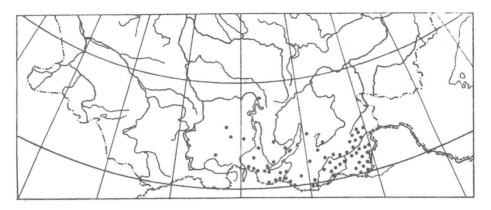

266. *Melica turczaninowiana.*

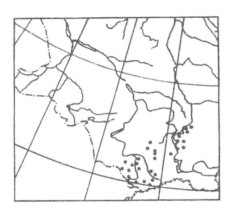

267. *Melica transsilvanica.*

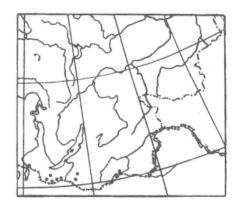

268. *Melica virgata.*

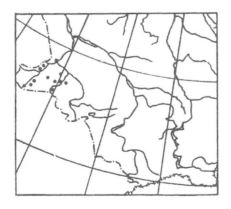

269. *Molinia caerulea.*

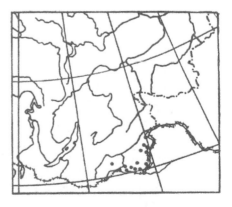

270. *Arundinella anomala.*

340

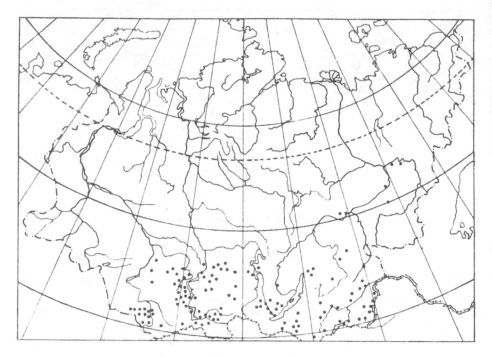

271. Achnatherum confusum.

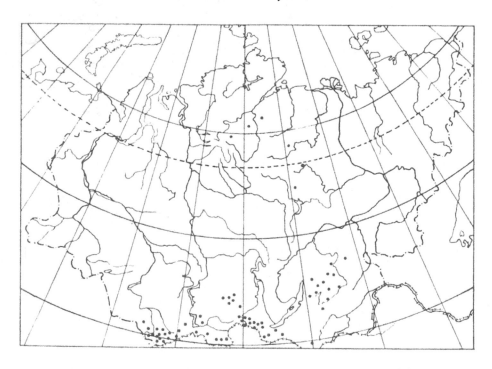

272. Ptilagrostis mongholica.

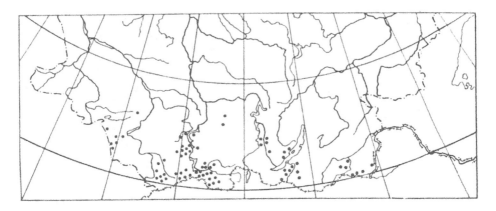

273. *Achnatherum splendens.*

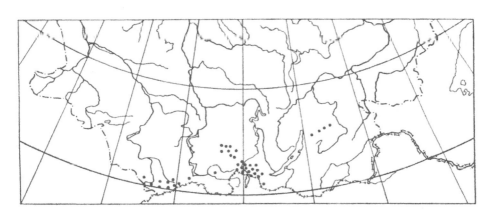

274. *Ptilagrostis junatovii.*

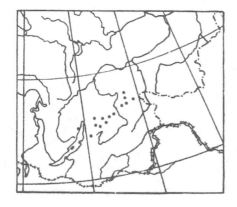

275. *Ptilagrostis alpina.*

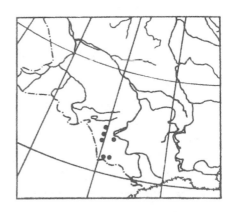

276. *Crypsis aculeata.*

342

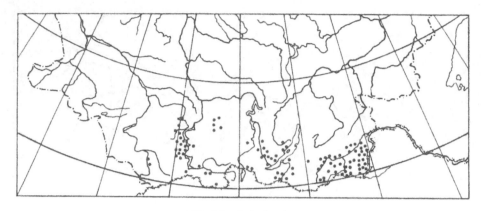

277. Stipa baicalensis.

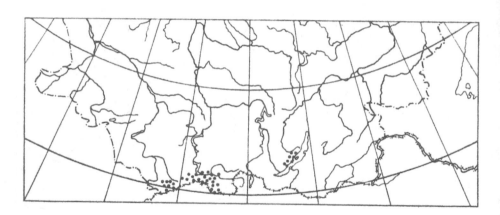

278. Stipa glareosa.

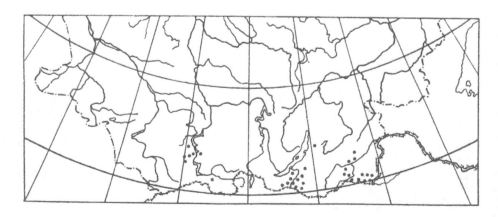

279. Stipa grandis.

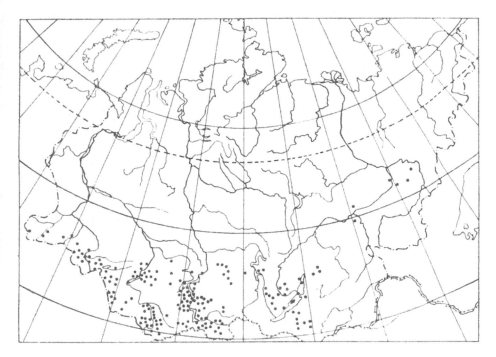

280. *Stipa capillata.*

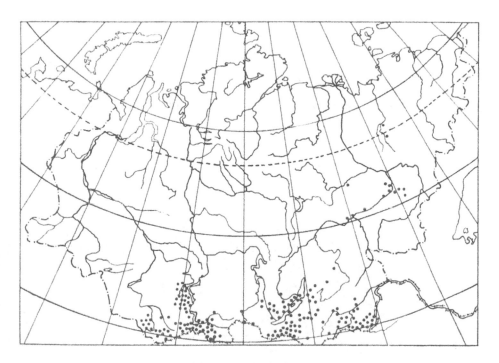

281. *Stipa krylovii.*

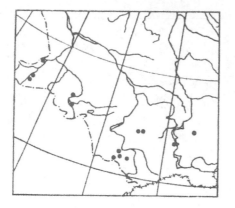

282. *Stipa dasyphylla.*

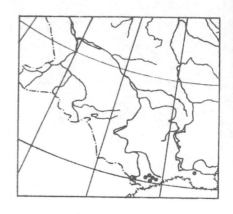

283. *Stipa kirghisorum.*

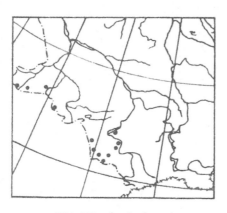

284. *Stipa lessingiana.*

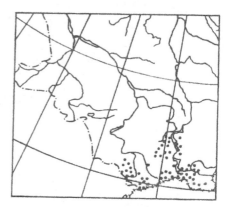

285. *Stipa orientalis.*

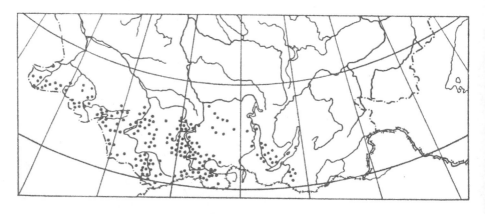

286. *Stipa pennata.*

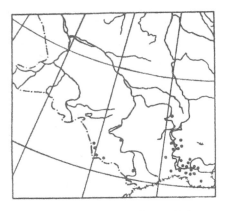

287. Stipa pennata subsp. *sabulosa.*

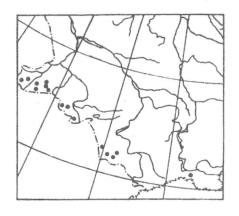

288. Stipa praecapillata.

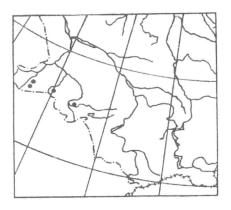

289. Stipa pulcherrima.

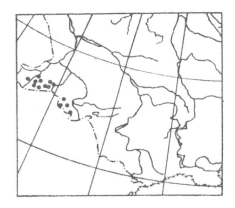

290. Stipa tirsa.

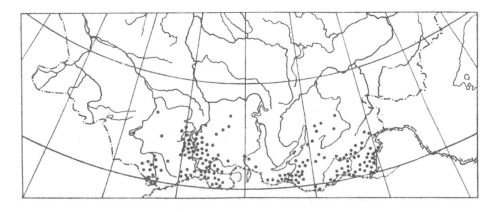

291. Achnatherum sibiricum.

346

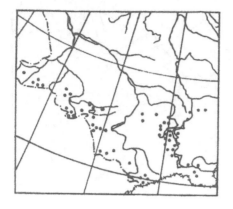

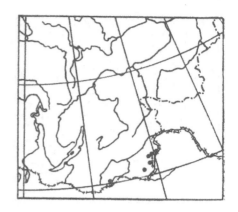

292. *Stipa zalesskii.*

293. *Tripogon chinensis.*

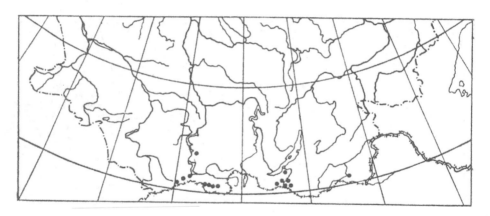

294. *Enneapogon borealis.*

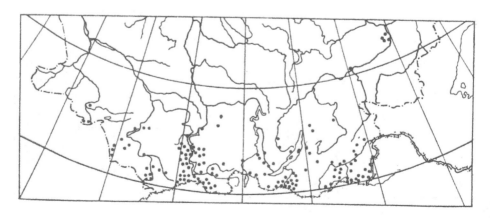

295. *Cleistogenes squarrosa.*

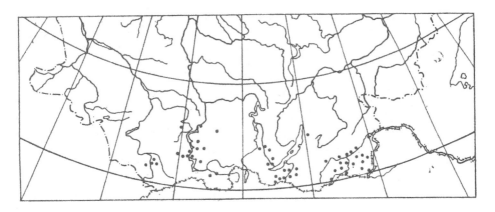

296. Cleistogenes kitagawae.

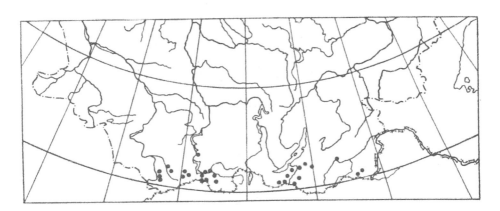

297. Eragrostis minor.

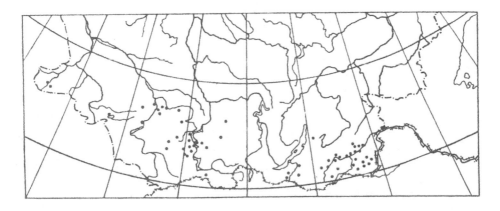

298. Eragrostis pilosa.

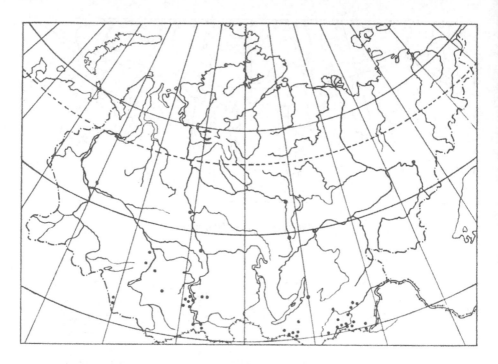

299. Eragrostis amurensis.

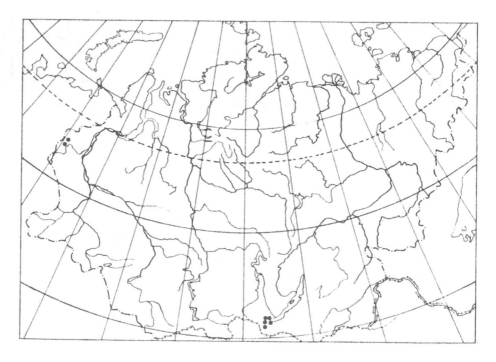

300. Nardus stricta.

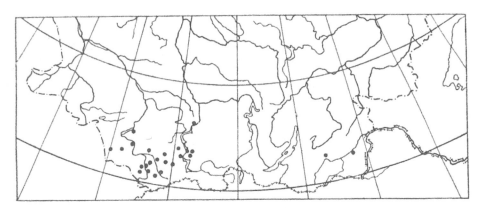

301. Setaria pumila.

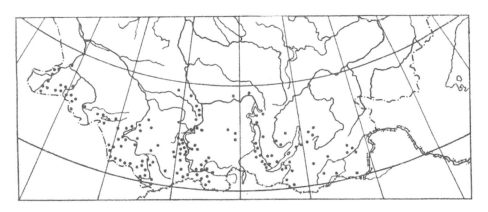

302. Setaria viridis

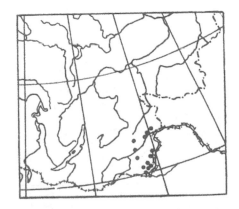

303. Setaria viridis subsp. *purpurascens.*

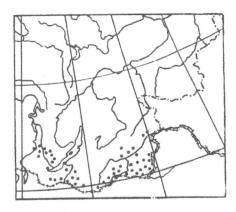

304. Spodiopogon sibiricus.

350

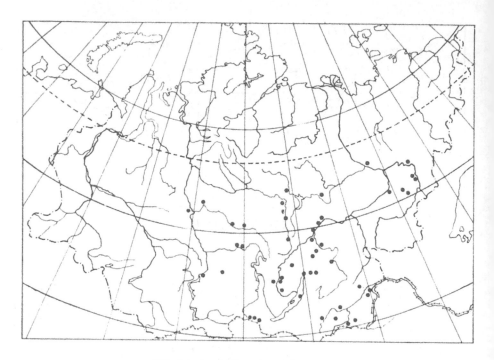

305. Setaria viridis subsp. *glareosa.*

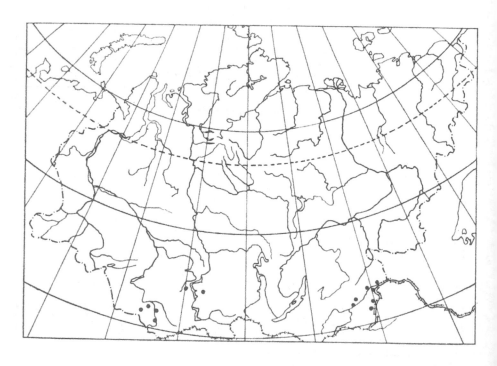

306. Digitaria ischaemum.

Index of Latin Names
of Plants

358

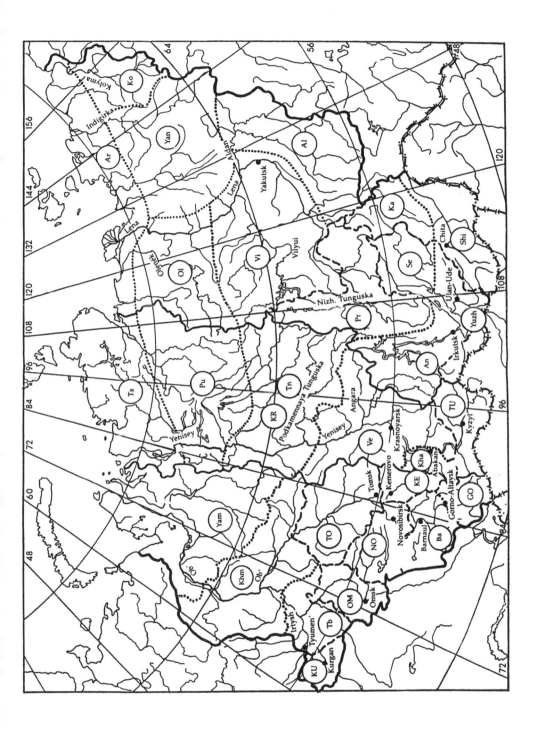